Structural Dynamic Analysis with Generalized Damping Models

To
Sonia Adhikari
Sunanda Adhikari
and
Tulsi Prasad Adhikari

Structural Dynamic Analysis with Generalized Damping Models

Analysis

Sondipon Adhikari

Series Editor
Noël Challamel

iSTE

WILEY

First published 2014 in Great Britain and the United States by ISTE Ltd and John Wiley & Sons, Inc.

ISTE Ltd
27-37 St George's Road
London SW19 4EU
UK

www.iste.co.uk

John Wiley & Sons, Inc.
111 River Street
Hoboken, NJ 07030
USA

www.wiley.com

Library of Congress Control Number: 2013948547

British Library Cataloguing-in-Publication Data
A CIP record for this book is available from the British Library
ISBN: 978-1-84821-521-4

Printed and bound in Great Britain by CPI Group (UK) Ltd., Croydon, Surrey CR0 4YY

Table of Contents

Preface

Among the various ingredients of structural dynamics, damping is one of the least understood topics. The main reason is that unlike the stiffness and inertia forces, damping forces cannot always be obtained from "first principles". The past two decades have seen significant developments in the modeling and analysis of damping in the context of engineering dynamic systems. Developments in composite materials including nanocomposites and their applications in advanced structures, such as new generation of aircrafts and large wind turbines, have led to the need for understanding damping in a better manner. Additionally, the rise of vibration energy harvesting technology using piezoelectric and electromagnetic principles further enhanced the importance of looking at damping more rigorously. The aim of this book is to systematically present the latest developments in the modeling and analysis of damping in the context of general linear dynamic systems with multiple degrees-of-freedom. The focus has been on the mathematical and computational aspects. This book will be relevant to aerospace, mechanical and civil engineering disciplines and various sub-disciplines within them. The intended readers of this book include senior undergraduate students and graduate students doing projects or doctoral research in the field of damped vibration. Researchers, professors and practicing engineers working in the field of advanced vibration will find this book useful. This book will also be useful for researchers working in the fields of aeroelasticity and hydroelasticity, where complex eigenvalue problems routinely arise due to fluid–structure interactions.

There are some excellent books which already exist in the field of damped vibration. The book by Nashif *et al.* [NAS 85] covers various material damping models and their applications in the design and analysis of dynamic systems. A valuable reference on dynamic analysis of damped structures is [SUN 95]. The book by Beards [BEA 96] takes a pedagogical approach toward structural vibration of damped systems. The handbook by Jones [JON 01] focuses on viscoelastic damping and analysis of structures with such damping models. These books represented the state of the art at the time of their publications. Since these publications, significant research works have gone into the dynamics of damped systems. The aim of this book is to cover some of these latest developments. The attention is mainly limited to theoretical and computational aspects, although some references to experimental works are given.

One of the key features of this book is the consideration of general non-viscous damping and how such general models can be seamlessly integrated into the framework of conventional structural dynamic analysis. New results are illustrated by numerical examples and, wherever possible, connections are made to well-known concepts of viscously damped systems. The related title, *Structural Dynamic Analysis with Generalized Damping Models: Identification* [ADH 14b], is complementary to this book, and, indeed, they could have been presented together. However, for practical reasons, it has proved more convenient to present the material separately. There are ten chapters and one appendix in the two books combined covering analysis and identification of dynamic systems with viscous and non-viscous damping.

This current book, *Structural Dynamic Analysis with Generalized Damping Models: Analysis*, deals with the analysis of linear systems with general damping models. Chapter 1 gives an introduction to the various damping models. Dynamics of viscously damped systems are discussed in Chapter 2. Chapter 3 considers dynamics of non-viscously damped single-degree-of-freedom systems in detail. Chapter 4 discusses non-viscously damped multiple degree-of-freedom systems. Linear systems with general non-viscous damping are studied in Chapter 5. Chapter 6 proposes reduced computational methods for damped systems. A method to deal with general asymmetric systems is described in the appendix.

The related book *Structural Dynamic Analysis with Generalized Damping Models: Identification* [ADH 14b] deals with the identification and quantification of damping. Chapter 1 describes parametric sensitivity of damped systems. Chapter 2 describes the problem of identification of viscous damping. The identification of non-viscous damping is detailed in Chapter 3. Chapter 4 gives some tools for the quantification of damping.

This book is the result of the last 15 years of research and teaching in the area of damped vibration problems. Initial chapters started taking shape when I offered a course on advanced vibration at the University of Bristol. The later chapters originated from the research works with numerous colleagues, students, collaborators and mentors. I am deeply indebted to all of them for numerous stimulating scientific discussions, exchanges of ideas and, on many occasions, direct contributions toward the intellectual content of the book. I am grateful to my teachers Professor C. S. Manohar (Indian Institute of Science, Bangalore), Professor R. S. Langley (University of Cambridge) and, in particular, Professor J. Woodhouse (University of Cambridge), who was heavily involved with the works reported in Chapters 2–4 of

[ADH 14b]. I am very thankful to my colleague Professor M. I. Friswell with whom I have a long-standing collaboration. Some joint works are directly related to the content of this book (Chapter 1 of [ADH 14b] in particular). I would also like to thank Professor D. J. Inman (University of Michigan) for various scientific discussions during his visits to Swansea. I am thankful to Professor A. Sarkar (Carleton University) and his doctoral student M. Khalil for joint research works. I am deeply grateful to Dr A. S. Phani (University of British Columbia) for various discussions related to damping identification and contributions toward Chapters 2 and 5 of this book and Chapter 2 of [ADH 14b]. Particular thanks go to Dr N. Wagner (Intes GmbH, Stuttgart) for joint works on non-viscously damped systems and contributions in Chapter 4 of this book. I am also grateful to Professor F. Papai for involving me in research works on damping identification. My former PhD students B. Pascual (contributed in Chapter 6), J. L. du Bois and F. A. Diaz De la O deserve particular thanks for various contributions throughout their time with me and putting up with my busy schedules. I am grateful to Dr Y. Lei (University of Defense Technology, Changsha) for carrying out joint research with me on non-viscously damped continuous systems. I am grateful to Professor A. W. Lees (Swansea University), Professor N. Lieven, Professor F. Scarpa (University of Bristol), Professor D. J. Wagg (University of Sheffield), Professor S. Narayanan (Indian Institute of Technology (IIT) Madras), Professor G. Litak (Lublin University), E. Jacquelin (Université Lyon), Dr A. Palmeri (Loughborough University), Professor S. Bhattacharya (University of Surrey), Dr S. F. Ali (IIT Madras), Dr R. Chowdhury (IIT Roorkee), Dr P. Duffour (University College London), and Dr P. Higino, Dr G. Caprio and Dr A. Prado (Embraer Aircraft) for their intellectual contributions and discussions at different times. Besides the names mentioned here, I am also thankful to many colleagues, fellow researchers and students working in this field of research around the world, whose names cannot be listed here due to page limitations. The lack of explicit mentions by no means implies that their contributions are any lesser. The opinions presented in the book are entirely mine, and none of my colleagues, students, collaborators and mentors have any responsibility for any shortcomings.

I have been fortunate to receive grants from various companies, charities and government organizations including an Advanced Research Fellowship from UK Engineering and Physical Sciences Research Council (EPSRC), the Wolfson Research Merit Award from the Royal Society and the Philip Leverhulme Prize from the Leverhulme Trust. Without these findings, it would have been impossible to have conducted the works leading to this book. Finally, I want to thank my colleagues at the College of Engineering at Swansea University. Their support proved to be a key factor in materializing the idea of writing this book.

Last, but by no means least, I wish to thank my wife Sonia and my parents for their constant support, encouragement and putting up with my ever-increasing long periods of "non-engagement" with them.

Sondipon ADHIKARI
October 2013

Nomenclature

C'_{jj}	diagonal element of the modal damping matrix
$\alpha_k^{(j)}$	terms in the expansion of approximate complex modes
α_1, α_2	proportional damping constants
α_j	coefficients in Caughey series, $j = 0, 1, 2, \cdots$
$\mathbf{0}_j$	a vector of j zeros
$\mathbf{A}$	state-space system matrix
$\mathbf{a}_j$	a coefficient vector for the expansion of jth complex mode
$\boldsymbol{\alpha}$	a vector containing the constants in Caughey series
$\bar{h}(\mathrm{i}\omega)$	frequency response function of an SDOF system
$\mathbf{B}$	state-space system matrix
$\mathbf{b}_j$	a vector for the expansion of jth complex mode
$\bar{\mathbf{f}}(s)$	forcing vector in the Laplace domain
$\bar{\mathbf{f}}'(s)$	modal forcing function in the Laplace domain
$\bar{\mathbf{p}}(s)$	effective forcing vector in the Laplace domain
$\bar{\mathbf{q}}(s)$	response vector in the Laplace domain
$\bar{\mathbf{u}}(s)$	Laplace transform of the state-vector of the first-order system
$\bar{\mathbf{y}}(s)$	modal coordinates in the Laplace domain
$\bar{\mathbf{y}}_k$	Laplace transform of the internal variable $\mathbf{y}_k(t)$
$\mathbb{R}^+$	positive real line
$\mathbf{C}$	viscous damping matrix
$\mathbf{C}'$	modal damping matrix
$\mathbf{C}_0$	viscous damping matrix (with a non-viscous model)
$\mathbf{C}_k$	coefficient matrices in the exponential model for $k = 0, ..., n$, where n is the number of kernels

$\mathcal{G}(t)$	non-viscous damping function matrix in the time domain
$\Delta\mathbf{K}$	error in the stiffness matrix
$\Delta\mathbf{M}$	error in the mass matrix
β	non-viscous damping factor
β_c	critical value of β for oscillatory motion, $\beta_c = \frac{1}{3\sqrt{3}}$
$\beta_i(\bullet)$	proportional damping functions (of a matrix)
$\beta_k(s)$	coefficients in the state-space modal expansion
β_{mU}	the value of β above which the frequency response function always has a maximum
$\mathbf{F}$	linear matrix pencil with time step in state-space, $\mathbf{F} = \mathbf{B} - \frac{h}{2}\mathbf{A}$
$\mathbf{F}_1, \mathbf{F}_2$	linear matrix pencils with time step in the configuration space
$\mathbf{F}_j$	regular linear matrix pencil for the jth mode
$\mathbf{f}'(t)$	forcing function in the modal coordinates
$\mathbf{f}(t)$	forcing function
$\mathbf{G}(s)$	non-viscous damping function matrix in the Laplace domain
$\mathbf{G}_0$	the matrix $\mathbf{G}(s)$ at $s \to 0$
$\mathbf{G}_\infty$	the matrix $\mathbf{G}(s)$ at $s \to \infty$
$\mathbf{H}(s)$	frequency response function matrix
$\hat{\mathbf{u}}_j$	real part of $\hat{\mathbf{z}}_j$
$\hat{\mathbf{v}}_j$	imaginary part of $\hat{\mathbf{z}}_j$
$\hat{\mathbf{z}}_j$	jth measured complex mode
$\mathbf{I}$	identity matrix
$\mathbf{K}$	stiffness matrix
$\mathbf{M}$	mass matrix
$\mathbf{O}_{ij}$	a null matrix of dimension $i \times j$
$\mathbf{\Omega}$	diagonal matrix containing the natural frequencies
$\mathbf{p}$	parameter vector (in [ADH 14b], Chapter 1)
$\mathbf{P}_j$	a diagonal matrix for the expansion of jth complex mode
$\boldsymbol{\phi}_j$	eigenvectors in the state-space
$\boldsymbol{\psi}_j$	left eigenvectors in the state-space
$\mathbf{q}(t)$	displacement response in the time domain
$\mathbf{q}_0$	vector of initial displacements
$\mathbf{Q}_j$	an off-diagonal matrix for the expansion of jth complex mode
$\mathbf{r}(t)$	forcing function in the state-space
$\mathbf{R}_k$	rectangular transformation matrices (in Chapter 4)

$\mathbf{R}_k$	residue matrix associated with pole s_k
$\mathbf{S}$	a diagonal matrix containing eigenvalues s_j
$\mathbf{T}$	a temporary matrix, $\mathbf{T} = \sqrt{\mathbf{M}^{-1}\mathbf{K}}$ ([ADH 14b], Chapter 2)
$\mathbf{T}_k$	Moore-Penrose generalized inverse of $\mathbf{R}_k$
$\mathbf{T}_k$	a transformation matrix for the optimal normalization of the kth complex mode
$\boldsymbol{\Theta}$	normalization matrix
$\mathbf{u}(t)$	the state-vector of the first-order system
$\mathbf{u}_0$	vector of initial conditions in the state-space
$\mathbf{u}_j$	displacement at the time step j
$\mathbf{v}(t)$	velocity vector $\mathbf{v}(t) = \dot{\mathbf{q}}(t)$
$\mathbf{v}_j$	a vector of the j-modal derivative in Nelson's methods (in [ADH 14b], Chapter 1)
$\mathbf{v}_j$	velocity at the time step j
ε_j	error vector associated with jth complex mode
$\varphi_k(s)$	eigenvectors of the dynamic stiffness matrix
$\mathbf{W}$	coefficient matrix associated with the constants in Caughey series
$\mathbf{X}$	matrix containing the undamped normal modes $\mathbf{x}_j$
$\mathbf{x}_j$	undamped eigenvectors, $j = 1, 2, \cdots, N$
$\mathbf{y}(t)$	modal coordinate vector (in Chapter 2)
$\mathbf{y}_k(t)$	vector of internal variables, $k = 1, 2, \cdots, n$
$\mathbf{y}_{k,j}$	internal variable $\mathbf{y}_k$ at the time step j
$\mathbf{Z}$	matrix containing the complex eigenvectors $\mathbf{z}_j$
$\mathbf{z}_j$	complex eigenvectors in the configuration space
ζ	diagonal matrix containing the modal damping factors
ζ_v	a vector containing the modal damping factors
χ	merit function of a complex mode for optimal normalization
χ_R, χ_I	merit functions for real and imaginary parts of a complex mode
Δ	perturbation in the real eigenvalues
δ	perturbation in complex conjugate eigenvalues
$\dot{q}_0$	initial velocity (SDOF systems)
ϵ	small error
η	ratio between the real and imaginary parts of a complex mode
$\mathcal{F}$	dissipation function
γ	non-dimensional characteristic time constant

γ_j	complex mode normalization constant
γ_R, γ_I	weights for the normalization of the real and imaginary parts of a complex mode
$\hat{\theta}(\omega)$	frequency-dependent estimated characteristic time constant
$\hat{\theta}_j$	estimated characteristic time constant for jth mode
$\hat{t}$	an arbitrary independent time variable
κ_j	real part of the complex optimal normalization constant for the jth mode
λ	complex eigenvalue corresponding to the oscillating mode (in Chapter 3)
λ_j	complex frequencies MDOF systems
$\mathcal{M}_r$	moment of the damping function
$\mathcal{D}$	dissipation energy
$\mathcal{G}(t)$	non-viscous damping kernel function in an SDOF system
$\mathcal{T}$	kinetic energy
$\mathcal{U}$	potential energy
μ	relaxation parameter
μ_k	relaxation parameters associated with coefficient matrix $\mathbf{C}_k$ in the exponential non-viscous damping model
ν	real eigenvalue corresponding to the overdamped mode
$\nu_k(s)$	eigenvalues of the dynamic stiffness matrix
ω	driving frequency
ω_d	damped natural frequency of SDOF systems
ω_j	undamped natural frequencies of MDOF systems, $j = 1, 2, \cdots, N$
ω_n	undamped natural frequency of SDOF systems
$\omega_{\max}$	frequency corresponding to the maximum amplitude of the response function
ω_{d_j}	damped natural frequency of MDOF systems
ρ	mass density
i	unit imaginary number, $\mathrm{i} = \sqrt{-1}$
τ	dummy time variable
θ_j	characteristic time constant for jth non-viscous model
$\tilde{\mathbf{f}}(t)$	forcing function in the modal domain
$\tilde{\omega}$	normalized frequency ω/ω_n
ς_j	imaginary part of the complex optimal normalization constant for the jth mode

ϑ	phase angle of the response of SDOF systems
ϑ_j	phase angle of the modal response
ψ	a trail complex eigenvector (in Chapter 2)
$\widehat{\mathbf{A}}$	asymmetric state-space system matrix
$\widehat{\mathbf{C}}$	fitted damping matrix
$\widehat{f}(\omega_j)$	fitted generalized proportional damping function (in [ADH 14b], Chapter 2)
$\widetilde{\mathbf{A}}$	state-space system matrix for rank-deficient systems
$\widetilde{\mathbf{B}}$	state-space system matrix for rank-deficient systems
$\widetilde{\mathbf{i}}_r$	integration of the forcing function in the state-space for rank-deficient systems
$\widetilde{\mathbf{i}}_r$	integration of the forcing function in the state-space
$\widetilde{\mathbf{\Phi}}$	matrix containing the state-space eigenvectors for rank-deficient systems
$\widetilde{\phi}_j$	eigenvectors in the state-space for rank-deficient systems
$\widetilde{\mathbf{r}}(t)$	forcing function in the state-space for rank-deficient systems
$\widetilde{\mathbf{u}}(t)$	the state vector for rank-deficient systems
$\widetilde{\mathbf{y}}_k(t)$	vector of internal variables for rank-deficient systems, $k = 1, 2, \cdots, n$
$\widetilde{\mathbf{y}}_{k,j}$	internal variable $\mathbf{y}_k$ at the time step j for rank-deficient systems
$\widetilde{\boldsymbol{y}}_{k_j}$	jth eigenvector corresponding to the kth the internal variable for rank-deficient systems
ξ	a function of ζ defined in equation [3.132]
ζ	viscous damping factor
ζ_c	critical value of ζ for oscillatory motion, $\zeta_c = \frac{4}{3\sqrt{3}}$
ζ_j	modal damping factors
ζ_L	lower critical damping factor
ζ_n	equivalent viscous damping factor
ζ_U	upper critical damping factor
ζ_{mL}	the value of ζ below which the frequency response function always has a maximum
a_k, b_k	non-viscous damping parameters in the exponential model
B	response amplitude of SDOF systems
B_j	modal response amplitude
c	viscous damping constant of an SDOF system
c_k	coefficients of exponential damping in an SDOF system
c_{cr}	critical damping factor

d_j	a constant of the j-modal derivative in Nelson's methods
E	Young's modulus
$f(t)$	forcing function (SDOF systems)
$f_d(t)$	non-viscous damping force
$G(\mathrm{i}\omega)$	non-dimensional frequency response function
$G(s)$	non-viscous damping kernel function in the Laplace domain (SDOF systems)
$g_{(i)}$	scalar damping functions, $i = 1, 2, \cdots$
h	constant time step
$h(t)$	impulse response function of SDOF systems
$h(t)$	impulse response function
I_k	non-proportionally indices, $k1 = 1, 2, 3, 4$
k	spring stiffness of an SDOF system
L	length of the rod
l_e	length of an element
m	dimension of the state-space for non-viscously damped MDOF systems
m	mass of an SDOF system
N	number of degrees of freedom
n	number of exponential kernels
n_d	number of divisions in the time axis
p	any element in the parameter vector **p** (in [ADH 14b], Chapter 1)
$q(t)$	displacement in the time domain
q_0	initial displacement (SDOF systems)
Q_{nc_k}	non-conservative forces
$R(\mathbf{x})$	Rayleigh quotient for a trail vector **x**
R_1, R_2, R_3	three new Rayleigh quotients
r_j	normalized eigenvalues of non-viscously damped SDOF systems (in Chapter 3)
r_k	rank of $\mathbf{C}_k$ matrices
s	Laplace domain parameter
s_j	eigenvalues of dynamic systems
t	time
T_n	natural time period of an undamped SDOF system
T_{min}	minimum time period for the system
$varrho_j$	complex optimal normalization constant for the jth mode

x	normalized frequency-squared, $x = \omega^2/\omega_n^2$ (in Chapter 3)
y_j	modal coordinates (in Chapter 2)
$\bar{f}(s)$	forcing function in the Laplace domain
$\bar{q}(s)$	displacement in the Laplace domain
$\hat{\mathbf{U}}$	matrix containing $\hat{\mathbf{u}}_j$
$\hat{\mathbf{V}}$	matrix containing $\hat{\mathbf{v}}_j$
$\mathbf{\Phi}$	matrix containing the eigenvectors $\boldsymbol{\phi}_j$
$\dot{\mathbf{q}}_0$	vector of initial velocities
$\mathcal{F}_i(\bullet, \bullet)$	non-viscous proportional damping functions (of a matrix)
$\boldsymbol{Y}_k$	a matrix of internal eigenvectors
$\boldsymbol{y}_{kj}$	jth eigenvector corresponding to the kth the internal variable
PSD	power spectral density
$\mathbf{0}$	a vector of zeros
$\mathcal{L}$	Lagrangian (in Chapter 3)
$\delta(t)$	Dirac-delta function
δ_{jk}	Kroneker-delta function
$\Gamma(\bullet)$	gamma function
γ	Lagrange multiplier (in Chapter 3)
$(\bullet)^*$	complex conjugate of $(\bullet)$
$(\bullet)^T$	matrix transpose
$(\bullet)^{-1}$	matrix inverse
$(\bullet)^{-T}$	matrix inverse transpose
$(\bullet)^H$	Hermitian transpose of $(\bullet)$
$(\bullet)_e$	elastic modes
$(\bullet)_{nv}$	non-viscous modes
$(\dot{\bullet})$	derivative with respect to time
$\mathbb{C}$	space of complex numbers
$\mathbb{R}$	space of real numbers
$\perp$	orthogonal to
$\mathcal{L}(\bullet)$	Laplace transform operator
$\mathcal{L}^{-1}(\bullet)$	inverse Laplace transform operator
$\det(\bullet)$	determinant of $(\bullet)$
$\text{diag}\,[\bullet]$	a diagonal matrix
$\forall$	for all
$\Im(\bullet)$	imaginary part of $(\bullet)$

$\in$	belongs to
$\notin$	does not belong to
$\otimes$	Kronecker product
$\overline{(\bullet)}$	Laplace transform of $(\bullet)$
$\Re(\bullet)$	real part of $(\bullet)$
vec	vector operation of a matrix
$O(\bullet)$	in the order of
ADF	anelastic displacement field model
$\mathrm{adj}(\bullet)$	adjoint matrix of $(\bullet)$
GHM	Golla, Hughes and McTavish model
MDOF	multiple-degree-of-freedom
SDOF	single-degree-of-freedom

Chapter 1

Introduction to Damping Models and Analysis Methods

It is true that the grasping of truth is not possible without empirical basis. However, the deeper we penetrate and the more extensive and embracing our theories become, the less empirical knowledge is needed to determine those theories.

Albert Einstein, December 1952

Problems involving vibration occur in many areas of mechanical, civil and aerospace engineering: wave loading of offshore platforms, cabin noise in aircrafts, earthquake and wind loading of cable-stayed bridges and high-rise buildings, performance of machine tools – to pick only few random examples. Quite often vibration is not desirable and the interest lies in reducing it by dissipation of vibration energy or *damping*. Characterization of damping forces in a vibrating structure has long been an active area of research in structural dynamics. Since the publication of Lord Rayleigh's classic monograph *Theory of Sound* (1877), a large body of literature can now be found on damping. Although the topic of damping is an age old problem, the demands of modern engineering have led to a steady increase of interest in recent years. Studies of damping have a major role in vibration isolation in automobiles under random loading due to surface irregularities and buildings subjected to earthquake loadings. The developments in the fields of robotics and active structures have provided impetus toward developing procedures for dealing with general dissipative forces in the context of structural dynamics. Beside these, in the last few decades, the sophistication of modern design methods together with the development of improved composite structural materials instilled a trend toward lighter structures. At the same time, there is also a constant demand for larger structures, capable of carrying more loads at higher speeds with minimum noise and vibration level as the safety/workability and environmental criteria become more stringent. Examples include very large wind turbines, which are being used increasingly for superior energy generation. Unfortunately, these two demands are conflicting and the problem cannot be solved without proper understanding of energy dissipation or damping behavior. Recent advances in vibration energy harvesting [ERT 11, ELV 13] demand further understanding of damping [LES 04, ALI 13], as it is crucial for the quantification of harvested energy. It is the aim of this book to provide fundamental techniques for the analysis and identification of damped structural systems.

In spite of a large amount of research, understanding of damping mechanisms is quite basic compared to the other aspects of modeling. A major reason for this is that, in contrast to inertia and stiffness forces, it is not, in general, clear which *state variables* are relevant to determining the damping forces. Moreover, it seems that in a realistic situation it is often the structural joints [SEG 06] which are more responsible for energy dissipation than the (solid) material. There have been detailed studies on the material damping (see [BER 73]) and also on energy dissipation mechanisms in the joints [EAR 66, BEA 77]. But here, difficulty lies in representing all these tiny mechanisms in different parts of the structure in a unified manner. Even in many cases, these mechanisms turn out to be locally nonlinear, requiring an equivalent linearization technique for a global analysis [BAN 83]. A well-known method for getting rid of all these problems is to use the so-called "viscous damping". This approach was first introduced by Lord Rayleigh [RAY 77] via his famous "dissipation function", a quadratic expression for the energy dissipation rate with a symmetric matrix of coefficients, the "damping matrix". A further idealization, also pointed out by Rayleigh, is to assume the damping matrix to be a linear combination of the mass and stiffness matrices. Since its introduction, this model has been used extensively and is now usually known as "Rayleigh damping", "proportional damping" or "classical damping". With such a damping model, the *modal analysis* procedure, originally developed for undamped systems, can be used to analyze damped systems in a very similar manner.

In this chapter, we review some existing works on damping and set the scene for the book. This book focuses on the mathematical analysis and identification of damped linear dynamic systems. We also look mainly at discrete or discretized continuous systems. This can be done by employing conventional finite element approximation to the original boundary value problem. This aspect is not discussed here as there are many excellent books which the readers can refer to [BAT 76, DAW 84, ZIE 91, BAT 95, FRI 95b, PET 98, HUG 00, COO 01]. Therefore, for the purpose of this book, we consider that the mass and stiffness matrices are available so that we mainly focus on the damping aspects.

Different mathematical models of damping used in literature are discussed in section 1.1. Damping models used for single-degree-of-freedom (SDOF), multiple-degrees-of-freedom (MDOF) and continuous systems have been included. The concepts of viscous and non-viscous damping are introduced. A brief review of modal analysis method for viscously damped systems is given in section 1.2. The state-space method and approximate methods based on the configuration space are reviewed. Analysis methods for non-viscously damped systems are discussed in section 1.3. State-space-based methods, time-domain-based methods and approximate methods in the configuration space are discussed. After the review of different analysis methods, we move to review various damping identification methods.

The methods for identification of viscous damping is discussed in section 1.4. Both SDOF and MDOF systems are considered. In section 1.5, the identification methods for non-viscous damping model in linear dynamic systems have been reviewed. For successful modeling and model updating of a dynamic system, it is necessary to know how much the eigenvalues and eigenvectors are sensitive to the parameters [MOT 93, FRI 95b, FRI 01]. Therefore, different methods for computing parametric sensitivity of eigenvalues and eigenvectors are reviewed in section 1.6. Sensitivity of undraped, viscously damped and non-viscously damped systems is discussed. Based on the review of existing works, the motivation behind this book is explained in section 1.7. Finally, in section 1.8, the scope of the book is outlined. Here, a summary of the topics that are covered in the following chapters is given.

1.1. Models of damping

Damping is the dissipation of energy from a vibrating structure. In this context, the term "dissipate" is used to mean the transformation of energy from one form into another and, therefore, a removal of energy from the vibrating system. The type of energy into which the mechanical energy is transformed is dependent on the system and the physical mechanisms that cause the dissipation. For most vibrating systems, a significant part of the energy is converted into heat.

The specific ways in which energy is dissipated in vibration are dependent on the physical mechanisms active in the structure. These physical mechanisms are complicated physical processes that are not totally understood. The types of damping that are present in the structure will depend on which mechanisms predominate in the given situation. Thus, any mathematical representation of the physical damping mechanisms in the equation of motion of a vibrating system will have to be a generalization and approximation of the true physical situation. As Scanlan [SCA 70] has observed, any mathematical damping model is really only a crutch that does not give a detailed explanation of the underlying mechanisms. The majority of vibration books, for example [MEI 67, MEI 80, PAZ 80, NEW 89, BAT 95, MEI 97, PET 98,

GÉR 97, INM 03, FRI 10b], consider the classical viscous damping model. However, there are some excellent books and papers which specifically focus on vibration damping in engineering structures. The books by Bland [BLA 60] and Lazan [LAZ 68] give a detailed account of earlier works on viscoelastic damping and damped systems. The book by Nashif *et al.* [NAS 85] covers various material damping models and their applications in the design and analysis of dynamic systems. A valuable reference on dynamics analysis of damped structures is the book by Sun and Lu [SUN 95]. The book by Beards [BEA 96] takes a pedagogical approach toward structural vibration of damped systems. The handbook by Jones [JON 01] focuses on viscoelastic damping and analysis of structures with such damping models. The important role of damping in the context of earthquake engineering has been analyzed in the book by Liang *et al.* [LIA 11]. The recent book by Veselic [VES 11] focuses on mathematical aspects of dynamics of MDOF damped systems. The paper by Gaul [GAU 99] gives a comprehensive overview of viscoelastically damped systems. The review paper by Mead [MEA 02] gives an overview of damping modeling structural dynamics. The two linked review papers by Vasques *et al.* [VAS 10a, VAS 10b] and the article by Vasques and Cardoso [VAS 11] discuss both mathematical aspects and experimental identification of linear dynamic systems with viscoelastic damping.

For our mathematical convenience, we divide the elements that dissipate energy into three classes: (1) damping in SDOF systems, (2) damping in continuous systems and (3) damping in MDOF systems. Elements such as dampers of a vehicle-suspension fall into the first class. Dissipation within a solid body, on the other hand, falls into the second class and demands a representation that accounts for both its intrinsic properties and its spatial distribution. Damping models for MDOF systems can be obtained by discretization of the equation of motion. There have been attempts to mathematically describe the damping in SDOF, continuous and MDOF systems.

1.1.1. *Single-degree-of-freedom systems*

Free oscillation of an undamped SDOF system never dies out and the simplest approach to introduce dissipation is to incorporate an ideal viscous dashpot in the model. The damping force (F_d) is assumed to be proportional to the instantaneous velocity, that is

$$F_d(t) = c\,\dot{q}(t) \qquad [1.1]$$

and the coefficient of proportionality c is known as the dashpot constant or viscous damping constant. The loss factor, which is the energy dissipation per radian to the peak potential energy in the cycle, is widely accepted as a basic measure of the damping. For an SDOF system, this loss factor can be given by

$$\eta = \frac{c|\omega|}{k} \qquad [1.2]$$

where k is the stiffness. The expression similar to this equation has been discussed in Ungar and Kerwin [UNG 62] in the context of viscoelastic systems. Equation [1.2] shows a linear dependence of the loss factor on the driving frequency. This dependence has been discussed by Crandall [CRA 70] where it has been pointed out that the frequency dependence, observed in practice, is usually not of this form. In such cases, we often resort to an equivalent ideal dashpot. Theoretical objections to the approximately constant value of damping over a range of frequency, as observed in aeroelasticity problems, have been raised by Naylor [NAY 70]. On the lines of equation [1.2], we are tempted to define the frequency-dependent dashpot as

$$c(\omega) = \frac{k\eta(\omega)}{|\omega|}. \qquad [1.3]$$

This representation, however, has some serious physical limitations. In references [CRA 70, CRA 91, NEW 89, SCA 70], it has been pointed out that such a representation violates causality, a principle which asserts that the states of a system at a given point of time can be affected only by the events in the past and not by those of the future.

Now for the SDOF system, the frequency-domain description of the equation of motion can be given by

$$\left[-m\omega^2 + \mathrm{i}\omega c(\omega) + k\right] \bar{q}(\mathrm{i}\omega) = \bar{f}(\mathrm{i}\omega) \qquad [1.4]$$

where $\bar{q}(\mathrm{i}\omega)$ and $\bar{f}(\mathrm{i}\omega)$ are the response and excitation, respectively, represented in the frequency domain. Note that the dashpot is now allowed to have frequency dependence. Inserting equation [1.3] into [1.4], we obtain

$$\left[-m\omega^2 + k\left\{1 + \mathrm{i}\eta(\omega)\mathrm{sgn}(\omega)\right\}\right] \bar{q}(\mathrm{i}\omega) = \bar{f}(\mathrm{i}\omega) \qquad [1.5]$$

where $\mathrm{sgn}(\bullet)$ represents the sign function. The "time-domain" representations of equations [1.4] and [1.5] are often taken as

$$m\ddot{q} + c(\omega)\dot{q} + kq = f \qquad [1.6]$$

and

$$m\ddot{q} + kq\left\{1 + \mathrm{i}\eta(\omega)\mathrm{sgn}(\omega)\right\} = f \qquad [1.7]$$

respectively. It has been pointed out by Crandall [CRA 70] that these are not the correct Fourier inverses of equations [1.4] and [1.5]. The reason is that the inertia, the stiffness and the forcing function are inverted properly, while the damping terms in equations [1.6] and [1.7] are obtained by mixing the frequency-domain and time-domain operations. Crandall [CRA 70] calls [1.6] and [1.7] the "non-equations" in the time domain. It has been pointed out by Newland [NEW 89] that only certain forms of frequency dependence for $\eta(\omega)$ are allowed in order to satisfy causality. Crandall [CRA 70] has shown that the impulse response function for the ideal hysteretic dashpot (η independent of frequency) is given by

$$h(t) = \frac{1}{\pi k \eta_0} . \frac{1}{t}, \quad -\infty < t < \infty. \qquad [1.8]$$

This response function is clearly non-causal because it states that the system responds before the excitation (or the cause) takes place. This non-physical behavior of the hysteretic damping model is a flaw, and further attempts have been made to address this problem. Bishop and Price [BIS 86] introduced the band-limited hysteretic damper and suggested that it might satisfy the causality requirement. However, Crandall [CRA 91] has further shown that the band-limited hysteretic dashpot is also non-causal. In view of this discussion, it can be said that most of the hysteretic damping models fail to satisfy the casualty condition. Based on the analyticity of the transfer function, Makris [MAK 99] has shown that for causal hysteretic damping the real and imaginary parts of the dynamic stiffness matrix must form a Hilbert transform pair. The Hilbert transform relation is also known as the Kramers–Kronig result. He has shown that the causal hysteretic damping model is the limiting case of a linear viscoelastic model with nearly frequency-independent dissipation that was proposed by Biot [BIO 58]. It was also shown that there is a continuous transition from the linear viscoelastic model to the ideally hysteretic damping model. More recently, Chen and Zhang [CHE 08] showed that ideal linear hysteretic damper possesses a non-causal impulse response precursor.

The physical mechanisms of damping, including various types of external friction, fluid viscosity and internal material friction, have been studied rather extensively in some detail and are complicated physical phenomena. However, a certain simplified mathematical formulation of damping forces and energy dissipation can be associated with a class of physical phenomena. Coulomb damping, for example, is used to represent dry friction present in sliding surfaces, such as structural joints. For this kind of damping, the force resisting the motion is assumed to be proportional to the normal force between the sliding surfaces and independent of the velocity except for the sign. The damping force is thus

$$F_d = \frac{\dot{q}}{|\dot{q}|} F_r = \text{sgn}(\dot{q}) F_r \qquad [1.9]$$

where F_r is the frictional force. In the context of finding an equivalent viscous damping, Bandstra [BAN 83] has reported several mathematical models of physical damping mechanisms in SDOF systems. For example, velocity-squared damping, which is present when a mass vibrates in a fluid or when fluid is forced rapidly through an orifice. The damping force in this case is

$$F_d = \text{sgn}(\dot{q}) a \dot{q}^2; \quad \text{or, more generally} \quad F_d = c\dot{q}|\dot{q}|^{n-1} \qquad [1.10]$$

where c is the damping proportionality constant. Viscous damping is a special case of this type of damping. If the fluid flow is relatively slow, i.e. laminar, then by letting $n = 1$ the above equation reduces to the case of viscous damping [1.1].

In the context of viscoelastically damped SDOF systems, there are several studies which analyze the dynamics in detail. Free and forced vibration of viscoelastic systems were considered in [MUR 98b, MUR 98a]. Muller [MUL 05] and Adhikari [ADH 05] considered the conditions of oscillatory motion for a viscoelastically damped SDOF system. Sieber *et al.* [SIE 08] considered exponential non-viscous damping with weak nonlinearities in a Duffing oscillator. Equation of motion of such a system can be given by

$$m\frac{\mathrm{d}^2 q}{\mathrm{d}\hat{t}^2} + c\int_{\hat{\tau}=0}^{\hat{\tau}=\hat{t}} \mu \mathrm{e}^{-\mu(\hat{t}-\hat{\tau})} \frac{\mathrm{d}q}{\mathrm{d}\hat{\tau}} \mathrm{d}\hat{\tau} + kq + \alpha k q^3 = A\cos(\Omega \hat{t}). \qquad [1.11]$$

Both hardening and softening type of nonlinearities were considered and the stability of the system was discussed. In [ADH 08], the dynamic response characteristics of a non-viscously damped oscillator was discussed in detail. Genta and Amati [GEN 10] considered dynamics of non-viscously damped SDOF system and proposed a general state-space approach. In [ADH 09a], some approximate methods (non-state-space approach) for the calculation of eigenvalues of non-viscously damped SDOF system were proposed. Palmeri and Muscolino [PAL 11b] proposed a Laguerre's polynomial approximation (LPA) technique for time-domain analysis of an oscillator with the generalized Maxwell's model. Some of the techniques can be extended to continuous and MDOF systems as discussed next.

1.1.2. *Continuous systems*

Construction of damping models becomes more difficult for continuous systems. Inman [INM 89] applied the GHM (Golla, Hughes and McTavish) approach to simple beams and used the separation of variables approach in conjunction with modal analysis. Banks and Inman [BAN 91] have considered four different damping models for a composite beam. These models of damping are:

1) *Viscous air damping:* For this model, the damping operator in the Euler–Bernoulli equation for beam vibration becomes

$$L_1 = \gamma \frac{\partial}{\partial t} \qquad [1.12]$$

where γ is the viscous damping constant.

2) *Kelvin–Voigt damping:* For this model, the damping operator becomes

$$L_1 = c_d I \frac{\partial^5}{\partial x^4 \partial t} \qquad [1.13]$$

where I is the moment of inertia and c_d is the strain-rate dependent damping coefficient. A similar damping model was also used in [MAN 98, ADH 99c] in the context of randomly parametered Euler–Bernoulli beams.

3) *Time hysteresis damping:* For this model, the damping operator is assumed as

$$L_1 = \int_{-\infty}^{t} g(\tau) q_{xx}(x, t+\tau) \mathrm{d}\tau \quad \text{where } g(\tau) = \frac{\alpha}{\sqrt{-\tau}} \exp(\beta \tau) \qquad [1.14]$$

where α and β are constants. Later, this model will be discussed in detail.

4) *Spatial hysteresis damping:*

$$L_1 = \frac{\partial}{\partial x} \left[\int_0^L h(x, \xi) \{ q_{xx}(x,t) - q_{xt}(\xi, t) \} \mathrm{d}\xi \right] \qquad [1.15]$$

The kernel function $h(x, \xi)$ is defined as

$$h(x, \xi) = \frac{a}{b\sqrt{\pi}} \exp[-(x-\xi)^2 / 2b^2] \qquad [1.16]$$

where b is some constant.

They observed that the spatial hysteresis model combined with a viscous air damping model results in the best quantitative agreement with the experimental time histories. Again, in the context of Euler–Bernoulli beams, Bandstra [BAN 83] has considered two damping models where the damping term is assumed to be of the forms $\{\text{sgn}\, q_t(x,t)\}\, b_1 q^2(x,t)$ and $\{\text{sgn}\, q_t(x,t)\}\, b_2 |q(x,t)|$.

Lesieutre [LES 92] considered the dynamics of uniaxial rods with frequency-dependent material properties. Lei *et al.* [LEI 06] proposed a Galerkin method for dynamics of beam with distributed non-viscous damping. Friswell *et al.* [FRI 07a, FRI 07b] considered dynamics of Euler–Bernoulli beams with non-local and non-viscous damping. They considered the following integro-partial-differential equation as the equation of motion for the beam

$$\frac{\partial^2}{\partial x^2}\left(EI(x)\frac{\partial^2 q(x,t)}{\partial x^2}\right) + \rho A(x)\frac{\partial^2 q(x,t)}{\partial t^2} + Q_N(x,t) = f\,(x,t) \qquad [1.17]$$

with the damping force $Q_N\,(x,t)$ given by

$$Q_N\,(x,t) = \frac{\partial^2}{\partial x^2}\left(\int_0^L \int_{-\infty}^t C\,(x,\xi,t-\tau)\,\frac{\partial^2 \dot{q}(\xi,\tau)}{\partial \xi^2} d\tau d\xi\right) \qquad [1.18]$$

Dynamic analysis of beams with general non-local and non-viscous damping [1.18] has been considered by several authors [ADH 07b, DI 13, FAI 13, GON 12, DI 11, CHE 11, TSA 09, CHI 09, PAN 13, POT 13, ABU 12]. Yuksel and Dalli [YUK 05] considered longitudinally vibrating elastic rods with locally and non-locally reacting viscous dampers. Cortes and Elejabarrieta [COR 06b] considered longitudinal vibration of a rod with exponential non-viscous damping model. They obtained expressions for complex natural frequencies and mode shapes. Damped vibration of spatially curved one-dimensional structures was considered by Otrin and Boltezar [OTR 07b, OTR 07a, OTR 09a]. Xue-Chuan *et al.* [XUE 08] studied axial vibration of non-local viscoelastic Kelvin bars in tension. Cortes *et al.* [COR 08] proposed a frequency-domain approach for the axial vibration problem of a uniform elastic rod with a viscoelastic end damper. They derived an analytical solution for the frequency response functions (FRFs). Calim [CAL 09] analyzed the dynamic behavior of Timoshenko beams on Pasternak-type viscoelastic foundation subjected to time-dependent loads. A Galerkin-type state-space approach for transverse vibrations of slender double-beam systems with viscoelastic inner layer was proposed by Palmeri and Adhikari [PAL 11a]. Calim and Akkurt [CAL 11] considered free vibration analysis of straight and circular Timoshenko beams on elastic foundation. Garcia-Barruetabena *et al.* [GAR 12] proposed both an analytical solution and finite element approach for axial vibration of rods with exponential non-viscous damping. A rotating Timoshenko beam with Maxwell–Weichert viscoelastic damping model is used in [SKA 12]. Lei *et al.* [LEI 13b, LEI 13a] studied free vibration of non-local Euler–Bernoulli and Timoshenko beams with non-viscous damping. A generalized one-dimensional elastoplastic model based on fractional calculus is presented in [MEN 12a]. Recently, Wang and Inman [WAN 13] considered GHM and anelastic displacement field (ADF) models of viscoelastic damping for the dynamics of constrained layer sandwich beams. A finite element approach and an experimental validation have been reported by the authors.

1.1.3. *Multiple-degrees-of-freedom systems*

The most popular approach to model damping in the context of MDOF systems is to assume viscous damping. This approach was first introduced by Lord Rayleigh [RAY 77]. By analogy with the potential energy and the kinetic energy, Rayleigh assumed the *dissipation function*, given by

$$\mathcal{F}(\mathbf{q}) = \frac{1}{2}\sum_{j=1}^{N}\sum_{k=1}^{N} C_{jk}\dot{q}_j\dot{q}_k = \frac{1}{2}\dot{\mathbf{q}}^T\mathbf{C}\dot{\mathbf{q}}. \qquad [1.19]$$

In the above expression, $\mathbf{C} \in \mathbb{R}^{N\times N}$ is a non-negative definite symmetric matrix known as the viscous damping matrix. It should be noted that not all forms of the viscous damping matrix can be handled within the scope of classical modal analysis. Based on the solution method, viscous damping matrices can be further divided into classical and non-classical dampings. Further discussions on viscous damping will follow in section 1.2.

It is important to avoid the widespread misconception that viscous damping is the *only* linear model of vibration damping in the context of MDOF systems. Any causal model which makes the energy dissipation functional non-negative is a possible candidate for a damping model. There have been several efforts to incorporate non-viscous damping models in MDOF systems. References [BAG 83, TOR 87, GAU 91, MAI 98] considered damping modeling in terms of fractional derivatives of the displacements. Following Maia *et al.* [MAI 98], the damping force using such models can be expressed by

$$\mathbf{F}_d = \sum_{j=1}^{l} \mathbf{g}_j D^{\nu_j}[\mathbf{q}(t)]. \qquad [1.20]$$

Here, $\mathbf{g}_j$ are complex constant matrices and the fractional derivative operator

$$D^{\nu_j}[\mathbf{q}(t)] = \frac{\mathrm{d}^{\nu_j}\mathbf{q}(t)}{\mathrm{d}t^{\nu_j}} = \frac{1}{\Gamma(1-\nu_j)}\frac{\mathrm{d}}{\mathrm{d}t}\int_0^t \frac{\mathbf{q}(t)}{(t-\tau)^{\nu_j}}\mathrm{d}\tau \qquad [1.21]$$

where ν_j is a fraction and $\Gamma(\bullet)$ is the gamma function. The familiar viscous damping appears as a special case when $\nu_j = 1$. We refer the readers to the review papers

[SLA 93, ROS 97, GAU 99] for further discussions on this topic. The physical justification for such models, however, may not always be very clear for engineering problems.

Possibly, the most general way to model damping within the linear range is to consider non-viscous damping models which depend on the past history of motion via convolution integrals over some kernel functions. A *modified dissipation function* for such damping model can be defined as

$$\mathcal{F}(\mathbf{q}) = \frac{1}{2}\sum_{j=1}^{N}\sum_{k=1}^{N} \dot{q}_k \int_0^t \mathcal{G}_{jk}(t-\tau)\dot{q}_j(\tau)\mathrm{d}\tau = \frac{1}{2}\dot{\mathbf{q}}^T \int_0^t \boldsymbol{\mathcal{G}}(t-\tau)\dot{\mathbf{q}}(\tau)\mathrm{d}\tau.[1.22]$$

Here, $\boldsymbol{\mathcal{G}}(t) \in \mathbb{R}^{N\times N}$ is a symmetric matrix of the damping kernel functions, $\mathcal{G}_{jk}(t)$. The kernel functions, or others closely related to them, are described under many different names in the literature of different subjects: for example, retardation functions, heredity functions, after-effect functions and relaxation functions. In the special case when $\boldsymbol{\mathcal{G}}(t-\tau) = \mathbf{C}\,\delta(t-\tau)$, where $\delta(t)$ is the Dirac-delta function, equation [1.22] reduces to the case of viscous damping as in equation [1.19]. The damping model of this kind is a further generalization of the familiar viscous damping. By choosing suitable kernel functions, it can also be shown that the fractional derivative model discussed before is also a special case of this damping model. Thus, as pointed by Woodhouse [WOO 98], this damping model is the most general damping model within the scope of a linear analysis.

Golla and Hughes [GOL 85] and McTavish and Hughes [MCT 93] have effectively used damping model of the form [1.22] in the context of viscoelastic structures. The damping kernel functions are commonly defined in the frequency/Laplace domain. Conditions which $\mathbf{G}(s)$, the Laplace transform of $\boldsymbol{\mathcal{G}}(t)$, must satisfy in order to produce dissipative motion were given by Golla and Hughes [GOL 85]. The approach pioneered by Lesieutre [LES 90, LES 92, LES 95, LES 96b, LES 96a] used a first-order state-space method called the anelastic displacement fields (ADF) method. A selection of different damping models proposed in literature is summarized in Table 1.1. Adhikari and Woodhouse [ADH 03b] proposed four indices to quantify non-viscous damping when the kernel function can have any form as given in Table 1.1.

1.1.4. *Other studies*

Another major source of damping in a vibrating structure is the structural joints, see [TAN 97] for a recent review. Here, a major part of the energy loss takes place through air-pumping. The air-pumping phenomenon is associated with damping when air is entrapped in pockets in the vicinity of a vibrating surface. In these situations, the entrapped air is "squeezed out" and "sucked in" through any available hole. Dissipation of energy takes place in the process of air flow and Coulomb-friction dominates around the joints. This damping behavior has been studied by many authors in some practical situations, for example by Cremer and Heckl [CRE 73]. Earls [EAR 66] has obtained the energy dissipation in a lap joint over a cycle under different clamping pressure. Beards and Williams [BEA 77] have noted that significant damping can be obtained by suitably choosing the fastening pressure at the interfacial slip in joints.

Model number	Damping functions	Author and reference
1	$G(s) = \sum_{k=1}^{n} \frac{a_k s}{s+b_k}$	Biot [BIO 55], [BIO 58]
2	$G(s) = as \int_0^\infty \frac{\gamma(\rho)}{s+\rho} \mathrm{d}\rho$ $\gamma(\rho) = \begin{cases} \frac{1}{\beta-\alpha} & \alpha \leq \gamma \leq \beta \\ 0 & \text{otherwise} \end{cases}$	Buhariwala [BUH 82]
3	$G(s) = \frac{E_1 s^\alpha - E_0 b s^\beta}{1+bs^\beta}$ $0 < \alpha < 1, \quad 0 < \beta < 1$	Bagley and Torvik [BAG 83]
4	$sG(s) = G^\infty \left[1 + \sum_k \alpha_k \frac{s^2+2\xi_k\omega_k s}{s^2+2\xi_k\omega_k s+\omega_k^2}\right]$	Golla and Hughes [GOL 85] and McTavish and Hughes [MCT 93]
5	$G(s) = 1 + \sum_{k=1}^{n} \frac{\Delta_k s}{s+\beta_k}$	Lesieutre and Mingori [LES 90]
6	$G(s) = c\frac{1-e^{-st_0}}{st_0}$	Adhikari [ADH 98]
7	$G(s) = c\frac{1+2(st_0/\pi)^2-e^{-st_0}}{1+2(st_0/\pi)^2}$	Adhikari [ADH 98]
8	$G(s) = c\, e^{s^2/4\mu} \left[1 - \mathrm{erf}\left(\frac{s}{2\sqrt{\mu}}\right)\right]$	Adhikari and Woodhouse [ADH 01c]

Table 1.1. *Summary of damping functions in the Laplace domain*

Energy dissipation within the material is attributed to a variety of mechanisms, such as thermoelasticity, grain-boundary viscosity and point-defect relaxation, see [LAZ 59, LAZ 68, BER 73]. Such effects are in general called material damping. In an imperfect elastic material, the stress-strain curve forms a closed hysteresis loop rather than a single line upon a cyclic loading. Much effort has been made by numerous investigators to develop models of hysteretic restoring forces and techniques to identify such systems. For a recent review on this literature, we refer the readers to [CHA 98]. Most of these studies are motivated by the observed fact that the energy dissipation from materials is only a weak function of frequency and almost directly proportional to q^n. The exponent on displacement for the energy dissipation of material damping ranges from 2 to 3, for example 2.3 for mild steel [BAN 83]. In this context, another large body of literature can be found on composite materials where many researchers have evaluated a material's specific damping capacity (SDC). Baburaj and Matsukai [BAB 94] and the references therein give an account of research that has been conducted in this area.

1.2. Modal analysis of viscously damped systems

Equation of motion of a viscously damped system can be obtained from the Lagrange's equation (see, for example, [MEI 67, GÉR 97, MEI 97] for further details). Using the Rayleigh's dissipation function given by [1.19], the damped forces can be obtained as

$$Q_{\mathrm{nc}_k} = -\frac{\partial \mathcal{F}}{\partial \dot{q}_k}, \quad k = 1, \cdots, N \qquad [1.23]$$

and, consequently, the equation of motion can expressed as

$$\mathbf{M}\ddot{\mathbf{q}}(t) + \mathbf{C}\dot{\mathbf{q}}(t) + \mathbf{K}\mathbf{q}(t) = \mathbf{f}(t). \qquad [1.24]$$

The aim is to solve this equation, together with the initial conditions, by modal analysis (which is described in detail in section 2.3.1). Using the modal transformation in [2.69], premultiplying equation [1.24] by the transpose of the modal matrix $\mathbf{X}^T$ and using the mode orthogonality relationships in [2.65] and [2.66], equation of motion of a damped system in the modal coordinates may be obtained as

$$\ddot{\mathbf{y}}(t) + \mathbf{X}^T\mathbf{C}\mathbf{X}\dot{\mathbf{y}}(t) + \boldsymbol{\Omega}^2\mathbf{y}(t) = \tilde{\mathbf{f}}(t). \qquad [1.25]$$

Clearly, unless $\mathbf{X}^T\mathbf{C}\mathbf{X}$ is a diagonal matrix, no advantage can be gained by employing modal analysis because the equations of motion will still be coupled. To solve this problem, it is common to assume *proportional damping*, that is $\mathbf{C}$ is simultaneously diagonalizable with $\mathbf{M}$ and $\mathbf{K}$. Such damping models allows us to analyze damped systems in very much the same manner as undamped systems. Later, Caughey and O'Kelly [CAU 65] have derived the condition which the system matrices must satisfy so that viscously damped linear systems possess classical normal modes. Adhikari [ADH 06a] introduced the concept of generalized proportional damping by which the damping matrix can be expressed as matrix functions of mass and stiffness matrices. This can significantly help in identifying the damping matrix from measured damping factors for multiple modes [ADH 09b, PAP 12]. Several authors have used a proportional damping modeling approach in wide-ranging applications [BIL 06, SUL 13, CAR 11, SUL 10, OTR 09b, LIN 09]. Phani [PHA 03] discussed the necessary and sufficient conditions for the existence of classical normal modes in damped linear dynamic systems. Recently, Chang [CHA 13] investigated the performance of proportional damping in the context of nonlinear MDOF systems. In Chapter 2, the concept of proportional damping or classical damping will be analyzed in more detail.

Modes of proportionally damped systems preserve the simplicity of the real normal modes as in the undamped case. Unfortunately, there is no physical reason why a general system should behave like this. In fact, practical experience in modal testing shows that most real-life structures do not do so, as they possess complex modes instead of real normal modes. This implies that, in general, linear systems are non-classically damped. When the system is non-classically damped, some or all of the N differential equations in [1.25] are coupled through the $\mathbf{X}^T\mathbf{C}\mathbf{X}$ term and cannot be reduced to N second-order uncoupled equations. This coupling brings several complication in the system dynamics – the eigenvalues and the eigenvectors no longer remain real and also the eigenvectors do not satisfy the classical orthogonality relationships. The methods for solving this kind of problem follow mainly two routes: the state-space method and the configuration space method or configuration space. A brief discussion of these two approaches is taken up in the following sections.

1.2.1. *The state-space method*

The state-space method is based on transforming the N second-order coupled equations into a set of $2N$ first-order coupled equations by augmenting the displacement response vectors with the velocities of the corresponding coordinates, see [NEW 89]. Equation [1.24] can be recast as

$$\dot{\mathbf{u}}(t) = \mathbf{A}\mathbf{u}(t) + \mathbf{r}(t) \qquad [1.26]$$

where $\mathbf{A} \in \mathbb{R}^{2N\times 2N}$ is the system matrix, $\mathbf{r}(t) \in \mathbb{R}^{2N}$ is the force vector and $\mathbf{u}(t) \in \mathbb{R}^{2N}$ is the response vector in the state-space given by

$$\mathbf{A} = \begin{bmatrix} \mathbf{O}_N & \mathbf{I}_N \\ -\mathbf{M}^{-1}\mathbf{K} & -\mathbf{M}^{-1}\mathbf{C} \end{bmatrix}, \quad \mathbf{u}(t) = \begin{Bmatrix} \mathbf{q}(t) \\ \dot{\mathbf{q}}(t) \end{Bmatrix}, \quad \text{and} \quad \mathbf{r}(t) = \begin{Bmatrix} \mathbf{0} \\ -\mathbf{M}^{-1}\mathbf{f}(t). \end{Bmatrix} \qquad [1.27]$$

In the above equation, $\mathbf{O}_N$ is the $N \times N$ null matrix and $\mathbf{I}_N$ is the $N \times N$ identity matrix. The eigenvalue problem associated with the above equation is an asymmetric matrix now. Uncoupling of equations in the state-space is again possible and has been considered by many authors, for example [MEI 80, NEW 89, VEL 86]. This analysis was further generalized by Newland [NEW 87] for the case of systems involving singular matrices. In the formulation of equation [1.26], the matrix $\mathbf{A}$ is no longer symmetric, and so eigenvectors are no longer orthogonal with respect to it. In fact, in this case, instead of an orthogonality relationship, we obtain a biorthogonality relationship, after solving the adjoint eigenvalue problem. The complete procedure for uncoupling the equations now involves solving two eigenvalue problems, each of which is double the size of an eigenvalue problem in the modal space. The details of the relevant algebra can be found in [MEI 80, MEI 97]. It should be noted that these solution procedures are exact in nature. One disadvantage of such an exact method is that it requires significant numerical effort to determine the eigensolutions. The effort required is evidently intensified by the fact that the eigensolutions of a non-classically damped system are complex. From the analyst's viewpoint, another disadvantage is the lack of physical insight afforded by this method which is intrinsically numerical in nature.

Another variation of the state-space method available in the literature is through the use of "Duncan form". This approach was introduced in [FOS 58], and later several authors, for example [BÉL 77, NEL 79, VIG 86, SUA 87, SUA 89, SES 94, REN 97, ELB 09], have used this approach to solve a wide range of interesting problems. The advantage of this approach is that the system matrices in the state-space retain symmetry as in the configuration space.

1.2.2. *Methods in the configuration space*

It has been pointed out that the state-space approach toward the solution of equation of motion in the context of linear structural dynamics is not only computationally expensive but also fails to provide the physical insight which modal analysis in configuration space or configuration space offers. The eigenvalue problem associated with equation [1.24] can be represented by the λ-matrix problem [LAN 66]

$$s_j^2\mathbf{M}\mathbf{z}_j + s_j\mathbf{C}\mathbf{z}_j + \mathbf{K}\mathbf{z}_j = \mathbf{0} \qquad [1.28]$$

where $s_j \in \mathbb{C}$ is the jth latent root (eigenvalue) and $\mathbf{z}_j \in \mathbb{C}^N$ is the jth latent vector (eigenvector). The eigenvalues s_j are the roots of the characteristic polynomial

$$\det\left[s^2\mathbf{M} + s\mathbf{C} + \mathbf{K}\right] = 0. \qquad [1.29]$$

The order of the polynomial is $2N$ and the roots appear in complex conjugate pairs. Several authors have studied non-classically damped linear systems using approximate methods. In this section, we briefly review the existing methods for this kind of analysis.

1.2.2.1. *Approximate decoupling method*

Consider the equation of motion of a general viscously damped system in the modal coordinates given by [1.25]. Earlier, it was mentioned that due to the non-classical nature of damping, this set of N differential equations is coupled through the $\mathbf{C}' = \mathbf{X}^T\mathbf{C}\mathbf{X}$ term. A usual approach in this case is simply to ignore the off-diagonal terms of the modal damping matrix $\mathbf{C}'$ which couple the equation of motion. This approach is termed the "decoupling approximation". For large-scale systems, the computational effort in adopting the decoupling approximation is an order of magnitude smaller than the methods of complex modes. The solution of the decoupled equation would be close to the exact solution of the coupled equations if the non-classical damping terms are sufficiently small. Analysis of this question goes back to Rayleigh [RAY 77]. A preliminary discussion on this topic can be found in [MEI 67, MEI 97]. Thomson *et al.* [THO 74] have studied the effect of neglecting off-diagonal entries of the modal damping matrix through numerical experiments and have proposed a method for improved accuracy. Warburton and Soni [WAR 77] have suggested a criterion for such a diagonalization so that the computed response is acceptable. Using the frequency-domain approach, Hasselsman [HAS 76] proposed a criterion for determining whether the equations of motion might be considered practically decoupled if non-classical damping exists. The criterion suggested by him was to have adequate frequency separation between the natural modes.

Using matrix norms, Shahruz and Ma [SHA 88] have tried to find an optimal diagonal matrix $\mathbf{C}_d$ in place of $\mathbf{C}'$. An important conclusion emerging from their study is that if $\mathbf{C}'$ is diagonally dominant, then among all approximating diagonal matrices $\mathbf{C}_d$, the one that minimizes the error bound is simply the diagonal matrix obtained by omitting the off-diagonal elements of $\mathbf{C}'$. Using a time-domain analysis, Shahruz [SHA 90] has rigorously proved that if $\mathbf{C}_d$ is obtained from $\mathbf{C}'$ by neglecting the off-diagonal elements of $\mathbf{C}'$, then the error in the solution of the approximately decoupled system will be small as long as the off-diagonal elements of $\mathbf{C}'$ are not too large. Udwadia [UDW 09] proved that, for systems with non-repeated eigenvalues, the best approximation of a diagonal modal damping matrix is simply to consider the diagonal of the $\mathbf{C}'$ matrix. Mentrasti [MEN 12b] considered complex modal analysis for a proportionally damped structure equipped with linear non-proportionally damped viscous elements.

Ibrahimbegovic and Wilson [IBR 89] have developed a procedure for analyzing non-proportionally damped systems using a subspace with a vector basis generated from the mass and stiffness matrices. Their approach avoids the use of complex eigensolutions. An iterative approach for solving the coupled equations is developed in [UDW 90] based on updating the forcing term appropriately. Felszeghy [FEL 93] presented a method that searches for another coordinate system in the neighborhood of the normal coordinate system so that in the new coordinate system removal of coupling terms in the equations of motion produces a minimum bound on the relative error introduced in the approximate solution. Hwang and Ma [HWA 93] have shown that the error due to the decoupling approximation can be decomposed into an infinite series and can be summed exactly in the Laplace domain. They also concluded that by solving a small number of additional coupled equations in an iterative manner, the accuracy of the approximate solution can be greatly improved. Felszeghy [FEL 94] developed a formulation based on biorthonormal eigenvectors for modal analysis of non-classically damped discrete systems. The analytical procedure takes advantage of the simplification that arises when the modal analysis of the motion separated into a classical and non-classical modal vector expansion.

From the above-mentioned studies, it has often been believed that either frequency separation between the normal modes [HAS 76], often known as "Hasselsman's criteria", or some form of diagonal dominance [SHA 88], in the modal damping matrix $\mathbf{C}'$, is sufficient for neglecting modal coupling. In contrast to these widely accepted beliefs, references [PAR 92a, PAR 92b, PAR 94] have shown, using Laplace transform methods, that within the practical range of engineering applications neither the diagonal dominance of the modal damping matrix nor the frequency separation between the normal modes would be sufficient for neglecting modal coupling. They have also given examples when the effect of modal coupling may even increase following the previous criterion. Phani and Adhikari [PHA 08] proposed three Rayleigh quotients for non-proportionally damped systems based on approximate complex modes. It was shown that the stationarity can only be obtained when the modal damping matrix is diagonally dominant.

In the context of approximate decoupling, Shahruz and Srimatsya [SHA 97] considered error vectors in modal and physical coordinates, say denoted by $\mathbf{e_N}(\bullet)$ and $\mathbf{e_P}(\bullet)$, respectively. They have shown that based on the norm (denoted here by $\| (\bullet) \|$) of these error vectors, three cases may arise:

1) $\| \mathbf{e_N}(\bullet) \|$ is small (respectively, large) and $\| \mathbf{e_P}(\bullet) \|$ is small (respectively, large);

2) $\| \mathbf{e_N}(\bullet) \|$ is large but $\| \mathbf{e_P}(\bullet) \|$ is small;

3) $\| \mathbf{e_N}(\bullet) \|$ is small but $\| \mathbf{e_P}(\bullet) \|$ is large.

From this study, especially in view of case (3), it is clear that the error norms based on the modal coordinates are not reliable to use in the actual physical coordinates. However, they have given conditions when $\| \mathbf{e_N}(\bullet) \|$ will lead to a reliable estimate of $\| \mathbf{e_P}(\bullet) \|$. For a flexible structure with light damping, it was shown [GAW 97] that neglecting off-diagonal terms of the modal damping matrix in most practical cases imposes negligible errors in the system dynamics. They also concluded that the requirement of diagonal dominance of the damping matrix is not necessary in the case of small damping, which relaxes the criterion earlier given by [SHA 88].

In order to quantify the extent of non-proportionality, several authors have proposed "non-proportionality indices". References [PAR 86, NAI 86] have developed several indices based on modal phase difference, modal polygon areas, relative magnitude of coupling terms in the modal damping matrix, system response and Nyquist plot. Based on the idea related to the modal polygon area, Bhaskar [BHA 99] has proposed two more indices of non-proportionality. Another index based on driving frequency and elements of the modal damping matrix is given in [BEL 90]. Bhaskar [BHA 95] has proposed a non-proportionality index based on the error introduced by ignoring the coupling terms in the modal damping matrix. An analytical index for the quantification of non-proportionality for discrete vibrating systems was developed in [TON 92, TON 94]. It has been shown that the fundamental nature of non-proportionality lies in finer decompositions of the damping matrix. Shahruz [SHA 95] has shown that the analytical index given by [TON 94] solely based on the damping matrix may lead to erroneous results when the driving frequency lies close to a system's natural frequency. He has suggested that a suitable index for non-proportionality should include the damping matrix and natural frequencies as well as the excitation vector. Prells and Friswell [PRE 00] have shown that the (complex) modal matrix of a non-proportionally damped system depends on an orthonormal matrix, which represents the phase between different degrees-of-freedom of the system. For proportionally damped systems, this matrix becomes an identity matrix and, consequently, they have used this orthonormal matrix as an indicator of non-proportionality. Three indices to measure the damping non-proportionality were proposed in [LIU 00]. The first index measures the correlation between the real and imaginary parts of the complex modes; the second index measures the magnitude of the imaginary parts of the complex modes; and the third index quantifies the degree of modal coupling. These indices are based on the fact that the complex modal matrix can be decomposed into a product of a real and complex matrix. Adhikari [ADH 04a] proposed the optimal normalization of complex modes and suggested a mode-by-mode non-proportionality index. Koruk and Sanliturk [KOR 13] quantified mode shape complexity based on conservation of energy principle when a structure is vibrating at a specific mode during a period of vibration.

In an another line of work, some researchers aimed at diagonalizing a linear dynamic system *exactly* using real transformations even when it is non-proportionally damped. Works by Garvey *et al.* [GAR 02b, GAR 02a, GAR 04, ABU 09, PRE 09, TIS 11] proposed the structure preserving transformation and for viscously damped systems and extended the idea to more general linear dynamical systems. In a series of papers, Ma *et al.* [KAW 11, MOR 11a, MOR 11b, MA 10, MA 09, MOR 09, MOR 08b, MOR 08a, MA 04] considered the possibility of decoupling the equation of motion using real modes. They showed that it is possible to diagonalize the **M**, **C**, **K** system using a real transformation even when these matrices are general in nature (i.e. not proportionally damped). These works have the potential to rethink the concept of proportional damping in linear dynamic systems in a new light.

1.2.2.2. *Complex modal analysis*

Other than the approximate decoupling methods, another approach toward the analysis of non-proportionally damped linear systems is to use complex modes. Since the original contribution of Caughey and O'Kelly [CAU 65], many papers have been written on complex modes. Several authors, for example [MIT 90, IMR 95, LAL 95], have given reviews on this subject. Placidi *et al.* [PLA 91] have used a series expansion of complex eigenvectors into the subspace of real modes, in order to identify normal modes from complex eigensolutions. In the context of modal analysis, Liang *et al.* [LIA 92] have proposed and analyzed the question of whether the existence of complex modes is an indicator of non-proportional damping and how a mode is influenced by damping. Analyzing the errors in the use of modal coordinates, references [SES 94, IBR 95] have concluded that the complex mode shapes are not necessarily the result of high damping. The complexity of the mode shapes is the result of particular damping distributions in the system and depends on the proximity of the mode shapes. Liu and Sneckenberger [LIU 94] have developed a complex mode theory for a linear vibrating deficient system based on the assumption that it has a complete set of eigenvectors. Complex mode superposition methods have been used by [OLI 96] in the context of soil structure interaction problems. Balmès [BAL 97] has proposed a method to find normal modes and the associated non-proportional damping matrix from the complex modes. He has also shown that a set of complex modes is complete if it verifies a defined properness condition which is used to find complete approximations of identified complex modes. Garvey *et al.* [GAR 95] have given a relationship between real and imaginary parts of complex modes for general systems whose mass, stiffness and damping can be expressed by real symmetric matrices. They have also observed that the relationship becomes most simple when all roots are complex and the real parts of all the roots have the same sign. Bhaskar [BHA 99] has analyzed complex modes in detail and addressed the problem of visualizing the deformed modes shapes when the motion is not synchronous.

While the above-mentioned works concentrate on the properties of the complex modes, several authors have considered the problem of determination of complex modes in the configuration space. Cronin [CRO 76] has obtained an approximate solution for a non-classically damped system under harmonic excitation by perturbation techniques. Clough and Mojtahedi [CLO 76] considered several methods of treating generally damped systems, and concluded that the proportional damping approximation may give unreliable results for many cases. Similarly, it was shown [DUN 79] that significant errors can be incurred when dynamic analysis of a non-proportionally damped system is based on a truncated set of modes, as is commonly done in modeling continuous systems. Meirovitch and Ryland [MEI 85] have used a perturbation approach to obtain the left and right eigenvectors of damped gyroscopic systems. Chung and Lee [CHU 86] applied perturbation techniques to obtain the eigensolutions of damped systems with weakly non-classical damping. Cronin [CRO 90] has developed an efficient perturbation-based series method to solve the eigenproblem for dynamic systems having a non-proportional damping matrix. To illustrate the general applicability of this method, Peres-Da-Silva *et al.* [PER 95] have applied it to determine the eigenvalues and eigenvectors of a damped gyroscopic system. In the context of non-proportionally damped gyroscopic systems, Malone *et al.* [MAL 97] have developed a perturbation method that uses an undamped gyroscopic system as the unperturbed system. Based on a small damping assumption, Woodhouse [WOO 98] has given the expression for complex natural frequencies and mode shapes of non-proportionally damped linear discrete systems with viscous and non-viscous damping.

Adhikari [ADH 99a] derived approximate expressions of complex modes using a Neumann expansion for each mode. A general expression of the FRF when the system matrices are asymmetric was derived. This is particularly useful when it is not possible to transform an asymmetric system [INM 83, AHM 87, SHA 89, AHM 84a, ADH 00c, LIU 05], into a symmetric system. Gallina [GAL 03] discussed the effect of damping on asymmetric system. Liu and Zheng [LIU 10] proposed a synthesis method for transient response of non-proportionally damped structures. An iterative approach to obtain complex modes for non-proportionally damped systems was proposed in [ADH 11a].

1.2.2.3. *Response bounds and frequency response*

Previously, we have mainly discussed the calculation of the eigensolutions of non-classically damped systems. Here, we briefly consider the problem of obtaining dynamic response of such systems. Nicholson [NIC 87b] and Nicholson and Baojiu [NIC 96] have reviewed the literature on stable response of non-classically damped mechanical systems. Nicholson [NIC 87a] gave upper bounds for the response of non-classically damped systems under impulsive loads and step loads. Yae and Inman [YAE 87] have obtained bound on the displacement response of non-proportionally damped discrete systems in terms of physical parameters of the system and input. They have also observed that the larger the deviation from proportional damping, the less accurate their results become.

Bellos and Inman [BEL 90] have given a procedure for computing the transfer functions of a non-proportionally damped discrete system. Their method was based on Laplace transformation of the equation of motion in modal coordinates. A fairly detailed survey of the previous research is made in [BEL 90]. Yang [YAN 93] has developed an iterative procedure for calculation of the transfer functions of non-proportionally damped systems. Bhaskar [BHA 95] has analyzed the behavior of errors in calculating FRF when the off-diagonal terms of modal damping matrix are neglected. It has been shown that the exact response can be expressed by an infinite Taylor series and the approximation of ignoring the off-diagonal terms of modal damping matrix is equivalent to retaining one term of the series.

Finally, it should be noted that frequency responses of viscously damped systems with non-proportional damping can be obtained *exactly* in terms of the complex frequencies and complex modes in the configuration space, see, for example, [LAN 66] (section 7.5) and [GÉR 97, pp. 126–128]. Similar expressions are also derived in [FAW 76, VIG 86, WOO 98]. This in turn requires determination of complex modes in the configuration space. This problem will be discussed in detail later in this book.

1.3. Analysis of non-viscously damped systems

In section 1.1.3, it was pointed out that the most general way to model (non-viscous) damping within the scope of linear theory is through the use of the modified dissipation function given by equation [1.22]. Equations of motion of such non-viscously damped systems can be obtained from Lagrange's equation (see, for example, [MEI 67, GÉR 97, MEI 97]). The damping forces can be obtained as

$$Q_{\mathrm{nc}_k} = -\frac{\partial \mathcal{F}}{\partial \dot{q}_k} = -\sum_{j=1}^{N} \int_0^t \mathcal{G}_{jk}(t-\tau)\dot{q}_j(\tau)\mathrm{d}\tau, \quad k = 1, \cdots, N \qquad [1.30]$$

and, consequently, the equation of motion can be expressed as

$$\mathbf{M}\ddot{\mathbf{q}}(t) + \int_0^t \boldsymbol{\mathcal{G}}(t-\tau)\dot{\mathbf{q}}(\tau)\mathrm{d}\tau + \mathbf{K}\mathbf{q}(t) = \mathbf{f}(t). \qquad [1.31]$$

This is a set of coupled second-order integro-differential equations. The presence of the "integral" term in the equations of motion complicates the analysis. Unlike the viscously damped systems, the concept of "proportional damping" cannot easily be formulated for such systems. The question of the existence of classical normal modes in such systems, i.e. if proportional damping can occur in such systems, will be discussed in Chapter 5.

Equations similar to [1.31] occur in many different subjects. Bishop and Price [BIS 79] have considered equations of motion similar to [1.31] in the context of ship dynamics. The convolution term appeared in order to represent the fluid forces and moments. They have discussed the eigenvalue problem associated with equation [1.31] and presented an orthogonality relationship for the right and left eigenvectors. They have also given an expression for the system response due to sinusoidal excitation. Their results were not very efficient because the orthogonality relationship of the eigenvectors was not utilized due to the difficulty associated with the form of the orthogonality equation, which itself became frequency dependent. Here, we briefly discuss different numerical methods proposed for linear dynamic systems with non-viscous damping.

1.3.1. *State-space-based methods*

Equation of motion like [1.31] arises in the dynamics of *viscoelastic structures*. A method to obtain such equations using a time-domain finite element formulation was proposed in [GOL 85, MCT 93]. Their approach (the GHM method), which introduces additional dissipation coordinates corresponding to the internal dampers, increases the size of the problem. Dynamic responses of the system were obtained by using the eigensolutions of the augmented problem in the state-space. A method to obtain the time- and frequency-domain descriptions of the response by introducing additional coordinates like the GHM method was proposed in [MUR 97a, MUR 98a]. To reduce the order of the problem, references [FRI 97, PAR 99, FRI 99] have proposed a state-space approach that employs a modal truncation and uses an iterative approach to obtain the eigensolutions. A state-space model reduction approach was considered by Yiu [YIU 93, YIU 94] using sub-structuring techniques for linear systems with exponential viscoelastic damping model. Trindade *et al.* [TRI 00] considered frequency-dependent viscoelastic material models for active–passive vibration damping and compared two widely used models. Adhikari [ADH 01a] derived the conditions for existence of proportional damping for non-viscously damped systems. Under such conditions, undamped normal modes can diagonalize the dynamic system. Palmeri [PAL 03] considered a state-space approach linear dynamic systems with memory. Time-domain approaches for viscoelastically damped systems were considered in [MUS 05, PAL 04]. Trindade [TRI 06] proposed a reduced order approach for viscoelastically damped beams through projection of the dissipative modes onto the structural modes. Wagner and Adhikari [WAG 03] proposed a symmetric extended state-space approach for exponentially damped systems using internal dissipation coordinates. Adhikari and Wagner [ADH 03a] considered general asymmetric non-viscously damped systems and explained the structures of the left and right eigenvectors. Zhang and Zheng [ZHA 07a] proposed a state-space approach for general linear dynamic systems with Biot viscoelastic model. Vasques *et al.* [VAS 06] considered finite element modeling and experimental validation of a beam with frequency-dependent viscoelastic damping. Vasques *et al.* [VAS 10a] discussed computational methods for viscoelastically damped systems using reduced approaches in the extended state-space. Genta and Amati [GEN 10] considered dynamics of non-viscously damped MDOF systems in the context of rotor dynamics using a state-space approach. Friswell *et al.* [FRI 10a] used internal variables for the time-domain analysis of rotors with frequency-dependent damping. The papers by de Lima *et al.* [DE 10, DE 09] proposed a novel component mode synthesis approach for general viscoelastic linear dynamic systems. Wang and Inman [WAN 13] used a symmetric state-space formulation linear system with GHM damping.

1.3.2. *Time-domain-based methods*

While the above methods often aimed at determining the eigensolutions of the system, few authors have considered the calculation of the dynamic response in the time domain. Adhikari and Wagner [ADH 04b] proposed a direct time-domain approach for exponentially damped systems which avoids the use of dissipation coordinates. Shen and Duan [SHE 09] proposed a Guass integration approach in conjunction with state-space representation of the equation of motion for linear MDOF systems with exponential damping. An efficient time-domain approach for linear dynamic systems with fractional damping was proposed by Trinks and Ruge [TRI 02]. Cortes and Elejabarrieta [COR 07a] proposed a time-domain integration approach for linear systems with fractional derivative damping model. Later, Cortes *et al.* [COR 09] proposed a direct integration formulation for linear dynamics systems with exponential non-viscous damping model. Pan and Wang [PAN 13] proposed a frequency as well as a time-domain approach using a discrete Fourier transform (DFT) method in combination with the fast Fourier transform (FFT) for exponentially damped systems.

1.3.3. *Approximate methods in the configuration space*

Computational cost for non-viscously damped systems can be prohibitive for large dimensional problems. To address this, several authors have proposed reduced approximate methods in the configuration space. Using a first-order perturbation approach, Woodhouse [WOO 98] has obtained expressions for the eigensolutions and transfer functions of system [1.31]. His method, although it avoids the state-space representations and additional dissipation coordinates, is valid for small damping terms only. Adhikari [ADH 02b] proposed an approximate method based on Neumann expansion for the eigenvectors of linear systems with general non-viscous damping. Several mathematical properties of the eigensolutions of such systems were derived in [ADH 01b]. Cortes and Elejabarrieta [COR 06a, COR 06c] proposed a new approximate method for the complex eigensolutions of a non-viscously damped system. The key idea proposed used the solution of the undamped system and approximated the complex eigensolutions by finite increments using the eigenvector derivatives and the Rayleigh quotient. Garcia-Barruetabena [GAR 11] demonstrated that non-viscous modes only contribute to the transient response in of a linear system with exponential non-viscous damping. Some approximate methods to obtain the eigensolutions of non-viscously damped systems using the eigensolutions of the underlying undamped systems were proposed in references [ADH 09a, ADH 10, ADH 11b]. Lázaro *et al.* [LAZ 12, LAZ 13a, LAZ 13b] proposed an approach for the computation of eigensolutions and dynamic response of MDOF system with exponential damping. The motivation was to approximate the response of the original viscoelastic system using the eigensolutions of the underlying undamped or proportionally damped system. Li *et al.* [LI 13a] approximated the FRF matrix without using the dissipation modes of the linear MDOF systems with viscoelastic hereditary terms. Pawlak and Lewandowski [PAW 13] proposed a reduced computational approach for nonlinear eigenvalue problems arising in non-viscously damped systems.

1.4. Identification of viscous damping

In section 1.2, we have discussed several methods for *analysis* of viscously damped linear dynamic systems. In this section, we focus our attention on the methodologies available for identification of viscous damping parameters from experimental measurements.

1.4.1. *Single-degree-of-freedom systems*

Several methods are available for identifying the viscous damping parameters for SDOF systems for linear and nonlinear damping models, see [NAS 85]. For linear damping models, these methods can be broadly described as:

1) *Methods based on transient response of the system:* This is also known as logarithmic decrement method: if q_i and q_{i+i} are heights of two subsequent peaks, then the damping ratio ζ can be obtained as

$$\delta = \log_e \left(\frac{q_i}{q_{i+i}} \right) \approx 2\pi\zeta \qquad [1.32]$$

For applicability of this method, the decay must be exponential.

2) *Methods based on harmonic response of the system:* These methods are based on calculating the half-power points and bandwidth from the frequency response curve. It can be shown that the damping factor ζ can be related to a peak of the normalized frequency response curve by

$$|H|_{\max} \approx \frac{1}{2\zeta} \qquad [1.33]$$

3) *Methods based on energy dissipation:* Consider the force-deflection behavior of a spring-mass-damper (equivalent to a block of material) under sinusoidal loading at some particular frequency. In steady state, considering conservation of energy, energy

loss per cycle (Δq_{cyc}) can be calculated by equating it with the input power. Here, it can be shown that the damping factor ζ can be related as

$$2\zeta = \frac{\Delta q_{\text{cyc}}}{2\pi U_{\max}} \qquad [1.34]$$

where $U_{\max}$ is maximum energy of the system.

The above-mentioned methods, although developed for SDOF systems, can be used for separate modes of MDOF systems, for example a cantilever beam vibrating in the first mode. Chassiakos *et al.* [CHA 98] proposed an online parameter identification technique for an SDOF hysteretic system. Some authors [KHA 09, KHA 10, SRA 11] have proposed methods to identify damping parameters in nonlinear systems.

1.4.2. *Multiple-degrees-of-freedom systems*

For MDOF systems, most of the common methods for experimental determination of the damping parameters use the proportional damping assumption. A typical procedure can be described as follows, see [EWI 84] for details:

1) Measure a set of transfer functions $H_{ij}(\omega)$ at a set of grid points on the structure.

2) Obtain the natural frequencies ω_k by a pole-fitting method.

3) Evaluate the modal half-power bandwidth $\Delta\omega_k$ from the FRFs, then the Q-factor $Q_k = \frac{\omega_k}{\Delta\omega_k}$ and the modal damping factor $\zeta_k = \frac{1}{2Q_k}$.

4) Determine the modal amplitude factors a_k to obtain the mode shapes, $\mathbf{x}_k$.

5) Finally, reconstruct some transfer functions to verify the accuracy of the evaluated parameters.

Such a procedure does not provide reliable information about the nature or spatial distribution of the damping, though the reconstructed transfer functions may match the measured ones well.

The next stage, followed by many researchers, is to attempt to obtain the full viscous damping matrix from the experimental measurements. Pilkey and Inman [PIL 98] have given a recent survey on methods of viscous damping identification. These methods can be divided into two basic categories [FAB 88]: (1) damping identification from modal testing and analysis and (2) direct damping identification from the forced response measurements.

The modal testing and analysis method seeks to determine the modal parameters, such as natural frequencies, damping ratio and mode shapes, from the measured transfer functions, and then fit a damping matrix to these data. In one of the earliest works, Lancaster [LAN 61] gave an expression from which the damping matrix can be constructed from complex modes and frequencies. Unfortunately, this expression relies on having all the modes, which is almost impossible in practice. For this reason, several authors have proposed identification methods by considering the modal data to be incomplete or noisy. Hasselsman [HAS 72] proposed a perturbation method to identify a non-proportional viscous damping matrix from complex modes and frequencies. Béliveau [BÉL 76] proposed a method that uses eigensolutions, phase angles and damping ratios to identify the parameters of viscous damping matrix. His method utilizes a Bayesian framework based on eigensolution perturbation and a Newton–Raphson scheme. Ibrahim [IBR 83b] uses the higher order analytical modes together with the experimental set of complex modes to compute improved mass, stiffness and damping matrices. Minas and Inman [MIN 91] proposed a method for viscous damping identification in which it is assumed that the mass and stiffness are *a priori* known and modal data, obtained from experiment, allowed to be incomplete. Starek and Inman [STA 97] proposed an inverse vibration problem approach in which it is assumed that the damping matrix has an *a priori* known structure. Their method yields a positive-definite damping matrix but requires the full set of complex modes. Pilkey and Inman [PIL 97] developed an iterative method for damping matrix identification by using Lancaster's [LAN 61] algorithm. This method requires experimentally identified complex eigensolutions and the mass matrix. Alvin *et al.* [ALV 97] proposed a method in which a correction was applied to the proportionally damped matrix by means of an error minimization approach. Halevi and Kenigsbuch [HAL 99] proposed a method for updating the damping matrix by using the reference basis approach in which error and incompleteness of the measured modal data were taken into account. As an intermediate step, their method corrects the imaginary parts of the measured complex modes which are more inaccurate than their corresponding real parts.

Direct damping identification methods attempt to fit the equations of motion to the measured forced response data at several time/frequency points. Caravani and Thomson [CAR 74] proposed a least-square error minimization approach to obtain the viscous damping matrix. Their method uses a measured frequency response at a set of chosen frequency points and utilizes an iterative method to successively improve the identified parameters. Fritzen [FRI 86] used the instrumental variable method for identification of the mass, damping and stiffness matrices. It was observed that the identified values are less sensitive to noise compared to what was obtained from the least-square approach. Fabunmi *et al.* [FAB 88] presented a damping matrix identification scheme that uses forced response data in the frequency domain and assumes that the mass and stiffness matrices are known. Mottershead [MOT 90] used the inverse of the FRFs to modify the system matrices so that the modified model varies minimally from an initial finite element model. Using a different approach, Roemer and Mook [ROE 92] developed methods in the time domain for simultaneous identification of the mass, damping and stiffness matrices. It was observed that the identified damping matrix has a larger relative error than that of the mass and stiffness matrices. Chen *et al.* [CHE 96a] proposed a frequency-domain technique for identification of the system matrices in which the damping matrix was determined independently. It was shown that separate identification of the damping matrix improves the result as relative magnitude of the damping matrix is less than those of the mass and stiffness matrices. Later, Baruch [BAR 97] proposed a similar approach in which the damping matrix was identified separately from the mass and stiffness matrices.

Adhikari and Woodhouse [ADH 01c, ADH 02e] proposed a complex mode based approach for the identification of viscous damping matrix. Later, a method to identify symmetric damping matrices [ADH 02d] was proposed. Li [LI 05] used modulations of the responses to identify damping. Damping identification in pneumatic tyres was discussed by Geng *et al.* [GEN 07] using complex modes. A pattern recognition approach was used to identify damping in sucker-rod pumping system [LIU 07]. Erlicher and Argoul [ERL 07] proposed a wavelet transform based method for damping identification. Lin and Zhu [LIN 06] and Phani and Woodhouse [PHA 07, PHA 09] discussed methods for damping matrix identification from measured FRFs. Khalil *et al.* [KHA 07] proposed a proper orthogonal decomposition approach for the identification of the damping matrix along with the mass and stiffness matrices. Arora *et al.* [ARO 10, ARO 09a, ARO 09d, ARO 09b, ARO 09c] considered updating the damping matrix using analytical and experimental approaches. Prandina *et al.* [PRA 09] discussed the philosophy and performance of different damping identification approaches and investigated the role of the first-order perturbation methods and modal truncation on damping identification. A system identification algorithm based on the free vibration response of structures was proposed to identify damping in references [ROY 09, CHA 10]. Cavacece *et al.* [CAV 09] used a Moore–Penrose pseudo-inverse for the identification of damping. Pradhan and Modak [PRA 12a, PRA 12b] considered the determination of damping matrices from the FRF data. They developed an updating formulation that seeks to separate updating of the damping matrix from the updating of the stiffness and the mass matrix. Holland *et al.* [HOL 12a, HOL 12b] considered identification of damping in bladed disks. In a series of works, Liu *et al.* [LIU 08c, LIU 08d, LIU 08b, LIU 08a, LIU 09] pioneered the Lie-group estimation method for the inverse problem and damping identification in linear structural dynamics. Li and Law [LI 09] proposed a time-domain approach for damping identification using the sensitivity of the acceleration response of the analytical model along with a model updating technique. Algorithms for the mass normalization of the mode shapes in the context of experimental modal analysis were proposed by Yang *et al.* [YAN 12]. A common-base proper orthogonal decomposition approach was used by Andrianne and Dimitriadis [AND 12]. A pattern recognition approach was proposed to identify damping in a sucker-rod pumping system [LIU 07]. Cheonhong *et al.* [MIN 12] discussed a direct method for the identification of a non-proportional damping matrix using modal parameters. A two-step model updating algorithm for parameter identification of linear elastic damped structures was proposed by García-Palencia and Santini-Bell [GAR 13]. More recently, Holland and Epureanu [HOL 13] suggested a technique to identify the overall damping matrix utilizing identified (simple) damping matrices from different components. They demonstrated the approach for a mistuned blisk with varying levels of measurement noise.

1.5. Identification of non-viscous damping

Unlike viscous damping, there is little available in the literature which discusses generic methodologies for identification of non-viscous damping. Most of the methods proposed in the literature are system specific. Banks and Inman [BAN 91] have considered the problem of estimating damping parameters in a non-proportionally damped beam. They have taken four different models of damping: viscous air damping, Kelvin–Voigt damping, time hysteresis damping and spatial hysteresis damping, and used a spline inverse procedure to form a least-square fit to the experimental data. A procedure for obtaining hysteretic damping parameters in free-hanging pipe systems is given by Fang and Lyons [FAN 94]. Assuming material damping is the only source of damping, they have given a theoretical expression for the loss factor of the n-th mode. Their theory predicts higher modal damping ratios in higher modes. Maia *et al.* [MAI 97b] have emphasized the need for development of identification methodologies of general damping models and indicated several difficulties that might arise. Dalenbring [DAL 99] has proposed a method for identification of (exponentially decaying) damping functions from the measured FRFs and finite element displacement modes. A limitation of this method is that it neglects the effect of modal coupling, that is the identified non-viscous damping model is effectively proportional.

Adhikari and Woodhouse [ADH 01d] proposed a complex mode based approach for the identification of exponential non-viscous damping model. Zhang and Zheng [ZHA 07a] considered the Biot model in the context of MDOF systems and experimentally identified the model parameters. Vasques *et al.* [VAS 10b] discussed experimental identification and model validation of viscoelastically damped systems. They have compared several viscoelastic models in their study. Adhikari [ADH 02c] used a modified Lancaster's method to identify viscous and non-viscous damping matrices from the FRF matrix. Cortes and Elejabarrieta [COR 07b] characterized the viscoelastic damping properties of cantilever beams using the seismic response. Parameter identification of dynamical systems with fractional derivative damping models using methods based on inverse sensitivity analysis of damped eigensolutions and FRFs was proposed by Sivaprasad *et al.* [SIV 09]. Ding and Law [DIN 11] proposed an iterative regularization method for the identification of structural damping. Wang and Inman [WAN 13] proposed methods to identify parameters of GHM and ADF models from vibration experiments.

1.6. Parametric sensitivity of eigenvalues and eigenvectors

As seen so far, the characterization of eigenvalues and eigenvectors constitutes a central role in the design, analysis and identification of damped dynamic systems. As a result, the study of the variation of the eigenvalues and eigenvectors due to variations in the system parameters, or more precisely the sensitivity of eigensolutions, has emerged as an important area of research. For physically representative damping modeling and model updating of a dynamic system, it is necessary to know how much the eigenvalues and eigenvectors might change due to the changes in the parameters [MOT 93, FRI 95b, FRI 01]. For generally damped systems, this tantamounts to computing sensitivity of complex eigenvalues and eigenvectors in general. Sensitivity of eigenvalues and eigenvectors with respect to some system parameters may be represented by their derivatives with respect to those parameters. We briefly review some of the existing works on sensitivity of eigensolutions of undamped and damped systems.

1.6.1. *Undamped systems*

In one of the earliest works, Fox and Kapoor [FOX 68] gave exact expressions for the first derivative of eigenvalues and eigenvectors with respect to any design variable. Their results were obtained in terms of changes in the system property matrices and the eigensolutions of the structure, and have been used extensively in a wide range of application areas of structural dynamics. The expressions derived in [FOX 68] are valid for symmetric undamped systems. In many problems in dynamics, the inertia, stiffness and damping properties of the system cannot be represented by symmetric matrices or self-adjoint differential operators. These kinds of problems typically arise in the dynamics of actively controlled structures and in many general damped dynamic systems, for example moving vehicles on roads, missiles following trajectories, ship motion in sea water or the study of aircraft flutter. The asymmetry of damping and stiffness terms is often addressed in the context of gyroscopic and follower forces. Many authors [ROG 70, PLA 73, GAR 73, RUD 74] have extended Fox and Kapoor's [FOX 68] approach to determine eigensolution derivatives for more general asymmetric *conservative* systems. For these kinds of systems, Nelson [NEL 76] proposed an efficient method to calculate the first-order derivative of eigenvectors which requires only the eigenvalue and eigenvector under consideration. Murthy and Haftka [MUR 88] have written an excellent review on calculating the derivatives of eigenvalues and eigenvectors associated with general (non-Hermitian) matrices. Eigensensitivity analysis of a defective matrix with zero first-order eigenvalue derivatives was considered in [ZHA 04]. A method for modal reanalysis due to topological modifications (which changes the degree-of-freedom of the system) of structures was discussed by Zhi *et al.* [ZHI 06]. A new eigensolution reanalysis method was developed by Chen *et al.* [CHE 06] based on the Neumann series expansion and epsilon-algorithm. A general approach for incorporating the eigenvector normalization condition in the computation of eigenvector design sensitivities was proposed in [SMI 06]. Cha and Sabater [CHA 11] considered eigenvalue sensitivities of a linear structure carrying lumped attachments.

First-order derivatives are useful for practical problems as long as the perturbations of the system parameters remain "small". To consider a wide range of changes in the design parameters, the linear approximation intrinsic to the first-order derivatives may not be sufficient. Apart from large perturbations of system parameters, Brandon [BRA 84] has shown that the second-order eigensolution derivatives are not negligible compared to the first-order derivatives when the system has closely spaced natural frequencies. Second-order eigensolution derivatives are also required in design optimization to calculate the so-called "Hessian Matrix". For these reasons, there has been considerable interest in obtaining the second-order derivatives of the eigensolutions. Plaut and Huseyin [PLA 73] gave an expression for the second derivative of the eigenvalues for asymmetric systems. Rudisill [RUD 74] suggested a similar expression for the second derivative of the eigenvalues and went on to derive the second derivative of the eigenvectors. Brandon [BRA 91] derived the second derivative of the eigenvalues and eigenvectors for the case when the system matrices are linear functions of the design variables. Chen *et al.* [CHE 94a, CHE 94b] derived the second-order derivative of eigenvectors in terms of a series in the eigenvectors. Friswell [FRI 95a] proposed a method, similar to [NEL 76], to obtain the second-order derivative of the eigenvectors which employs only the eigensolutions of interest. Most of the methods discussed so far do not explicitly consider damped systems. In order to apply these results to obtain the second derivatives of the eigensolutions of general (non-proportionally) damped systems, the state-space formalism is required.

1.6.2. *Damped systems*

The work discussed so far does not explicitly consider the damping present in the system. In order to apply these results to systems with general non-proportional damping, it is required to convert the equations of motion into state-space form (see [ZEN 95], for example). Although exact in nature, the state-space methods require significant numerical effort as the size of the problem doubles. Moreover, these methods also lack some of the intuitive simplicity of the analysis based on configuration space. For these reasons, the determination of the derivatives of eigenvalues and eigenvectors in the configuration space for damped systems is very desirable. Unlike undamped systems, in damped systems the eigenvalues and eigenvectors, and consequently their derivatives, become complex in general. Some authors have considered the problem of the calculation of first-order derivatives of eigensolutions of viscously damped symmetric systems. Lee *et al.* [LEE 99a, LEE 99b] have proposed first-order formalism to determine natural frequency and mode shape sensitivities of damped systems. Adhikari [ADH 99b] derived an exact expression for the first-order derivative of complex eigenvalues and eigenvectors. The results were expressed in terms of the complex eigenvalues and eigenvectors of the second-order system and the first-order representation of the equation of motion was avoided. Later, Adhikari [ADH 00a] suggested an approximate method for calculating the first derivative of complex modes using a modal series involving only classical normal modes. An expression for the derivatives of eigenvalues and eigenvectors of non-conservative systems is presented by Choi *et al.* [CHO 04] in the configuration space. Moon *et al.* [MOO 04] proposed modified modal methods for calculating eigenpair sensitivity of an asymmetric damped system. They have used few lowest sets of modes to reduce the computational time. Guedria *et al.* [GUE 06] presented a new approach for simultaneously calculating the derivatives of the eigenvalues and their associated derivatives of the left and right eigenvectors for asymmetric damped systems. Friswell and Adhikari [FRI 00] extended Nelson's method to symmetric and asymmetric viscously damped systems. Guedria *et al.* [GUE 07] considered the computation of the second-order derivatives of the eigenvalues and eigenvectors of symmetric and asymmetric damped systems using Nelson's method. A modal approach for efficient calculation of complex eigenvector derivatives was proposed by Zhang-Ping and Jin-Wu [ZHA 07b]. Derivatives of repeated complex eigenvalues and corresponding eigenvectors of non-proportionally damped systems were considered in [HUI 07]. Chouchane *et al.* [CHO 07] proposed an algebraic approach for the calculation of eigensensitivity of asymmetric damped systems. Calculation of derivatives of multiple eigenvalues and eigenvectors of general unsymmetrical quadratic eigenvalue problems was considered in [XIE 08]. Abuazoum and Garvey [ABU 09] used structure-preserving equivalences to obtain eigenvalue and eigenvector derivatives of general second-order systems. Burchett [BUR 09] proposed a QZ-based algorithm for calculating derivatives of the system pole, transmission zero and residues. For the case when the system matrices are defective, efficient approaches for calculating the sensitivity of the eigensolutions were proposed in [XU 10, ZHA 11]. In the context of bridge deck flutter problems, Omenzetter [OME 12] considered sensitivity analysis of the eigenvalues for general dynamic systems. Some iterative methods for the derivatives of eigenvectors of quadratic eigenvalue problems arising in damped systems were suggested by Xie [XIE 12, XIE 13]. Li *et al.* [LI 12b, LI 13e, LI 13b] proposed efficient computational methods for the problem of eigensensitivity analysis of damped systems using both distinct and repeated eigenvalues.

Most of the above studies consider viscously damped systems. Adhikari [ADH 02a] proposed a modal approach for the eigensensitivity of linear systems with general non-viscous and non-proportional damping. It was shown that the eigenvector derivative can be expressed as a linear combination of other eigenvectors even when they do not satisfy any simple orthogonality relationships. Later, Adhikari and Friswell [ADH 06b] used Nelson's method to calculate the eigenvector derivatives of general non-viscously damped systems. Li *et al.* [LI 12a] proposed an algebraic method to compute the eigensolution derivatives for non-viscously damped systems. Later, they extended the formulation to asymmetric non-viscous systems [LI 13d]. More recently, Li *et al.* [LI 13c] discussed sensitivity analysis for general nonlinear eigenproblems arising in non-proportional and non-viscously damped systems.

1.7. Motivation behind this book

From the discussions so far in this chapter, it emerges that significant developments in the analysis of damped systems have taken place in the past two decades. This is fueled by the emergence of new materials such as composite and nanocomposite materials and the need to predict the system response ever more accurately in an efficient way. Based on the existing literature, it is clear that there are some pressing questions of general interest. These questions include, but are not limited to:

1) What damping model has to be used for a given structure, i.e. viscous or non-viscous, and if non-viscous then what kind of model should it be?

2) How can conventional modal analysis be extended to systems with non-viscous damping?

3) Can we physically understand the role of non-viscous damping in structural dynamics, as we do for viscous damping?

4) How is it possible to determine the damping parameters by conventional modal testing if a system is non-viscous?

5) How can we efficiently calculate the dynamic response of a large complex system in an efficient manner if the damping is non-proportional and non-viscous?

6) How sensitive is the dynamics of a system to the damping parameters? Does it matter if we get errors in some damping parameters?

7) How can we quantify damping in a system? What measures and tools can we use when the damping is, in general, non-proportional and non-viscous?

This book is motivated by these types of questions. We do not necessarily provide precise answers to these questions. The aim is to develop mathematical tools so that we can at least appreciate and investigate these types of questions for practically relevant engineering problems.

The first question is a major issue, and in the context of general vibration analysis, has been "settled" by assuming viscous damping, although it has been pointed out in the literature that, in general, it will not be the correct model. The next three questions are related to each other in the sense that for the identification of non-viscous damping parameters, a reliable method of modal analysis is also required. The fifth question on computational efficiency is becoming an issue as structural dynamic finite element models are getting larger. The consideration of parametric sensitivity of dynamic systems in the sixth question is important due to the recent drive toward model validation and uncertainty quantification of computational models. Finally, the last question regarding the quantification of damping is related to conceptual and intuitive feeling about how much damping there is in a system provided by certain parametric models.

Most of the techniques for detecting damping in a structure either consider the structure to be viscously damped or *a priori* assume some particular non-viscous model of damping and try to fit its parameters with regard to some specific structure. This *a priori* selection of damping no doubt hides the mechanisms of the system and there has not been any indication in the literature on how to find a damping model by doing conventional vibration testing. However, another relevant question in this context is whether this *a priori* selection of a damping model matters from an engineering point of view: it may be possible that a pre-assumed damping model with a "correct" set of parameters may represent the system response quite well, although the actual physical mechanism behind the damping may be different. These issues will be discussed in this book. Next, the scope of the book is discussed together with a brief overview of the chapters.

1.8. Scope of the book

Motivated by the pressing questions identified in the last section, a systematic study on the *analysis* and *identification* of damped discrete linear dynamic systems has been carried out. This book deals with analysis of linear systems with general damping models. The related book [ADH 14b] deals with identification and quantification of damping. The focus of these books is toward theoretical and computational aspects. However, some limited experimental results are given to support the theoretical developments. In section 1.1, it has been brought forward that the convolution integral model is the most general damping model for MDOF linear systems. Attention is specifically focused on this kind of general damping model. However, for comparing and establishing the relationship with current practice, viscously damped systems are also discussed. The book is divided into 10 chapters and an Appendix.

In Chapter 2, we begin by reviewing the theory of dynamics of SDOF undamped systems. The concept of resonance frequency is explained and methods to calculate dynamic response with initial conditions are discussed. Next, viscously damped SDOF systems are considered. Fundamental ideas such as damped natural frequency, damping ratio, FRF and impulse response function are discussed. A general expression of the forced dynamic response with non-zero initial condition is derived. Undamped vibration of MDOF system is discussed next. Classical concepts of eigenfrequencies, eigenmodes and mode orthogonality are introduced. Expressions of the dynamic response in the frequency domain and time domain are derived using the eigensolutions. The discussions are then extended to viscously damped MDOF systems. The idea of classical damping or proportional damping is critically reviewed and generalized proportional damping is introduced. Expressions of the dynamic response of proportionally damped systems are derived in terms of the classical normal modes and modal damping factors. General non-proportionally damped MDOF systems are discussed within the scope of the state-space method. Expressions of the dynamic response in the frequency and time domain due to general forcing and initial conditions are derived. The idea of Rayleigh quotient for damped systems is discussed. Stationarity properties for systems with proportional damping and non-proportional damping are derived. Numerical examples are provided to illustrate the theoretical developments.

Dynamics of SDOF non-viscously damped oscillators is considered in Chapter 3. It is assumed that the non-viscous damping force depends on the history of velocity via a convolution integral over an exponentially decaying kernel function. Classical qualitative dynamic properties known for viscously damped oscillators have been generalized to such non-viscously damped oscillators. The following questions of fundamental interest have been addressed: (1) under what conditions can a non-viscously damped oscillator sustain oscillatory motions? (2) How does the natural frequency of a non-viscously damped oscillator compare with that of an equivalent undamped oscillator? And (3) How does the decay rate compare with that of an equivalent viscously damped oscillator? Next, the characteristics of the FRF are discussed. The classical dynamic response properties known for viscously damped oscillators have been generalized to such non-viscously damped oscillators. The following questions of wide interest have been investigated: (1) under what conditions can the amplitude of the FRF reach a maximum value? (2) At what frequency will it occur? And (3) what will be the value of the maximum amplitude of the FRF? Introducing two non-dimensional factors, namely the viscous damping factor and the non-viscous damping factor, answers to these questions are provided. Wherever possible, attempts have been made to relate the new results with equivalent classical results for a viscously damped oscillator. It is shown that the classical concepts based on viscously damped systems can be extended to a non-viscously damped system only under certain conditions. Such conditions have been explicitly determined and illustrated numerically.

Chapter 4 extends the study in Chapter 3 to MDOF systems. Possible choices of non-viscous kernel functions are discussed. A general non-proportionally damped

MDOF system with exponential non-viscous damping is considered. The traditional state-space approach, well-known for viscously damped systems, is extended to such non-viscously damped systems using a set of internal variables. Two physically realistic cases: (1) when all the damping coefficient matrices are of full rank and (2) when the damping coefficient matrices have rank deficiency, are presented. For both cases, the equation of motion has been represented in terms of two symmetric matrices. The eigenvalues and the corresponding eigenvectors of the system are obtained by solving the state-space eigenvalue problem. It is shown that, unlike viscously damped systems, the number of eigensolutions is more than $2N$ and depends on the rank of the damping coefficient matrices. The ideas of elastic modes and non-viscous modes are introduced. The nature of these eigensolutions in the extended state-space has been explored. Some useful results relating the modal matrix in the extended state-space to the modal matrix in the original space are derived. Closed-form expressions of the dynamic response in the time domain and frequency domain due to arbitrary forcing and initial conditions are derived. It is shown that, even for general non-viscously damped systems, the response can be obtained using an approach similar to classical modal superposition method. A direct time-domain analysis of linear systems with exponentially decaying damping memory kernels is also considered. The method is based on the extended state-space representation of the equations of motion. Numerical examples are provided to illustrate the theoretical expressions.

Chapter 5 is aimed at extending classical modal analysis to treat lumped-parameter *general* non-viscously damped linear dynamic systems. This chapter extends the results of the last chapter where the *special case* of exponential damping was considered. The analytical approach adopted here is very different as the state-space approach has not been used. The nature of the eigenvalues and eigenvectors is discussed under certain simplified but physically realistic assumptions concerning the system matrices and the damping kernel functions. A numerical method based on the Neumann series expansion for the calculation of the eigenvectors is suggested. The transfer function matrix of the system is derived in terms of the eigenvectors of the second-order system. Exact closed-form expressions for the dynamic response due to general forces and initial conditions are derived. The mode-orthogonality relationships, known for undamped or viscously damped systems, have been generalized to such non-viscously damped systems. Some expressions are suggested for the normalization of the complex eigenvectors. A number of useful results which relate the system matrices with the eigensolutions are established. The approach taken in this chapter uses neither the state-space approach nor employs additional dissipation coordinates. The concept of the Rayleigh quotient for non-viscously damped systems is discussed. Three new Rayleigh quotients are proposed and their stationary properties are investigated. Suitable examples are given throughout the chapter to illustrate the derived analytical results.

Chapter 6 is devoted to reduced computational methods for damped dynamic systems. First, non-proportionally damped system with viscous model is considered. An iterative method to calculate complex modes from classical normal modes is proposed. A simple numerical algorithm is given to implement the iterative method. The calculation of eigenvalues of SDOF linear non-viscously damped systems with exponential model is considered next. An approximate non-state-space based approach is proposed for this type of problem. The proposed approximations are based on certain physical assumptions which simplify the underlying characteristic equation to be solved. Closed-form approximate expressions of the complex and real eigenvalues of the system are derived. These approximate expressions are obtained as functions of the undamped eigenvalues only. The methods are then extended to exponentially damped MDOF systems. This technique enables us to approximately calculate the eigenvalues of non-viscously damped systems by simple post-processing of the undamped eigenvalues. Beside these reduced modal methods, another model reduction approach based on an equivalent second-order form is discussed. This method is applicable to any general non-viscous model and *not* only the exponential model. The proposed approximation utilizes the idea of generalized proportional damping and expressions of approximate eigenvalues of the system. A closed-form expression of the equivalent second-order system has been derived. The new expression is obtained by elementary operations involving the mass, stiffness and the kernel function matrix only. This enables us to approximately calculate the dynamic response of general non-viscously damped systems using the standard tools for conventional second-order viscously damped systems. Representative numerical examples are given throughout to verify the accuracy of the derived expressions.

The ability to efficiently calculate the eigensolutions and dynamic response of general non-viscously damped MDOF systems is useful. However, it is also necessary to understand how different parameters in the model impact the dynamics of the system. [ADH 14b], Chapter 1 is devoted to calculating the parametric sensitivity of the eigensolutions of damped systems with respect to the system parameters. Sensitivity of eigenvalues and eigenvectors of MDOF undamped systems is considered first. Then, we consider MDOF systems with a non-proportional viscous damping model. Because of the non-proportional nature of the damping, the mode shapes and natural frequencies become complex, and as a consequence the sensitivities of eigenvalues and eigenvectors are also complex. The results are presented in terms of the complex modes and frequencies of the second-order system and avoid the use of rather undesirable state-space representation. The eigenvector derivatives are obtained using two approaches, namely via the superposition of complex modes and Nelson's method. Next, the parametric sensitivity of the eigensolutions of MDOF systems with general non-viscous damping is considered. Because of the general nature of the damping, eigensolutions are generally complex valued and eigenvectors do not satisfy any orthogonality relationship. It is shown that even under such general conditions, the derivative of the eigensolutions can still be expressed in a way similar to that of undamped or viscously damped systems. All the results derived in the chapter are explained by many numerical examples.

The above-mentioned studies give a firm basis for the modal analysis of non-viscously damped systems. Motivated by these results, in [ADH 14b], Chapter 2, studies on damping identification are considered. First, the identification of a viscous damping matrix from the modal damping factors using the generalized proportional damping model is discussed. The theoretical formulation is presented and a step-by-step procedure is given. The practical utility of the proposed identification scheme is illustrated on three representative structures: (1) a free-free beam in flexural vibration, (2) a quasi-periodic three-cantilever structure model by inserting slots in a plate, in out-of-plane flexural vibration and (3) a point-coupled-beam system. Then, the identification of a non-proportional viscous damping matrix under the circumstances when the actual damping model in the structure can be non-viscous is considered. A method is presented to obtain a full (non-proportional) viscous damping matrix from complex modes and complex natural frequencies. It is assumed that the damping is "small" so that a first-order perturbation method is applicable. The proposed method and several related issues are discussed by considering numerical examples based on a linear array of damped spring-mass oscillators. It is shown that the method can predict the spatial location of damping with good accuracy, and also gives some indication of the correct mechanism of damping. In some cases, the identified damping matrix becomes non-symmetric, which in a way is a non-physical result because the original system is reciprocal. Through an error analysis, how the identified damping matrix is influenced by errors in the identified modal quantities is discussed. Methods are also developed to identify damping models which preserve the symmetry of the system. This procedure is based on a constrained error minimization approach. Another approach based on the Lancaster method is presented. Here, the damping matrix is obtained directly from the poles and residues of the measured transfer functions. The proposed methods are supported by suitable numerical examples.

From the studies in [ADH 14b], Chapter 2, it is observed that when a system is non-viscously damped, an identified equivalent viscous damping model does not accurately represent the damping behavior. This demands new methodologies to identify non-viscous damping models. [ADH 14b], Chapter 3, takes a first step, by outlining a procedure for identifying a damping model involving an exponentially decaying relaxation function. The method uses experimentally identified complex modes and complex natural frequencies, together with the knowledge of the mass matrix of the system. The proposed method and several related issues are discussed by considering numerical examples of a linear array of damped spring-mass oscillators. It is shown that good estimates can be obtained for the exponential time constant and the spatial distribution of the damping. However, when the fitted model does not match with the "true" model, the identification method results in an asymmetric coefficient matrix. A symmetry preserving damping identification method is presented, which guarantees a symmetric coefficient matrix. This method is based on a matrix variate constrained optimization method. Finally, another approach for damping identification directly from the poles and residues of the measured transfer functions is presented. This approach bypasses the need to obtain complex modes and frequencies from experiments. Numerical examples with simulated data are given to illustrate the damping identification methods.

Chapter 4 of [ADH 14b] is inspired by a simple question – given a damping model, how much damping is there in a system? In the context of a proportionally damped system, the answer to this question is straightforward as the damping is completely quantified by the modal damping factors. However, for non-proportionally damped systems and non-viscously damped systems, the answer to this question is less obvious. Here, numerical tools are developed to measure damping. First, a simple method is proposed to normalize the complex modes of a non-proportionally damped system so that they are closest to their corresponding classical normal modes. Based on these "optimal complex modes", an index of damping non-proportionality is proposed. The methodologies developed here are applicable to both viscous and non-viscously damped systems. Later, indices to quantify the extent of any departures of a non-viscous model from a viscous model are developed. In other words, the amount of "non-viscosity" of damping in discrete linear systems is quantified. Four indices are proposed. Two of these indices are based on the non-viscous damping matrix of the system. A third index is based on the residue matrices of the system transfer functions and the fourth is based on the (measured) complex modes of the system. The performance of the proposed indices is examined by considering numerical examples.

Theory of dynamics of MDOF *symmetric* systems has been studied in this book. However, dynamical behavior of some systems encountered in practice can be asymmetric in nature. In the Appendix methods are proposed by which asymmetric dynamic systems can be transformed into symmetric systems. In this way, the methods proposed in the book can in turn be applied to asymmetric systems as well. The conditions for transforming multiple-degrees-of-freedom linear asymmetric dynamical systems to equivalent symmetric systems by non-singular linear transformations are discussed. An approach is proposed to transform asymmetric systems into symmetric systems by an equivalence transformation. The existing approach of symmetrization by similarity transformation is the "first kind" and the proposed approach by equivalence transformation is the "second kind". Because equivalence transformations are the most general non-singular linear transformations, conditions of symmetrizability obtained here are more "liberal" compared to the first kind and numerical calculations also become more straightforward. Several examples are provided to illustrate this approach.

The intended readers of this book are primarily senior undergraduate students, graduate students and practicing engineers working in the field of advanced vibration. Limited examples are provided to support of the theoretical developments. The book is written with the aim of being self-contained. However, a recommended prerequisite is an undergraduate-level vibration course. There are many excellent books which cover the fundamentals of the theory of vibration, for example references [MEI 67, MEI 80, PAZ 80, NEW 89, CLO 93, BAT 95, MEI 97, PET 98, GÉR 97, INM 03, RAO 11]. Readers will greatly benefit by familiarizing themselves with the basics of the theory of vibration.

In spite of the attempt at being exhaustive at the time of writing, clearly many relevant and possibly important bibliographic references are missed. This is inevitable as a huge amount of literature was published recently due to the significant rise in the interest in this topic. However, the author expects that the book covers the necessary background so that at least the readers will appreciate existing publications and future research works and developments in the field of damping. It is hoped that the readers will not only gain an understanding of the material presented in the book, but will also be able do their personal research and take this field forward.

Chapter 2

Dynamics of Undamped and Viscously Damped Systems

In this chapter, we discuss dynamics of single- and multiple-degrees-of-freedom (SDOF and MDOF) undamped and viscously damped systems. There are many excellent books [MEI 67, MEI 80, PAZ 80, NEW 89, MEI 97, GÉR 97, INM 03, RAO 11], which discuss these topics very well. Here, we give only the details as a necessary background for further developments in the following chapters.

The outline of this chapter is as follows. Dynamic analysis of SDOF undamped systems is discussed in section 2.1. SDOF viscously damped systems are considered in section 2.2. Dynamics of MDOF undamped systems are discussed in section 2.3. The concepts of natural frequencies (eigenvalues) and mode-shapes (eigenvectors) are introduced. Proportionally damped systems are investigated in detail in section 2.4. General non-proportionally damped systems are considered in section 2.5. The idea of Rayleigh quotient for damped systems is proposed in section 2.6. Stationarity properties for systems with proportional damping and non-proportional damping are discussed. Finally, a summary of the topics discussed in this chapter is given in section 2.7.

2.1. Single-degree-of-freedom undamped systems

An undamped SDOF system is possibly the simplest system and is of fundamental interest in structural dynamics. A schematic diagram of the system is shown in Figure 2.1. The equation of motion of the system can be expressed as

$$m\ddot{q}(t) + kq(t) = f(t). \quad [2.1]$$

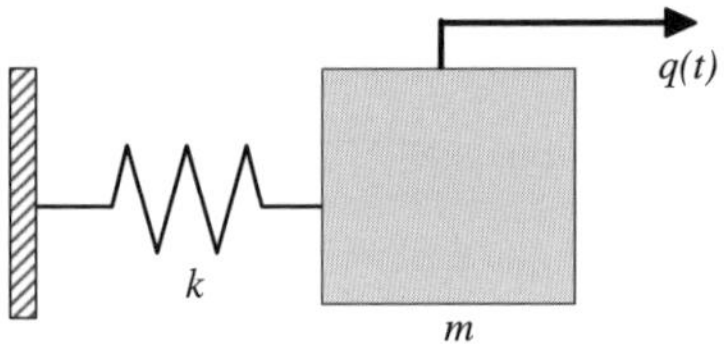

Figure 2.1. *A single degree-of-freedom undamped system*

Here m is the mass, k is the stiffness and $f(t)$ is the applied force on the system. The term $\ddot{q}(t)$ denotes the acceleration of the system. The initial conditions associated with the equation of motion can be given by

$$q(0) = q_0 \quad \text{and} \quad \dot{q}(0) = \dot{q}_0. \qquad [2.2]$$

It is shown that the response of the system [2.1] subject to the initial conditions [2.2] can be expressed in terms of the natural frequency of the system.

2.1.1. *Natural frequency*

First, consider the harmonic vibration of the system. For this, the forcing function can be expressed as $f(t) = \bar{f}e^{st}$. Assume that the solution $q(t)$ is of the form $q(t) = \bar{q}e^{st}$, $\bar{q}, s \neq 0$. Substituting this in the equation of motion [2.1] as assuming the free vibration (i.e. $\bar{f} = 0$), we obtain

$$\bar{q}\left(ms^2 + k\right) = 0 \quad \text{or} \quad s_{1,2} = \pm \mathrm{i}\sqrt{\frac{k}{m}} = \pm \mathrm{i}\omega_n. \qquad [2.3]$$

Here, $\mathrm{i} = \sqrt{-1}$ and ω_n, the natural frequency of the system, is given by

$$\omega_n = \sqrt{\frac{k}{m}}. \qquad [2.4]$$

The roots of the characteristic equation [2.3] are purely imaginary in nature and they appear in complex conjugate pairs. The natural frequency given by equation [2.4] is the frequency of natural oscillation of the system. The natural time period of the system is given by

$$T_n = \frac{2\pi}{\omega_n}. \qquad [2.5]$$

When the excitation frequency is close to the resonance frequency, we say that the SDOF system is in resonance. Near the resonance frequency, the response of the system can amplify significantly as we will see in the next section.

2.1.2. *Dynamic response*

We consider the steady-sate response of the dynamic system. Taking the Laplace transform of the equation of motion [2.1] and considering the initial conditions in [2.2], we have

$$s^2 m\overline{q}(s) - smq_0 - m\dot{q}_0 + k\overline{q}(s) = \overline{f}(s) \tag{2.6}$$

$$\text{or} \quad \left(s^2 m + k\right)\overline{q}(s) = \overline{f}(s) + m\dot{q}_0 + smq_0. \tag{2.7}$$

Here, $\overline{q}(s)$ and $\overline{f}(s)$ are Laplace transforms of $q(t)$ and $f(t)$, respectively. The response in the Laplace domain can be obtained as

$$\overline{q}(s) = \frac{\overline{f}(s) + m\dot{q}_0 + smq_0}{s^2 m + k} = \frac{1}{m}\frac{\overline{f}(s) + m\dot{q}_0 + smq_0}{s^2 + \omega_n^2} \tag{2.8}$$

$$= \left\{\frac{1}{m}\frac{\overline{f}(s)}{s^2 + \omega_n^2} + \frac{\dot{q}_0}{s^2 + \omega_n^2} + \frac{s}{s^2 + \omega_n^2} q_0\right\}. \tag{2.9}$$

To obtain the vibration response in the time domain, it is required to consider the inverse Laplace transform. Taking the inverse Laplace transform of $\overline{q}(s)$, we have

$$q(t) = \mathcal{L}^{-1}\left[\overline{q}(s)\right] = \mathcal{L}^{-1}\left[\frac{1}{m}\frac{\overline{f}(s)}{s^2 + \omega_n^2}\right] + \mathcal{L}^{-1}\left[\frac{1}{s^2 + \omega_n^2}\right]\dot{q}_0 + \mathcal{L}^{-1}\left[\frac{s}{s^2 + \omega_n^2}\right] q_0. \tag{2.10}$$

The inverse Laplace transform of the second and third parts can be obtained from the table of Laplace transforms (see, for example, [KRE 99]) as

$$\mathcal{L}^{-1}\left[\frac{1}{s^2 + \omega_n^2}\right] = \frac{\sin(\omega_n t)}{\omega_n} \tag{2.11}$$

$$\text{and} \quad \mathcal{L}^{-1}\left[\frac{s}{s^2 + \omega_n^2}\right] = \cos(\omega_n t). \tag{2.12}$$

The inverse Laplace transform of the first part can be obtained using the "convolution theorem" [KRE 99], which says that the inverse Laplace transform of product of two functions can be expressed as

$$\mathcal{L}^{-1}\left[\bar{f}(s)\bar{g}(s)\right] = \int_0^t f(\tau)g(t-\tau)d\tau. \qquad [2.13]$$

Considering $\bar{g}(s) = \frac{1}{m}\frac{1}{s^2+\omega_n^2}$, the inverse Laplace transform of the first part can be obtained as

$$\mathcal{L}^{-1}\left[\bar{f}(s)\frac{1}{s^2+\omega_n^2}\right] = \int_0^t \frac{1}{m\omega_n} f(\tau)\sin\left(\omega_n(t-\tau)\right)d\tau. \qquad [2.14]$$

Combining [2.14], [2.11] and [2.12], from equation [2.10], we have

$$q(t) = \int_0^t \frac{1}{m\omega_n} f(\tau)\sin\left(\omega_n(t-\tau)\right)d\tau + \frac{1}{\omega_n}\sin(\omega_n t)\dot{q}_0 + \cos(\omega_n t)q_0. \qquad [2.15]$$

Collecting the terms associated with $\sin(\omega_n t)$ and $\cos(\omega_n t)$ this expression can be simplified as

$$q(t) = \frac{1}{m\omega_n}\int_0^t f(\tau)\sin\left(\omega_n(t-\tau)\right)d\tau + B\cos\left(\omega_n t - \vartheta\right) \qquad [2.16]$$

where the amplitude and the phase angle can be obtained from

$$B = \sqrt{q_0^2 + \left(\frac{\dot{q}_0}{\omega_n}\right)^2} \qquad [2.17]$$

$$\text{and} \quad \tan\vartheta = \frac{\dot{q}_0}{\omega_n q_0}. \qquad [2.18]$$

The second part of equation [2.16], i.e. the term $B\cos(\omega_n t - \vartheta)$ only depends on the initial conditions and independent of the applied forcing. The first part, on the other hand, only depends on the applied forcing. The complete response of the SDOF system is given by equation [2.15].

2.1.2.1. *Impulse response function*

In the special case when all the initial conditions are zero and the forcing function is a Dirac delta function, that is $f(\tau) = \delta(\tau)$, the response is known as the impulse response function. This can be obtained from equation [2.15] by substituting $q_0 = 0, \dot{q}_0 = 0$ and $f(\tau) = \delta(\tau)$ as

$$h(t) = \int_0^t \frac{1}{m\omega_n}\delta(\tau)\sin\left(\omega_n(t-\tau)\right)d\tau = \frac{1}{m\omega_n}\sin\omega_n t. \qquad [2.19]$$

This is a fundamental property of the system. If this function is known (e.g. from experiments), then the response to any forcing can be obtained from the convolution integral expressions in [2.15] as

$$q(t) = \int_0^t f(\tau)h(t-\tau)d\tau. \qquad [2.20]$$

In the following, we consider viscously damped SDOF systems.

2.2. Single-degree-of-freedom viscously damped systems

An SDOF system with viscous damping is shown in Figure 2.2. The equation of motion of the system can be expressed as

$$m\ddot{q}(t) + c\dot{q}(t) + kq(t) = f(t). \qquad [2.21]$$

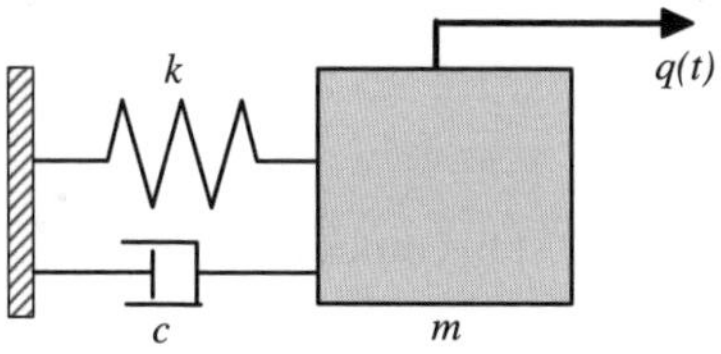

Figure 2.2. *A single degree-of-freedom system with viscous damping*

Here, m is the mass of the system, c is the damping, k is the spring stiffness and $f(t)$ is the applied forcing. As before, the symbol $(\dot{\bullet})$ represents derivative with respect to time, so that $\dot{q}(t)$ denotes the velocity and $\ddot{q}(t)$ denotes the acceleration of the system. The initial conditions associated with the equation of motion can be given by [2.2]. It is shown that the response of the system [2.21] subject to the initial conditions [2.2] can be expressed in terms of the undamped and damped natural frequencies of the system.

2.2.1. *Natural frequency*

Like the undamped system, first we consider harmonic vibration of the system. For this, both the forcing function and the solution can be expressed as $f(t) = \bar{f}e^{st}$ and $q(t) = \bar{q}e^{st}$, $\bar{q}, s \neq 0$. Substituting these in the equation of motion [2.21] and assuming the free vibration (i.e. $\bar{f} = 0$), we obtain the characteristic equation

$$\bar{q}\left(ms^2 + +sc + k\right) = 0 \quad \text{or} \quad \frac{\bar{q}}{m}\left(s^2 + +s\frac{c}{m} + \frac{k}{m}\right) = 0. \qquad [2.22]$$

We introduce the viscous damping factor as

$$\zeta = \frac{c}{2\sqrt{k\,m}}. \qquad [2.23]$$

Using the definition of the natural frequency in equation [2.4], it can be shown that

$$\frac{c}{m} = 2\zeta\omega_n \qquad [2.24]$$

Using this, from equation [2.22], we have the characteristic equation

$$s^2 + +2\zeta\omega_n\, s + \omega_n^2 = 0 \quad \text{or} \quad s_{1,2} = -\zeta\omega_n + \pm \mathrm{i}\omega_n\sqrt{1-\zeta^2}. \qquad [2.25]$$

The nature of the dynamic response of the system depends on the characteristics of the two roots in [2.25]. It turns out that the value of ζ controls the characteristics of both the roots. It is useful to define the critical damping factor

$$c_{cr} = 2\sqrt{km} \qquad [2.26]$$

so that the damping factor in [2.23] becomes the ratio

$$\zeta = \frac{c}{c_{cr}}. \qquad [2.27]$$

Three possible cases arise based on the value of ζ:

1) Overdamped motion, $\zeta > 1$: in this case, both roots in [2.25] are real and negative as

$$s_1 = -\omega_n\left(\zeta + \sqrt{\zeta^2 - 1}\right) \quad \text{and} \quad s_2 = -\omega_n\left(\zeta - \sqrt{\zeta^2 - 1}\right). \qquad [2.28]$$

The motion of the system is therefore non-oscillatory and exponentially decaying in nature.

2) Critically damped motion, $\zeta = 1$: in this case, both roots in [2.25] are real and negative, and identical in nature. That is, the system have repeated roots as

$$s_1 = -\omega_n \quad \text{and} \quad s_2 = -\omega_n. \tag{2.29}$$

The motion of the system can be expressed as $q(t) = (a_1 + a_2 t)e^{-\omega_n t}$ for some real constants a_1 and a_2. This motion is also non-oscillatory and exponentially decaying in nature. The motion of a critically damped system can be viewed as the borderline between non-oscillatory and oscillatory motions.

3) Underdamped motion, $\zeta < 1$: in this case, the roots of the characteristic equation [2.25] appear in complex conjugate pairs as

$$s_1 = -\zeta\omega_n + \mathrm{i}\omega_d \quad \text{and} \quad s_2 = -\zeta\omega_n - \mathrm{i}\omega_d. \tag{2.30}$$

Here, the damped natural frequency of the system is given by

$$\omega_d = \omega_n\sqrt{1-\zeta^2} = \sqrt{\frac{k}{m}}\sqrt{1-\zeta^2}. \tag{2.31}$$

As a result, the damped time period of the system can be defined as

$$T_d = \frac{2\pi}{\omega_d}. \tag{2.32}$$

In view of the roots in [2.30], the motion of the system can be expressed as $q(t) = Be^{-\zeta\omega_n t}\sin(\omega_d t\vartheta)$. The motion of the system is therefore oscillatory and exponentially decaying in nature.

2.2.2. *Dynamic response*

We consider the steady-sate response of the dynamic system. Taking the Laplace transform of the equation of motion [2.21] and considering the initial conditions in [2.2], we have

$$s^2 m\overline{q}(s) - smq_0 - m\dot{q}_0 + sc\overline{q}(s) - cq_0 + k\overline{q}(s) = \overline{f}(s) \tag{2.33}$$

$$\text{or} \quad \left(s^2 m + sc + k\right)\overline{q}(s) = \overline{f}(s) + m\dot{q}_0 + smq_0 + cq_0. \tag{2.34}$$

The response in the Laplace domain can be obtained as

$$\overline{q}(s) = \frac{\overline{f}(s) + m\dot{q}_0 + smq_0 + cq_0}{s^2 m + sc + k} = \frac{1}{m}\frac{\overline{f}(s) + m\dot{q}_0 + smq_0 + cq_0}{s^2 + 2s\zeta\omega_n + \omega_n^2} \tag{2.35}$$

$$= \left\{\frac{1}{m}\frac{\overline{f}(s)}{s^2 + 2s\zeta\omega_n + \omega_n^2} + \frac{\dot{q}_0 + 2\zeta\omega_n q_0}{s^2 + 2s\zeta\omega_n + \omega_n^2} + \frac{s}{s^2 + 2s\zeta\omega_n + \omega_n^2}q_0\right\}. \tag{2.36}$$

Using the definition of the damped natural frequency from equation [2.31], we can reorganize the denominator as

$$s^2 + 2s\zeta\omega_n + \omega_n^2 = (s + \zeta\omega_n)^2 - (\zeta\omega_n)^2 + \omega_n^2 = (s + \zeta\omega_n)^2 + \omega_d^2. \quad [2.37]$$

To obtain the vibration response in the time domain, it is required to consider the inverse Laplace transform. Taking the inverse Laplace transform of $\overline{q}(s)$ in [2.36], we have

$$\begin{aligned} q(t) &= \mathcal{L}^{-1}\left[\overline{q}(s)\right] = \mathcal{L}^{-1}\left[\frac{1}{m}\frac{\overline{f}(s)}{(s+\zeta\omega_n)^2+\omega_d^2}\right] \\ &+\mathcal{L}^{-1}\left[\frac{1}{(s+\zeta\omega_n)^2+\omega_d^2}\right](\dot{q}_0 + 2\zeta\omega_n q_0) + \mathcal{L}^{-1}\left[\frac{s}{(s+\zeta\omega_n)^2+\omega_d^2}\right]q_0. \end{aligned} \quad [2.38]$$

The inverse Laplace transform of the second and third parts can be obtained from the table of Laplace transforms (see, for example, [KRE 99]) as

$$\mathcal{L}^{-1}\left[\frac{1}{(s+\alpha)^2+\beta^2}\right] = \frac{e^{-\alpha t}\sin(\beta t)}{\beta} \quad [2.39]$$

$$\text{and} \quad \mathcal{L}^{-1}\left[\frac{s}{(s+\alpha)^2+\beta^2}\right] = e^{-\alpha t}\cos(\beta t) - \frac{\alpha e^{-\alpha t}\sin(\beta t)}{\beta}. \quad [2.40]$$

Here, α and β stand for $\zeta\omega_n$ and ω_d, respectively.

The inverse Laplace transform of the first part can be obtained using the convolution theorem [2.13]. Considering $\overline{g}(s) = \frac{1}{m}\frac{1}{(s+\zeta\omega_n)^2+\omega_d^2}$, the inverse Laplace transform of the first part can be obtained as

$$\mathcal{L}^{-1}\left[\overline{f}(s)\frac{1}{(s+\zeta\omega_n)^2+\omega_d^2}\right] = \int_0^t \frac{1}{m\omega_d} f(\tau)e^{-\zeta\omega_n(t-\tau)}\sin(\omega_d(t-\tau))\,d\tau. \quad [2.41]$$

Combining [2.39], [2.40] and [2.41], from equation [2.38], we have

$$\begin{aligned} q(t) &= \int_0^t \frac{1}{m\omega_d} f(\tau) e^{-\zeta\omega_n(t-\tau)} \sin\left(\omega_d(t-\tau)\right) d\tau + \frac{e^{-\zeta\omega_n t}}{\omega_d} \sin(\omega_d t) \\ &\quad \times (\dot{q}_0 + 2\zeta\omega_n q_0) + \left\{ e^{-\zeta\omega_n t} \cos(\omega_d t) - \frac{\zeta\omega_n\, e^{-\zeta\omega_n t} \sin(\omega_d\, t)}{\omega_d} \right\} q_0 \\ &= \int_0^t \frac{1}{m\omega_d} f(\tau) e^{-\zeta\omega_n(t-\tau)} \sin\left(\omega_d(t-\tau)\right) d\tau + \sin(\omega_n t) \left(\frac{\dot{q}_0 + \zeta\omega_n q_0}{\omega_d} \right) \\ &\quad + \cos(\omega_n t) q_0. \end{aligned} \quad [2.42]$$

Collecting the terms associated with $\sin(\omega_d\, t)$ and $\cos(\omega_d\, t)$ this expression can be simplified as

$$q(t) = \int_0^t \frac{1}{m\omega_d} f(\tau) e^{-\zeta\omega_n(t-\tau)} \sin\left(\omega_d(t-\tau)\right) d\tau + B\cos(\omega_d t - \vartheta) \quad [2.43]$$

where the amplitude and the phase angle can be obtained from

$$B = \sqrt{q_0^2 + \left(\frac{\dot{q}_0 + \zeta\omega_n q_0}{\omega_d} \right)^2} \quad [2.44]$$

$$\text{and} \quad \tan\vartheta = \frac{\dot{q}_0 + \zeta\omega_n q_0}{\omega_d q_0}. \quad [2.45]$$

The second part of equation [2.43], i.e. the term $B\cos(\omega_n t - \vartheta)$ only depends on the initial conditions and independent of the applied forcing. The first part, on the other hand, only depends on the applied forcing. The complete response of a viscously damped SDOF system is given by equation [2.42]. In the special case when the system is undamped, by substituting $\zeta = 0$ we can obtain the results derived in the previous section.

2.2.2.1. *Impulse response and frequency response function*

When all the initial conditions are zero and the forcing function is a Dirac delta function, that is $f(\tau) = \delta(\tau)$, the response is known as the impulse response function.

This can be obtained from equation [2.42] by substituting $q_0 = 0, \dot{q}_0 = 0$ and $f(\tau) = \delta(\tau)$ as

$$h(t) = \int_0^t \frac{1}{m\omega_d}\delta(\tau)e^{-\zeta\omega_n(t-\tau)}\sin\left(\omega_d(t-\tau)\right)d\tau = \frac{1}{m\omega_d}e^{-\zeta\omega_n t}\sin\omega_n t. \quad [2.46]$$

This is a fundamental property of the damped system and depends only on the natural frequency and the damping factor. If this function is known (e.g. from experiments), then the response to any forcing can be obtained from the convolution integral expressions in [2.42] as $q(t) = \int_0^t f(\tau)h(t-\tau)d\tau$.

The frequency response function can be obtained directly from equation [2.34] by substituting $q_0 = 0, \dot{q}_0 = 0, \bar{f}(s) = 1$ and $s = \mathrm{i}\omega$ as

$$\bar{h}(\mathrm{i}\omega) = \frac{1}{-\omega^2 m + \mathrm{i}\omega c + k} = \frac{1}{m}\frac{1}{\left(-\omega^2 + 2\mathrm{i}\zeta\omega\omega_n + \omega_n^2\right)}. \quad [2.47]$$

Here, ω is the driving frequency.

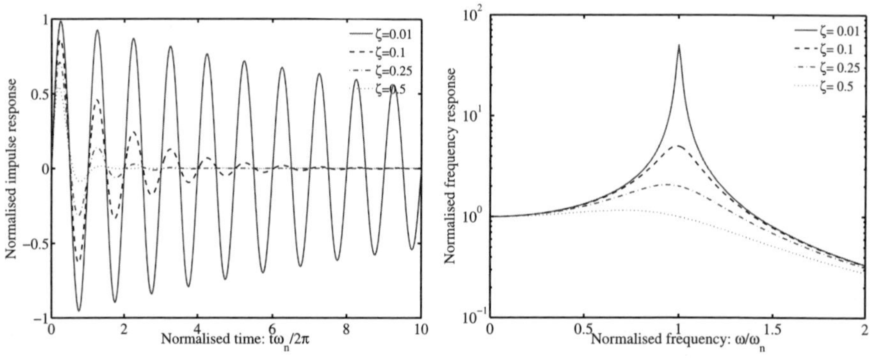

(a) Impulse response function

(b) Amplitude of the frequency response function

Figure 2.3. *Normalized impulse response function and amplitude of the frequency response function of a viscously damped SDOF system for various values of the damping factor ζ*

For convenience we introduce a normalized time $t' = t/T_n = \omega_n t/2\pi$. Using this, the normalized impulse response function is shown in Figure 2.3(a) for four values of the damping factor, i.e. $\zeta = 0.01, 0.1, 0.25$ and 0.5. It can be seen that with increasing damping (1) the oscillations diminish to almost zero very quickly, and (2) the amplitude of successive peaks also reduce significantly. For example when $\zeta = 0.5$, the vibration ceases after about three period of oscillations.

Dividing the expression for the frequency response function in [2.47] by ω_n and introducing the non-dimensional frequency parameter $\tilde{\omega} = \omega/\omega_n$, we have

$$\begin{aligned}\bar{h}(\mathrm{i}\omega) &= \frac{1}{m}\frac{1}{(-\omega^2 + 2\mathrm{i}\zeta\omega\omega_n + \omega_n^2)} = \frac{1}{m\omega_n^2}\frac{1}{(-\tilde{\omega}^2 + 2\mathrm{i}\zeta\tilde{\omega} + 1)}\\ &= \frac{1}{k}\frac{1}{(-\tilde{\omega}^2 + 2\mathrm{i}\zeta\tilde{\omega} + 1)}.\end{aligned} \quad [2.48]$$

The amplitude of the normalized frequency response function $(\bar{h}(\mathrm{i}\tilde{\omega})/(1/k))$ can be obtained as

$$\left|\frac{\bar{h}(\mathrm{i}\tilde{\omega})}{(1/k)}\right| = \frac{1}{\sqrt{(1-\tilde{\omega}^2)^2 + 4\zeta^2\tilde{\omega}^2}}. \quad [2.49]$$

This function is shown in Figure 2.3(b) for the same four values of the damping factor considered to plot the impulse response function. The following observations can be made:

– For lower values of the damping factor, the response amplitude has a peak near $\omega = \omega_n$.

– The higher the damping, the lower the value of the peak response.

– The frequency of the peak response shifts to the left for increasing damping values.

These observations can be explained by analytically obtaining the "peak frequency". Differentiating the amplitude of the normalized frequency response function with respect to $\tilde{\omega}$ and setting it to zero, we have

$$\frac{\partial}{\partial\tilde{\omega}}\left|\frac{\bar{h}(\mathrm{i}\tilde{\omega})}{(1/k)}\right| = 0 \quad \text{or} \quad \frac{\partial}{\partial\tilde{\omega}}\left[(1-\tilde{\omega}^2)^2 + 4\zeta^2\tilde{\omega}^2\right]^{-1/2}. \quad [2.50]$$

Solving this equation, we have

$$\tilde{\omega}_{\max} = \frac{\omega_{\max}}{\omega_n} = \sqrt{1-2\zeta^2}. \quad [2.51]$$

For real values of $\tilde{\omega}_{\max}$, we must have $1 - 2\zeta^2 \geq 0$, or $\zeta < 1/\sqrt{2}$. Provided the value of the damping factor is lower than $1/\sqrt{2} = 0.7071$, there will be a peak in the frequency response function. The maximum amplitude can be obtained by substituting the $\tilde{\omega}_{\max}$ in the expression of the normalized frequency response [2.49] as

$$\left|\frac{\bar{h}(\mathrm{i}\tilde{\omega})}{(1/k)}\right|_{\max} = \frac{1}{2\zeta\sqrt{1-\zeta^2}}. \qquad [2.52]$$

This shows that the maximum amplitude will decrease with increasing value of the damping factor. The influence of the damping factor on the dynamics of a viscously damped SDOF system is summarized in Table 2.1.

Parameter values	Dynamic characteristics
$\zeta > 1$	Overdamped motion – the motion is non-oscillatory and exponentially decaying
$\zeta = 1$	Critically damped motion – the motion is non-oscillatory and exponentially decaying
$\zeta < 1$	Underdamped motion – the motion is oscillatory and exponentially decaying
$1 < \zeta \leq 1/\sqrt{2}$	Oscillatory motion, but no peak in the frequency response function
$1/\sqrt{2} < \zeta$	Oscillatory motion, the frequency response function has peak at a normalized frequency $\sqrt{1-2\zeta^2}$ with a normalized peak value of $\frac{1}{2\zeta\sqrt{1-\zeta^2}}$

Table 2.1. *The influence of the damping factor on the dynamics of a viscously damped single-degree-of-freedom system*

2.3. Multiple-degree-of-freedom undamped systems

The equation of motion of an undamped non-gyroscopic system with N degrees of freedom can be given by

$$\mathbf{M}\ddot{\mathbf{q}}(t) + \mathbf{K}\mathbf{q}(t) = \mathbf{f}(t) \qquad [2.53]$$

where $\mathbf{M} \in \mathbb{R}^{N\times N}$ is the mass matrix, $\mathbf{K} \in \mathbb{R}^{N\times N}$ is the stiffness matrix, $\mathbf{q}(t) \in \mathbb{R}^N$ is the vector of generalized coordinates and $\mathbf{f}(t) \in \mathbb{R}^N$ is the forcing vector. Equation [2.53] represents a set of coupled second-order ordinary differential equations. The solution of this equation also requires the knowledge of the initial conditions in terms of displacements and velocities of all the coordinates. The initial conditions can be specified as

$$\mathbf{q}(0) = \mathbf{q}_0 \in \mathbb{R}^N \quad \text{and} \quad \dot{\mathbf{q}}(0) = \dot{\mathbf{q}}_0 \in \mathbb{R}^N. \qquad [2.54]$$

We aim to solve equation [2.53] together with the initial conditions in [2.54] using modal analysis.

2.3.1. *Modal analysis*

Lord Rayleigh [RAY 77] showed that undamped linear systems, equations of motion of which are given by [2.53], are capable of the so-called *natural motions*. This essentially implies that all the system coordinates execute harmonic oscillation at a given frequency and form a certain displacement pattern. The oscillation frequency and displacement pattern are called *natural frequencies* and *normal modes*, respectively. The natural frequencies (ω_j) and the mode shapes ($\mathbf{x}_j$) are intrinsic characteristics of a system and can be obtained by solving the associated matrix eigenvalue problem

$$\mathbf{K}\mathbf{x}_j = \omega_j^2 \mathbf{M}\mathbf{x}_j, \quad \forall j = 1, \cdots, N. \tag{2.55}$$

Since the above eigenvalue problem is in terms of real symmetric non-negative definite matrices $\mathbf{M}$ and $\mathbf{K}$, the eigenvalues and consequently the eigenvectors are real, that is $\omega_j \in \mathbb{R}$ and $\mathbf{x}_j \in \mathbb{R}^N$. Premultiplying equation [2.55] by $\mathbf{x}_k^T$, we have

$$\mathbf{x}_k^T \mathbf{K}\mathbf{x}_j = \omega_j^2 \mathbf{x}_k^T \mathbf{M}\mathbf{x}_j. \tag{2.56}$$

Transposing the above equation and noting that $\mathbf{M}$ and $\mathbf{K}$ are symmetric matrices, we have

$$\mathbf{x}_j^T \mathbf{K}\mathbf{x}_k = \omega_j^2 \mathbf{x}_j^T \mathbf{M}\mathbf{x}_k. \tag{2.57}$$

Now consider the eigenvalue equation for the kth mode

$$\mathbf{K}\mathbf{x}_k = \omega_k^2 \mathbf{M}\mathbf{x}_k. \tag{2.58}$$

Premultiplying equation [2.58] by $\mathbf{x}_j^T$, we have

$$\mathbf{x}_j^T \mathbf{K}\mathbf{x}_k = \omega_k^2 \mathbf{x}_j^T \mathbf{M}\mathbf{x}_k. \tag{2.59}$$

Subtracting equation [2.57] from [2.59], we have

$$\left(\omega_k^2 - \omega_j^2\right) \mathbf{x}_j^T \mathbf{M}\mathbf{x}_k = 0. \tag{2.60}$$

Since we assumed that the natural frequencies are not repeated, when $j \neq k$, $\omega_j \neq \omega_k$. Therefore, from equation [2.60] it follows that

$$\mathbf{x}_k^T \mathbf{M} \mathbf{x}_j = 0. \quad [2.61]$$

Using this in equation [2.57], we can also obtain

$$\mathbf{x}_k^T \mathbf{K} \mathbf{x}_j = 0. \quad [2.62]$$

These two relationships are often known as the mode orthogonality relationships. If we normalize $\mathbf{x}_j$ such that $\mathbf{x}_j \mathbf{M} \mathbf{x}_j = 1$, then from equation [2.57] it follows that $\mathbf{x}_j \mathbf{K} \mathbf{x}_j = \omega_j^2$. This normalization is known as the unity mass normalization, a convention often used in practice. Equations [2.61] and [2.62] are known as the orthogonality relationships. These equations combined with the normalization relationships can be concisely written in terms of Kroneker delta function δ_{lj} as

$$\mathbf{x}_l^T \mathbf{M} \mathbf{x}_j = \delta_{lj} \quad [2.63]$$

$$\text{and} \quad \mathbf{x}_l^T \mathbf{K} \mathbf{x}_j = \omega_j^2 \delta_{lj}, \quad \forall\, l, j = 1, \cdots, N. \quad [2.64]$$

Note that $\delta_{lj} = 1$ for $l = j$ and $\delta_{lj} = 0$ otherwise. The property of the eigenvectors in [2.63] is also known as the mass orthonormality relationship. The solution of undamped eigenvalue problem is now standard in many finite element packages. There are various efficient algorithms available for this purpose, see, for example, [BAT 95, WIL 88]. We refer to these publications [BAI 02, BAI 05, CHR 13] for Krylov subspace-based techniques for further numerical efficiency.

This orthogonality property of the undamped modes is very powerful as it allows us to transform a set of *coupled* differential equations into a set of *independent* equations. For convenience, we construct the matrices

$$\boldsymbol{\Omega} = \text{diag}\left[\omega_1, \omega_2, \cdots, \omega_N\right] \in \mathbb{R}^{N \times N} \quad [2.65]$$

$$\text{and} \quad \mathbf{X} = \left[\mathbf{x}_1, \mathbf{x}_2, \cdots, \mathbf{x}_N\right] \in \mathbb{R}^{N \times N} \quad [2.66]$$

where the eigenvalues are arranged such that $\omega_1 < \omega_2$, $\omega_2 < \omega_3, \cdots, \omega_k < \omega_{k+1}$. The matrix $\mathbf{X}$ is known as the undamped modal matrix. Using these matrix notations, the orthogonality relationships [2.63] and [2.64] can be rewritten as

$$\mathbf{X}^T \mathbf{M} \mathbf{X} = \mathbf{I} \quad [2.67]$$

$$\text{and} \quad \mathbf{X}^T \mathbf{K} \mathbf{X} = \boldsymbol{\Omega}^2 \quad [2.68]$$

where $\mathbf{I}$ is an $N \times N$ identity matrix. We use the following coordinate transformation (known as the modal transformation)

$$\mathbf{q}(t) = \mathbf{X}\mathbf{y}(t). \tag{2.69}$$

Substituting $\mathbf{q}(t)$ in equation [2.53], premultiplying by $\mathbf{X}^T$ and using the orthogonality relationships in [2.67] and [2.68], the equation of motion in the modal coordinates may be obtained as

$$\begin{aligned} \ddot{\mathbf{y}}(t) + \boldsymbol{\Omega}^2\mathbf{y}(t) &= \mathbf{f}'(t) \\ \text{or} \quad \ddot{y}_j(t) + \omega_j^2 y_j(t) &= f'_j(t) \quad \forall j = 1, \cdots, N \end{aligned} \tag{2.70}$$

where $\overline{\mathbf{f}}'(t) = \mathbf{X}^T\mathbf{f}(t)$ is the forcing function in the modal coordinates. This method significantly simplifies the dynamic analysis because complex MDOF systems can be treated as collections of SDOF oscillators. This approach of analyzing linear undamped systems is known as the *modal analysis*, possibly the most efficient tool for vibration analysis of complex engineering structures. Beside the numerical efficiency, the natural frequencies and the mode shapes provide valuable physical insights into the nature of vibration of the system.

2.3.2. *Dynamic response*

In this section, we outline how modal analysis can be used to obtain expressions for dynamic response of a system. Analysis in the frequency domain is described first, followed by the time-domain analysis.

2.3.2.1. *Frequency-domain analysis*

We consider the steady-state response of the dynamic system. Taking the Laplace transform of [2.53] and considering the initial conditions in [2.54], we have

$$s^2\mathbf{M}\bar{\mathbf{q}}(s) - s\mathbf{M}\mathbf{q}_0 - \mathbf{M}\dot{\mathbf{q}}_0 + \mathbf{K}\bar{\mathbf{q}}(s) = \bar{\mathbf{f}}(s) \tag{2.71}$$

$$\text{or} \quad \left[s^2\mathbf{M} + \mathbf{K}\right]\bar{\mathbf{q}}(s) = \bar{\mathbf{f}}(s) + \mathbf{M}\dot{\mathbf{q}}_0 + s\mathbf{M}\mathbf{q}_0 = \bar{\mathbf{p}}(s) \text{ (say).} \tag{2.72}$$

Using the modal transformation

$$\bar{\mathbf{q}}(s) = \mathbf{X}\bar{\mathbf{y}}(s) \tag{2.73}$$

and premultiplying [2.72] by $\mathbf{X}^T$, we have

$$\left[s^2\mathbf{M} + \mathbf{K}\right]\mathbf{X}\bar{\mathbf{y}}(s) = \bar{\mathbf{p}}(s) \quad \text{or} \quad \left\{\mathbf{X}^T\left[s^2\mathbf{M} + \mathbf{K}\right]\mathbf{X}\right\}\bar{\mathbf{y}}(s) = \mathbf{X}^T\bar{\mathbf{p}}(s). \tag{2.74}$$

Using the orthogonality relationships in [2.67] and [2.68], this equation reduces to

$$\left[s^2\mathbf{I}+\boldsymbol{\Omega}^2\right]\bar{\mathbf{y}}(s) = \mathbf{X}^T\bar{\mathbf{p}}(s) \quad [2.75]$$

$$\text{or } \bar{\mathbf{y}}(s) = \left[s^2\mathbf{I}+\boldsymbol{\Omega}^2\right]^{-1}\mathbf{X}^T\bar{\mathbf{p}}(s) \quad [2.76]$$

$$\text{or } \mathbf{X}\bar{\mathbf{y}}(s) = \mathbf{X}\left[s^2\mathbf{I}+\boldsymbol{\Omega}^2\right]^{-1}\mathbf{X}^T\bar{\mathbf{p}}(s) \ \text{(premultiplying by } \mathbf{X}\text{)} \quad [2.77]$$

$$\text{or } \bar{\mathbf{q}}(s) = \mathbf{X}\left[s^2\mathbf{I}+\boldsymbol{\Omega}^2\right]^{-1}\mathbf{X}^T\bar{\mathbf{p}}(s) \ \text{(using [2.73])} \quad [2.78]$$

$$\text{or } \bar{\mathbf{q}}(s) = \mathbf{X}[s^2\mathbf{I}+\boldsymbol{\Omega}^2]^{-1}\mathbf{X}^T\{\bar{\mathbf{f}}(s)+\mathbf{M}\dot{\mathbf{q}}_0+s\mathbf{M}\mathbf{q}_0\} \ \text{(using [2.72])}. \quad [2.79]$$

Equation [2.79] is the complete solution of the undamped dynamic response using modal analysis. In structural dynamics, often frequency-domain analysis is used. The dynamic response in the frequency domain can be obtained by substituting $s = \mathrm{i}\omega$ as

$$\begin{aligned}\bar{\mathbf{q}}(\mathrm{i}\omega) &= \mathbf{X}\left[-\omega^2\mathbf{I}+\boldsymbol{\Omega}^2\right]^{-1}\mathbf{X}^T\left\{\bar{\mathbf{f}}(\mathrm{i}\omega)+\mathbf{M}\dot{\mathbf{q}}_0+\mathrm{i}\omega\mathbf{M}\mathbf{q}_0\right\}\\ &= \mathbf{H}(\mathrm{i}\omega)\left\{\bar{\mathbf{f}}(\mathrm{i}\omega)+\mathbf{M}\dot{\mathbf{q}}_0+\mathrm{i}\omega\mathbf{M}\mathbf{q}_0\right\}.\end{aligned} \quad [2.80]$$

The term appearing in the preceding equation

$$\mathbf{H}(\mathrm{i}\omega) = \mathbf{X}\left[-\omega^2\mathbf{I}+\boldsymbol{\Omega}^2\right]^{-1}\mathbf{X}^T \quad [2.81]$$

is often known as the transfer function matrix. Note that $\left[-\omega^2\mathbf{I}+\boldsymbol{\Omega}^2\right]$ is a diagonal matrix and therefore its inverse is easy to obtain since

$$\left[-\omega^2\mathbf{I}+\boldsymbol{\Omega}^2\right]^{-1} = \mathrm{diag}\left[\frac{1}{\omega_1^2-\omega^2}, \frac{1}{\omega_2^2-\omega^2}, \cdots, \frac{1}{\omega_N^2-\omega^2}\right]. \quad [2.82]$$

The product $\mathbf{X}\left[-\omega^2\mathbf{I}+\boldsymbol{\Omega}^2\right]^{-1}\mathbf{X}^T$ can be expressed as

$$\begin{aligned}\mathbf{X}\left[-\omega^2\mathbf{I}+\boldsymbol{\Omega}^2\right]^{-1}\mathbf{X}^T &= [\mathbf{x}_1, \mathbf{x}_2, \cdots, \mathbf{x}_N]\\ &\quad\times \mathrm{diag}\left[\frac{1}{\omega_1^2-\omega^2}, \frac{1}{\omega_2^2-\omega^2}, \cdots, \frac{1}{\omega_N^2-\omega^2}\right]\begin{bmatrix}\mathbf{x}_1^T\\ \mathbf{x}_2^T\\ \vdots\\ \mathbf{x}_N^T\end{bmatrix}\end{aligned} \quad [2.83]$$

$$= [\mathbf{x}_1, \mathbf{x}_2, \cdots, \mathbf{x}_N]\begin{bmatrix}\frac{\mathbf{x}_1^T}{\omega_1^2-\omega^2}\\ \frac{\mathbf{x}_2^T}{\omega_2^2-\omega^2}\\ \vdots\\ \frac{\mathbf{x}_N^T}{\omega_N^2-\omega^2}\end{bmatrix} = \left[\frac{\mathbf{x}_1\mathbf{x}_1^T}{\omega_1^2-\omega^2}+\frac{\mathbf{x}_2\mathbf{x}_2^T}{\omega_2^2-\omega^2}+\cdots+\frac{\mathbf{x}_N\mathbf{x}_N^T}{\omega_N^2-\omega^2}\right]. \quad [2.84]$$

From this, we obtain the familiar expression of the transfer function matrix as

$$\mathbf{H}(\mathrm{i}\omega) = \sum_{j=1}^{N} \frac{\mathbf{x}_j \mathbf{x}_j^T}{\omega_j^2 - \omega^2}. \qquad [2.85]$$

Substituting $\mathbf{H}(\mathrm{i}\omega)$ in the expression of dynamic response [2.80], we have

$$\begin{aligned} \bar{\mathbf{q}}(\mathrm{i}\omega) &= \sum_{j=1}^{N} \frac{\mathbf{x}_j \mathbf{x}_j^T \left\{ \bar{\mathbf{f}}(\mathrm{i}\omega) + \mathbf{M}\dot{\mathbf{q}}_0 + \mathrm{i}\omega \mathbf{M}\mathbf{q}_0 \right\}}{\omega_j^2 - \omega^2} \\ &= \sum_{j=1}^{N} \frac{\mathbf{x}_j^T \bar{\mathbf{f}}(\mathrm{i}\omega) + \mathbf{x}_j^T \mathbf{M}\dot{\mathbf{q}}_0 + \mathrm{i}\omega \mathbf{x}_j^T \mathbf{M}\mathbf{q}_0}{\omega_j^2 - \omega^2} \mathbf{x}_j. \end{aligned} \qquad [2.86]$$

This expression shows that the dynamic response of the system is a linear combination of the mode shapes.

2.3.2.2. *Time-domain analysis*

Rewriting the expression of the frequency-domain response in equation [2.86] in the Laplace domain, we have

$$\bar{\mathbf{q}}(s) = \sum_{j=1}^{N} \left\{ \frac{\mathbf{x}_j^T \bar{\mathbf{f}}(s)}{s^2 + \omega_j^2} + \frac{\mathbf{x}_j^T \mathbf{M}\dot{\mathbf{q}}_0}{s^2 + \omega_j^2} + \frac{s}{s^2 + \omega_j^2} \mathbf{x}_j^T \mathbf{M}\mathbf{q}_0 \right\} \mathbf{x}_j. \qquad [2.87]$$

To obtain the vibration response in the time domain it is required to consider the inverse Laplace transform. Taking the inverse Laplace transform of $\bar{\mathbf{q}}(s)$, we have

$$\mathbf{q}(t) = \mathcal{L}^{-1} \left[\bar{\mathbf{q}}(s) \right] = \sum_{j=1}^{N} a_j(t) \mathbf{x}_j \qquad [2.88]$$

where the time-dependent constants are given by

$$a_j(t) = \mathcal{L}^{-1} \left[\frac{\mathbf{x}_j^T \bar{\mathbf{f}}(s)}{s^2 + \omega_j^2} \right] + \mathcal{L}^{-1} \left[\frac{1}{s^2 + \omega_j^2} \right] \mathbf{x}_j^T \mathbf{M}\dot{\mathbf{q}}_0 + \mathcal{L}^{-1} \left[\frac{s}{s^2 + \omega_j^2} \right] \mathbf{x}_j^T \mathbf{M}\mathbf{q}_0. \qquad [2.89]$$

Following the inverse Laplace transform approach described in section 2.1.2, the time-dependent coefficients a_j in equation [2.89] can be expressed as

$$a_j(t) = \int_0^t \frac{1}{\omega_j} \mathbf{x}_j^T \mathbf{f}(\tau) \sin\left(\omega_j (t-\tau)\right) d\tau + \frac{1}{\omega_j} \sin(\omega_j t) \mathbf{x}_j^T \mathbf{M} \dot{\mathbf{q}}_0 + \cos(\omega_j t) \mathbf{x}_j^T \mathbf{M} \mathbf{q}_0. \tag{2.90}$$

Collecting the terms associated with $\sin\left(\omega_{d_j} t\right)$ and $\cos\left(\omega_{d_j} t\right)$ this expression can be simplified as

$$a_j(t) = \int_0^t \frac{1}{\omega_j} \mathbf{x}_j^T \mathbf{f}(\tau) \sin\left(\omega_j (t-\tau)\right) d\tau + B_j \cos\left(\omega_j t - \vartheta_j\right) \tag{2.91}$$

where

$$B_j = \sqrt{\left(\mathbf{x}_j^T \mathbf{M} \mathbf{q}_0\right)^2 + \left(\frac{\mathbf{x}_j^T \mathbf{M} \dot{\mathbf{q}}_0}{\omega_j}\right)^2} \tag{2.92}$$

$$\text{and} \quad \tan \vartheta_j = \frac{\mathbf{x}_j^T \mathbf{M} \dot{\mathbf{q}}_0}{\omega_j \left(\mathbf{x}_j^T \mathbf{M} \mathbf{q}_0\right)}. \tag{2.93}$$

Observe that the second part of equation [2.91], i.e. the term $B_j \cos\left(\omega_j t - \vartheta_j\right)$ only depends on the initial conditions and independent of the applied loading.

2.4. Proportionally damped systems

Equation of motion of a viscously damped system can be obtained from Lagrange's equation and using the Rayleigh dissipation function given by [1.19]. The non-conservative forces can be obtained as

$$Q_{\mathrm{nc}_k} = -\frac{\partial \mathcal{F}}{\partial \dot{q}_k}, \quad k = 1, \cdots, N \tag{2.94}$$

and consequently the equation of motion can be expressed as

$$\mathbf{M}\ddot{\mathbf{q}}(t) + \mathbf{C}\dot{\mathbf{q}}(t) + \mathbf{K}\mathbf{q}(t) = \mathbf{f}(t). \tag{2.95}$$

The aim is to solve this equation, together with the initial conditions, by classical modal analysis as described in section 2.3.1. Using the modal transformation in [2.69], premultiplying equation [2.95] by $\mathbf{X}^T$ and using the orthogonality relationships in [2.67] and [2.68], equation of motion of a damped system in the modal coordinates may be obtained as

$$\ddot{\mathbf{y}}(t) + \mathbf{X}^T\mathbf{C}\mathbf{X}\dot{\mathbf{y}}(t) + \boldsymbol{\Omega}^2\mathbf{y}(t) = \tilde{\mathbf{f}}(t). \quad [2.96]$$

Clearly, unless $\mathbf{X}^T\mathbf{C}\mathbf{X}$ is a diagonal matrix, no advantage can be gained by employing modal analysis because the equations of motion will still be coupled. To solve this problem, it is common to assume *proportional damping* which we will discuss in some detail.

With the proportional damping assumption, the damping matrix $\mathbf{C}$ is simultaneously diagonalizable with $\mathbf{M}$ and $\mathbf{K}$. This implies that the damping matrix in the modal coordinate

$$\mathbf{C}' = \mathbf{X}^T\mathbf{C}\mathbf{X} \quad [2.97]$$

is a diagonal matrix. This matrix is also known as the modal damping matrix. The damping factors ζ_j are defined from the diagonal elements of the modal damping matrix as

$$C'_{jj} = 2\zeta_j\omega_j \quad \forall j = 1, \cdots, N. \quad [2.98]$$

Such a damping model, introduced by Lord Rayleigh [RAY 77], allows us to analyze damped systems in very much the same manner as undamped systems since the equation of motion in the modal coordinates can be decoupled as

$$\ddot{y}_j(t) + 2\zeta_j\omega_j\dot{y}_j(t) + \omega_j^2 y_j(t) = \tilde{f}_j(t) \quad \forall j = 1, \cdots, N. \quad [2.99]$$

The proportional damping model expresses the damping matrix as a linear combination of the mass and stiffness matrices, that is

$$\mathbf{C} = \alpha_1\mathbf{M} + \alpha_2\mathbf{K} \quad [2.100]$$

where α_1, α_2 are real scalars. This damping model is also known as "Rayleigh damping" or "classical damping". Modes of classically damped systems preserve the simplicity of the real normal modes as in the undamped case. Later, in a significant paper, Caughey and O'Kelly [CAU 65] have shown that the classical damping can exist in a more general situation.

2.4.1. *Condition for proportional damping*

Caughey and O'Kelly [CAU 65] have derived the condition that the system matrices must satisfy so that viscously damped linear systems possess classical normal modes. Their result can be described in the following theorem.

THEOREM 2.1.– A viscously damped system [2.95] possesses classical normal modes if and only if $\mathbf{CM}^{-1}\mathbf{K} = \mathbf{KM}^{-1}\mathbf{C}$.

OUTLINE OF THE PROOF.– See the original paper by Caughey and O'Kelly [CAU 65] for the detailed proof. Assuming $\mathbf{M}$ is not singular, premultiplying equation [2.95] by $\mathbf{M}^{-1}$ we have

$$\mathbf{I}\ddot{\mathbf{q}}(t) + [\mathbf{M}^{-1}\mathbf{C}]\dot{\mathbf{q}}(t) + [\mathbf{M}^{-1}\mathbf{K}]\mathbf{q}(t) = \mathbf{M}^{-1}\mathbf{f}(t). \qquad [2.101]$$

For classical normal modes, [2.101] must be diagonalized by an orthogonal transformation. Two matrices $\mathbf{A}$ and $\mathbf{B}$ can be diagonalized by an orthogonal transformation if and only if the commute in product [BEL 60], i.e. $\mathbf{AB} = \mathbf{BA}$. Using this condition in [2.101], we have

$$\begin{aligned} & [\mathbf{M}^{-1}\mathbf{C}][\mathbf{M}^{-1}\mathbf{K}] = [\mathbf{M}^{-1}\mathbf{K}][\mathbf{M}^{-1}\mathbf{C}], \\ \text{or} \quad & \mathbf{CM}^{-1}\mathbf{K} = \mathbf{KM}^{-1}\mathbf{C} \quad (\text{premultiplying both sides by } \mathbf{M}). \end{aligned} \qquad [2.102]$$

A modified and more general version of this theorem was proved in [ADH 01a].

EXAMPLE 2.1.– Assume that a system's mass, stiffness and damping matrices are given by

$$\mathbf{M} = \begin{bmatrix} 1.0 & 1.0 & 1.0 \\ 1.0 & 2.0 & 2.0 \\ 1.0 & 2.0 & 3.0 \end{bmatrix}, \quad \mathbf{K} = \begin{bmatrix} 2 & -1 & 0.5 \\ -1 & 1.2 & 0.4 \\ 0.5 & 0.4 & 1.8 \end{bmatrix} \quad \text{and} \quad \mathbf{C} = \begin{bmatrix} 15.25 & -9.8 & 3.4 \\ -9.8 & 6.48 & -1.84 \\ 3.4 & -1.84 & 2.22 \end{bmatrix}. \qquad [2.103]$$

It may be verified that all the system matrices are positive definite. The mass-normalized undamped modal matrix is obtained as

$$\mathbf{X} = \begin{bmatrix} 0.4027 & -0.5221 & -1.2511 \\ 0.5845 & -0.4888 & 1.1914 \\ -0.1127 & 0.9036 & -0.4134 \end{bmatrix}. \qquad [2.104]$$

Since Caughey and O'Kelly's [CAU 65], condition

$$\mathbf{K}\mathbf{M}^{-1}\mathbf{C} = \mathbf{C}\mathbf{M}^{-1}\mathbf{K} = \begin{bmatrix} 125.45 & -80.92 & 28.61 \\ -80.92 & 52.272 & -18.176 \\ 28.61 & -18.176 & 7.908 \end{bmatrix} \qquad [2.105]$$

is satisfied, the system possesses classical normal modes and that **X** given in equation [2.104] is the modal matrix. Because the system is positive definite, two other equivalent conditions, i.e.

$$\mathbf{M}\mathbf{K}^{-1}\mathbf{C} = \mathbf{C}\mathbf{K}^{-1}\mathbf{M} = \begin{bmatrix} 2.0 & -1.0 & 0.5 \\ -1.0 & 1.2 & 0.4 \\ 0.5 & 0.4 & 1.8 \end{bmatrix} \qquad [2.106]$$

and

$$\mathbf{M}\mathbf{C}^{-1}\mathbf{K} = \mathbf{K}\mathbf{C}^{-1}\mathbf{M} = \begin{bmatrix} 4.1 & 6.2 & 5.6 \\ 6.2 & 9.73 & 9.2 \\ 5.6 & 9.2 & 9.6 \end{bmatrix} \qquad [2.107]$$

are also satisfied.

2.4.2. *Generalized proportional damping*

Obtaining a damping matrix from "first principles" as with the mass and stiffness matrices is not possible for most systems. For this reason, assuming **M** and **K** are known, we often want to find **C** in terms of **M** and **K** such that the system still possesses classical normal modes. Of course, the earliest work along this line is the proportional damping shown in equation [2.100] by Rayleigh [RAY 77]. It can be verified that, for positive definite systems, expressing **C** in such a way will always satisfy the condition given by theorem 2.1. Caughey [CAU 60] proposed that a *sufficient* condition for the existence of classical normal modes is: if $\mathbf{M}^{-1}\mathbf{C}$ can be expressed in a series involving powers of $\mathbf{M}^{-1}\mathbf{K}$. His result generalized Rayleigh's result, which turns out to be the first two terms of the series. Later, Caughey and O'Kelly [CAU 65] proved that the series representation of damping

$$\mathbf{C} = \mathbf{M}\sum_{j=0}^{N-1} \alpha_j \left[\mathbf{M}^{-1}\mathbf{K}\right]^j \qquad [2.108]$$

is the *necessary and sufficient* condition for existence of classical normal modes for systems without any repeated roots. This series is now known as the "Caughey series".

A further generalized and useful form of proportional damping was proposed in [ADH 01a]. It can be shown that we can express the damping matrix in the form

$$\mathbf{C} = \mathbf{M}\,\beta_1\left(\mathbf{M}^{-1}\mathbf{K}\right) + \mathbf{K}\,\beta_2\left(\mathbf{K}^{-1}\mathbf{M}\right) \qquad [2.109]$$

$$\text{or} \quad \mathbf{C} = \beta_3\left(\mathbf{K}\mathbf{M}^{-1}\right)\mathbf{M} + \beta_4\left(\mathbf{M}\mathbf{K}^{-1}\right)\mathbf{K} \qquad [2.110]$$

The functions $\beta_i(\bullet)$, $i = 1, \cdots, 4$ can have very general forms – they may consist of an arbitrary number of multiplications, divisions, summations, subtractions or powers of any other functions or can even be functional compositions. Thus, any conceivable form of analytic functions that are valid for scalars can be used in equations [2.109] and [2.110]. In a natural way, common restrictions applicable to scalar functions are also valid, for example logarithm of a negative number is not permitted. Although the functions $\beta_i(\bullet)$, $i = 1, \cdots, 4$ are general, the expression of $\mathbf{C}$ in [2.109] or [2.110] gets restricted because of the special nature of the *arguments* in the functions. As a result, $\mathbf{C}$ represented in [2.109] or [2.110] does not cover the whole $\mathbb{R}^{N\times N}$, it is well known that many damped systems do not possess classical normal modes.

Rayleigh's result [2.100] can be obtained directly from equation [2.109] or [2.110] as a very special case by choosing each matrix function $\beta_i(\bullet)$ as real scalar times an identity matrix, that is

$$\beta_i(\bullet) = \alpha_i \mathbf{I}. \qquad [2.111]$$

The damping matrix expressed in equation [2.109] or [2.110] provides a new way of interpreting the "Rayleigh damping" or "proportional damping" where the identity matrices (always) associated in the right or left side of $\mathbf{M}$ and $\mathbf{K}$ are replaced by arbitrary matrix functions $\beta_i(\bullet)$ with proper arguments. This kind of damping model is called *generalized proportional damping*. We call the representation in equation [2.109] *right-functional form* and that in equation [2.110] *left-functional form*. Caughey series [2.108] is an example of right-functional form. Note that if $\mathbf{M}$ or $\mathbf{K}$ is singular then the argument involving its corresponding inverse has to be removed from the functions. We call the functions $\beta_i(\bullet)$ as *proportional damping functions* which are consistent with the definition of proportional damping constants (α_i) in Rayleigh's model. From this discussion, we have the following general result for damped linear systems.

THEOREM 2.2.– A viscously damped positive definite linear system possesses classical normal modes if and only if $\mathbf{C}$ can be represented by

1) $\mathbf{C} = \mathbf{M}\,\beta_1\left(\mathbf{M}^{-1}\mathbf{K}\right) + \mathbf{K}\,\beta_2\left(\mathbf{K}^{-1}\mathbf{M}\right)$, or

2) $\mathbf{C} = \beta_3\left(\mathbf{K}\mathbf{M}^{-1}\right)\mathbf{M} + \beta_4\left(\mathbf{M}\mathbf{K}^{-1}\right)\mathbf{K}$

for any $\beta_i(\bullet), i = 1, \cdots, 4$.

PROOF.– Consider the "if" part first. Suppose $\mathbf{X}$ is the mass normalized modal matrix and $\boldsymbol{\Omega}$ is the diagonal matrix containing the undamped natural frequencies. By definitions of these quantities, we have

$$\mathbf{X}^T\mathbf{M}\mathbf{X} = \mathbf{I} \qquad [2.112]$$

$$\text{and} \quad \mathbf{X}^T\mathbf{K}\mathbf{X} = \boldsymbol{\Omega}^2. \qquad [2.113]$$

From these equations, we obtain

$$\mathbf{M} = \mathbf{X}^{-T}\mathbf{X}^{-1}, \quad \mathbf{K} = \mathbf{X}^{-T}\boldsymbol{\Omega}^2\mathbf{X}^{-1} \qquad [2.114]$$

$$\mathbf{M}^{-1}\mathbf{K} = \mathbf{X}\boldsymbol{\Omega}^2\mathbf{X}^{-1} \quad \text{and} \quad \mathbf{K}^{-1}\mathbf{M} = \mathbf{X}\boldsymbol{\Omega}^{-2}\mathbf{X}^{-1}. \qquad [2.115]$$

Because the functions $\beta_1(\bullet)$ and $\beta_2(\bullet)$ are assumed to be analytic in the neighborhood of all the eigenvalues of $\mathbf{M}^{-1}\mathbf{K}$ and $\mathbf{K}^{-1}\mathbf{M}$, respectively, they can be expressed in polynomial forms using the Taylor series expansion. Following Bellman [BEL 60] (Chapter 6), we may obtain

$$\beta_1\left(\mathbf{M}^{-1}\mathbf{K}\right) = \mathbf{X}\,\beta_1\left(\boldsymbol{\Omega}^2\right)\mathbf{X}^{-1} \qquad [2.116]$$

$$\text{and} \quad \beta_2\left(\mathbf{K}^{-1}\mathbf{M}\right) = \mathbf{X}\,\beta_2\left(\boldsymbol{\Omega}^{-2}\right)\mathbf{X}^{-1}. \qquad [2.117]$$

A viscously damped system will possess classical normal modes if $\mathbf{X}^T\mathbf{C}\mathbf{X}$ is a diagonal matrix. Considering expression (1) in the theorem and using equations [2.114] and [2.115], we have

$$\begin{aligned}\mathbf{X}^T\mathbf{C}\mathbf{X} &= \mathbf{X}^T\left[\mathbf{M}\,\beta_1\left(\mathbf{M}^{-1}\mathbf{K}\right) + \mathbf{K}\,\beta_2\left(\mathbf{K}^{-1}\mathbf{M}\right)\right]\mathbf{X}\\ &= \mathbf{X}^T\left[\mathbf{X}^{-T}\mathbf{X}^{-1}\,\beta_1\left(\mathbf{M}^{-1}\mathbf{K}\right) + \mathbf{X}^{-T}\boldsymbol{\Omega}^2\mathbf{X}^{-1}\,\beta_2\left(\mathbf{K}^{-1}\mathbf{M}\right)\right]\mathbf{X}.\end{aligned} \qquad [2.118]$$

Utilizing equations [2.116] and [2.117] and carrying out the matrix multiplications, equation [2.118] reduces to

$$\begin{aligned}\mathbf{X}^T\mathbf{C}\mathbf{X} &= \left[\mathbf{X}^{-1}\mathbf{X}\,\beta_1\left(\boldsymbol{\Omega}^2\right)\mathbf{X}^{-1} + \boldsymbol{\Omega}^2\mathbf{X}^{-1}\mathbf{X}\,\beta_2\left(\boldsymbol{\Omega}^{-2}\right)\mathbf{X}^{-1}\right]\mathbf{X}\\ &= \beta_1\left(\boldsymbol{\Omega}^2\right) + \boldsymbol{\Omega}^2\beta_2\left(\boldsymbol{\Omega}^{-2}\right).\end{aligned} \qquad [2.119]$$

Equation [2.119] clearly shows that $\mathbf{X}^T\mathbf{C}\mathbf{X}$ is a diagonal matrix.

To prove the "only if" part, suppose

$$\mathbf{P} = \mathbf{X}^T\mathbf{C}\mathbf{X} \qquad [2.120]$$

is a general matrix (not necessary diagonal). Then there exists a non-zero matrix $\mathbf{S}$ such that (similarity transform)

$$\mathbf{S}^{-1}\mathbf{P}\mathbf{S} = \mathcal{D} \qquad [2.121]$$

where $\mathcal{D}$ is a diagonal matrix. Using equations [2.119] and [2.120], we have

$$\mathbf{S}^{-1}\mathcal{D}_1\mathbf{S} = \mathcal{D} \tag{2.122}$$

where $\mathcal{D}_1$ is another diagonal matrix. Equation [2.122] indicates that two diagonal matrices are related by a similarity transformation. This can only happen when they are the same and the transformation matrix is an identity matrix, that is $\mathbf{S} = \mathbf{I}$. Using this in equation [2.121] proves that $\mathbf{P}$ must be a diagonal matrix.

EXAMPLE 2.2.– To investigate the applicability of this result, it is shown that the linear dynamic system satisfying the following equations of free vibration

$$\begin{aligned}\mathbf{M}\ddot{\mathbf{q}} + \Big[&\mathbf{M}e^{-\left(\mathbf{M}^{-1}\mathbf{K}\right)^2/2}\sinh(\mathbf{K}^{-1}\mathbf{M}\ln(\mathbf{M}^{-1}\mathbf{K})^{2/3}) \\ &+ \mathbf{K}\cos^2(\mathbf{K}^{-1}\mathbf{M})\sqrt[4]{\mathbf{K}^{-1}\mathbf{M}}\tan^{-1}\frac{\sqrt{\mathbf{M}^{-1}\mathbf{K}}}{\pi}\Big]\dot{\mathbf{q}} + \mathbf{K}\mathbf{q} = \mathbf{0}\end{aligned} \tag{2.123}$$

possesses classical normal modes and can be analyzed using modal analysis. Here, $\mathbf{M}$ and $\mathbf{K}$ are the same as example 2.1.

Direct calculation shows

$$\mathbf{C} = \begin{bmatrix} -67.9188 & -104.8208 & -95.9566 \\ -104.8208 & -161.1897 & -147.7378 \\ -95.9566 & -147.7378 & -135.2643 \end{bmatrix}. \tag{2.124}$$

Using the modal matrix calculated before in equation [2.104], we obtain

$$\mathbf{X}^T\mathbf{C}\mathbf{X} = \begin{bmatrix} -88.9682 & 0.0 & 0.0 \\ 0.0 & 0.0748 & 0.0 \\ 0.0 & 0.0 & 0.5293 \end{bmatrix}, \tag{2.125}$$

which is a diagonal matrix. Analytically, the modal damping factors can be obtained as

$$2\zeta_j\omega_j = e^{-\omega_j^4/2}\sinh\left(\frac{1}{\omega_j^2}\ln\frac{4}{3}\omega_j\right) + \omega_j^2\cos^2\left(\frac{1}{\omega_j^2}\right)\frac{1}{\sqrt{\omega_j}}\tan^{-1}\frac{\omega_j}{\pi}. \tag{2.126}$$

A natural question that arises in the context of the generalized proportional damping is how to obtain the damping functions from experimental modal analysis. This will be addressed later in [ADH 14b], Chapter 2.

2.4.3. *Dynamic response*

Dynamic response of proportionally damped systems can be obtained in a similar way to that of undamped systems discussed before in section 2.3.2. Both frequency-domain and time-domain approaches are shown here.

2.4.3.1. *Frequency-domain analysis*

Taking the Laplace transform of [2.95] and considering the initial conditions in [2.54], we have

$$s^2\mathbf{M}\bar{\mathbf{q}} - s\mathbf{M}\mathbf{q}_0 - \mathbf{M}\dot{\mathbf{q}}_0 + s\mathbf{C}\bar{\mathbf{q}} - \mathbf{C}\mathbf{q}_0 + \mathbf{K}\bar{\mathbf{q}} = \bar{\mathbf{f}}(s) \qquad [2.127]$$

$$\text{or} \quad \left[s^2\mathbf{M} + s\mathbf{C} + \mathbf{K}\right]\bar{\mathbf{q}} = \bar{\mathbf{f}}(s) + \mathbf{M}\dot{\mathbf{q}}_0 + \mathbf{C}\mathbf{q}_0 + s\mathbf{M}\mathbf{q}_0. \qquad [2.128]$$

Consider the modal damping matrix

$$\mathbf{C}' = \mathbf{X}^T\mathbf{C}\mathbf{X} = 2\boldsymbol{\zeta}\boldsymbol{\Omega} \qquad [2.129]$$

where

$$\boldsymbol{\zeta} = \text{diag}\left[\zeta_1, \zeta_2, \cdots, \zeta_N\right] \in \mathbb{R}^{N\times N} \qquad [2.130]$$

is the diagonal matrix containing the modal damping factors. Using the mode orthogonality relationships and following the procedure similar to undamped systems, it is easy to show that

$$\bar{\mathbf{q}}(s) = \mathbf{X}\left[s^2\mathbf{I} + 2s\boldsymbol{\zeta}\boldsymbol{\Omega} + \boldsymbol{\Omega}^2\right]^{-1}\mathbf{X}^T\left\{\bar{\mathbf{f}}(s) + \mathbf{M}\dot{\mathbf{q}}_0 + \mathbf{C}\mathbf{q}_0 + s\mathbf{M}\mathbf{q}_0\right\}. \qquad [2.131]$$

The dynamic response in the frequency domain can be obtained by substituting $s = \mathrm{i}\omega$. Note that $\left[s^2\mathbf{I} + 2s\boldsymbol{\zeta}\boldsymbol{\Omega} + \boldsymbol{\Omega}^2\right]$ is a diagonal matrix and therefore its inverse is easy to obtain. Following the procedure similar to undamped systems, the transfer function matrix can be obtained as

$$\mathbf{H}(\mathrm{i}\omega) = \mathbf{X}\left[-\omega^2\mathbf{I} + 2\mathrm{i}\omega\boldsymbol{\zeta}\boldsymbol{\Omega} + \boldsymbol{\Omega}^2\right]^{-1}\mathbf{X}^T = \sum_{j=1}^{N} \frac{\mathbf{x}_j\mathbf{x}_j^T}{-\omega^2 + 2\mathrm{i}\omega\zeta_j\omega_j + \omega_j^2}. \qquad [2.132]$$

Using this, the dynamic response in the frequency domain can be conveniently represented from equation [2.131] as

$$\bar{\mathbf{q}}(\mathrm{i}\omega) = \sum_{j=1}^{N} \frac{\mathbf{x}_j^T \bar{\mathbf{f}}(\mathrm{i}\omega) + \mathbf{x}_j^T \mathbf{M}\dot{\mathbf{q}}_0 + \mathbf{x}_j^T \mathbf{C}\mathbf{q}_0 + \mathrm{i}\omega \mathbf{x}_j^T \mathbf{M}\mathbf{q}_0}{-\omega^2 + 2\mathrm{i}\omega\zeta_j\omega_j + \omega_j^2} \mathbf{x}_j. \qquad [2.133]$$

Therefore, like undamped systems, the dynamic response of proportionally damped system can also be expressed as a linear combination of the undamped mode shapes. This shows how classical modal analysis can be used for proportionally damped systems.

2.4.3.2. *Time-domain analysis*

Rewrite the frequency-domain response in equation [2.133] in the Laplace domain as

$$\begin{aligned} \bar{\mathbf{q}}(s) &= \sum_{j=1}^{N} \left\{ \frac{\mathbf{x}_j^T \bar{\mathbf{f}}(s)}{s^2 + 2s\zeta_j\omega_j + \omega_j^2} + \frac{\mathbf{x}_j^T \mathbf{M}\dot{\mathbf{q}}_0 + \mathbf{x}_j^T \mathbf{C}\mathbf{q}_0}{s^2 + 2s\zeta_j\omega_j + \omega_j^2} \right. \\ &\quad \left. + \frac{s}{s^2 + 2s\zeta_j\omega_j + \omega_j^2} \mathbf{x}_j^T \mathbf{M}\mathbf{q}_0 \right\} \mathbf{x}_j. \end{aligned} \qquad [2.134]$$

We can reorganize the denominator as

$$s^2 + 2s\zeta_j\omega_j + \omega_j^2 = (s + \zeta_j\omega_j)^2 - (\zeta_j\omega_j)^2 + \omega_j^2 = (s + \zeta_j\omega_j)^2 + \omega_{d_j}^2 \qquad [2.135]$$

where the damped natural frequency for the jth mode is given by

$$\omega_{d_j} = \omega_j \sqrt{1 - \zeta_j^2} \qquad [2.136]$$

Now we use the expressions of the inverse Laplace transforms in equations [2.39] and [2.40] with α and β standing for $\zeta\omega_j$ and ω_{d_j}, respectively. Taking the inverse Laplace transform of the dynamic response [2.134], we have

$$\mathbf{q}(t) = \mathcal{L}^{-1}\left[\bar{\mathbf{q}}(s)\right] = \sum_{j=1}^{N} a_j(t)\mathbf{x}_j \qquad [2.137]$$

where the time-dependent constants are given by

$$
\begin{aligned}
a_j(t) &= \mathcal{L}^{-1}\left[\frac{\mathbf{x}_j^T\bar{\mathbf{f}}(s)}{(s+\zeta_j\omega_j)^2+\omega_{d_j}^2}\right] \\
&\quad +\mathcal{L}^{-1}\left[\frac{1}{(s+\zeta_j\omega_j)^2+\omega_{d_j}^2}\right]\left(\mathbf{x}_j^T\mathbf{M}\dot{\mathbf{q}}_0+\mathbf{x}_j^T\mathbf{C}\mathbf{q}_0\right) \\
&\quad +\mathcal{L}^{-1}\left[\frac{s}{(s+\zeta_j\omega_j)^2+\omega_{d_j}^2}\right]\mathbf{x}_j^T\mathbf{M}\mathbf{q}_0 \\
&= \int_0^t \frac{1}{\omega_{d_j}}\mathbf{x}_j^T\mathbf{f}(\tau)e^{-\zeta_j\omega_j(t-\tau)}\sin\left(\omega_{d_j}(t-\tau)\right)d\tau \\
&\quad +\frac{e^{-\zeta_j\omega_j t}}{\omega_{d_j}}\sin(\omega_{d_j}t)\left(\mathbf{x}_j^T\mathbf{M}\dot{\mathbf{q}}_0+\mathbf{x}_j^T\mathbf{C}\mathbf{q}_0\right) \\
&\quad +\left\{e^{-\zeta_j\omega_j t}\cos\left(\omega_{d_j}t\right)-\frac{\zeta_j\omega_j\, e^{-\zeta_j\omega_j t}\sin\left(\omega_{d_j}t\right)}{\omega_{d_j}}\right\}\mathbf{x}_j^T\mathbf{M}\mathbf{q}_0 \\
&= \int_0^t \frac{1}{\omega_{d_j}}\mathbf{x}_j^T\mathbf{f}(\tau)e^{-\zeta_j\omega_j(t-\tau)}\sin\left(\omega_{d_j}(t-\tau)\right)d\tau \\
&\quad +e^{-\zeta_j\omega_j t}\left\{\frac{\sin(\omega_{d_j}t)}{\omega_{d_j}}\left(\mathbf{x}_j^T\mathbf{M}\dot{\mathbf{q}}_0+\mathbf{x}_j^T\mathbf{C}\mathbf{q}_0-\zeta_j\omega_j\mathbf{x}_j^T\mathbf{M}\mathbf{q}_0\right)\right. \\
&\quad \left.+\cos\left(\omega_{d_j}t\right)\left(\mathbf{x}_j^T\mathbf{M}\mathbf{q}_0\right)\right\}
\end{aligned}
\quad [2.138]
$$

The convolution theorem is used to obtain the inverse Laplace transform of the first part. The formulas shown in [2.39] and [2.40] are used to obtain the inverse Laplace transform of the second and third parts. The coefficients associated with $\sin(\omega_{d_j}t)$ in the preceding equation can be simplified. Suppose the initial displacement in the modal coordinate is $\mathbf{y}_0$ so that $\mathbf{q}_0 = \mathbf{X}\mathbf{y}_0$. Therefore

$$
\mathbf{x}_j^T\mathbf{M}\mathbf{q}_0 = \mathbf{x}_j^T\mathbf{M}\mathbf{X}\mathbf{y}_0 = \mathbf{x}_j^T\mathbf{M}\left[\mathbf{x}_1, \mathbf{x}_2, \cdots, \mathbf{x}_N\right]\mathbf{y}_0 = y_{0_j} \quad [2.139]
$$

which is the initial displacement of the jth modal coordinate. We have used the mass orthonormality properties of the eigenvectors in [2.63]. Using this and recalling the modal damping matrix in equation [2.129], we have

$$
\begin{aligned}
\mathbf{x}_j^T\mathbf{C}\mathbf{q}_0-\zeta_j\omega_j\mathbf{x}_j^T\mathbf{M}\mathbf{q}_0 &= \mathbf{x}_j^T\mathbf{C}\mathbf{X}\mathbf{y}_0-\zeta_j\omega_j\mathbf{x}_j^T\mathbf{M}\mathbf{X}\mathbf{y}_0 = 2\zeta_j\omega_j y_{0_j}-\zeta_j\omega_j y_{0_j} \\
&= \zeta_j\omega_j y_{0_j}
\end{aligned}
\quad [2.140]
$$

Using equation [2.139], we can establish

$$\mathbf{x}_j^T\mathbf{C}\mathbf{q}_0 - \zeta_j\omega_j\mathbf{x}_j^T\mathbf{M}\mathbf{q}_0 = \zeta_j\omega_j\mathbf{x}_j^T\mathbf{M}\mathbf{q}_0 \quad \text{or} \quad \mathbf{x}_j^T\mathbf{C}\mathbf{q}_0 = 2\zeta_j\omega_j\mathbf{x}_j^T\mathbf{M}\mathbf{q}_0 \qquad [2.141]$$

Substituting this simplification in equation [2.138] and collecting the terms associated with $\sin\left(\omega_{d_j}\,t\right)$ and $\cos\left(\omega_{d_j}\,t\right)$, the resulting expression can be simplified as

$$\begin{aligned} a_j(t) &= \int_0^t \frac{1}{\omega_{d_j}}\mathbf{x}_j^T\mathbf{f}(\tau)e^{-\zeta_j\omega_j(t-\tau)}\sin\left(\omega_{d_j}(t-\tau)\right)d\tau \\ &\quad + e^{-\zeta_j\omega_j t}B_j\cos\left(\omega_{d_j}t-\vartheta_j\right) \end{aligned} \qquad [2.142]$$

where

$$B_j = \sqrt{\left(\mathbf{x}_j^T\mathbf{M}\mathbf{q}_0\right)^2 + \left(\frac{\mathbf{x}_j^T\mathbf{M}\dot{\mathbf{q}}_0 + \frac{1}{2}\mathbf{x}_j^T\mathbf{C}\mathbf{q}_0}{\omega_{d_j}}\right)^2} \qquad [2.143]$$

$$\text{and} \quad \tan\vartheta_j = \frac{\mathbf{x}_j^T\mathbf{M}\dot{\mathbf{q}}_0 + \frac{1}{2}\mathbf{x}_j^T\mathbf{C}\mathbf{q}_0}{\omega_{d_j}\left(\mathbf{x}_j^T\mathbf{M}\mathbf{q}_0\right)} \qquad [2.144]$$

All the terms appearing in the preceding equations can be obtained from the natural frequencies, modal damping factors, forcing function and initial conditions. The modal amplitudes and phase angles in equations [2.143] and [2.144] can also be expressed in terms of the initial conditions in the modal coordinates as

$$B_j = \sqrt{y_{0_j}^2 + \left(\frac{\dot{y}_{0_j} + \zeta_j\omega_j y_{0_j}}{\omega_{d_j}}\right)^2} \quad \text{and} \quad \tan\vartheta_j = \frac{\dot{y}_{0_j} + \zeta_j\omega_j y_{0_j}}{\omega_{d_j} y_{0_j}}. \qquad [2.145]$$

This can be computationally efficient as $\mathbf{y}_0$ and $\dot{\mathbf{y}}_0$ can be stored *a priori*. In practice, often only the first few modes are used in the modal superposition expression in equation [2.137]. In the special case when there is no damping, by substituting $\mathbf{C} = \mathbf{O}$ it can be seen that the results derived here reduce to the ones derived in section 2.3.2.2.

EXAMPLE 2.3.– Figure 2.4 shows a three DOF spring-mass system. The mass of each block is m kg and the stiffness of each spring is k N/m. The viscous damping constant of the damper associated with each block is c Ns/m. The aim is to obtain the dynamic response for the following load cases:

1) When only the first mass (DOF 1) is subjected to a unit step input (see Figure 2.5) so that $\mathbf{f}(t) = \{f(t), 0, 0\}^T$ and $f(t) = 1 - U(t - t_0)$ with $t_0 = \frac{2\pi}{\omega_1}$ where ω_1 is the first undamped natural frequency of the system and $U(\bullet)$ is the unit step function.

2) When only the second mass (DOF 2) is subjected to unit initial displacement, i.e. $\mathbf{q}_0 = \{0, 1, 0\}^T$.

3) When only the second and the third masses (DOF 3) are subjected to unit initial velocities, i.e. $\dot{\mathbf{q}}_0 = \{0, 1, 1\}^T$.

4) When all three of the above loading are acting together on the system.

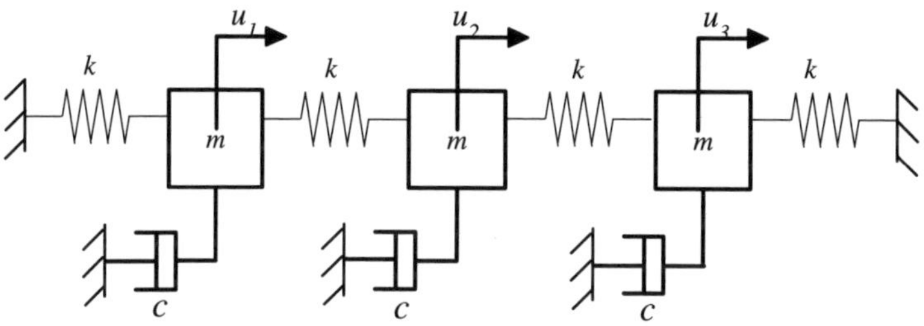

Figure 2.4. *Three DOF damped spring-mass system with dampers attached to the ground*

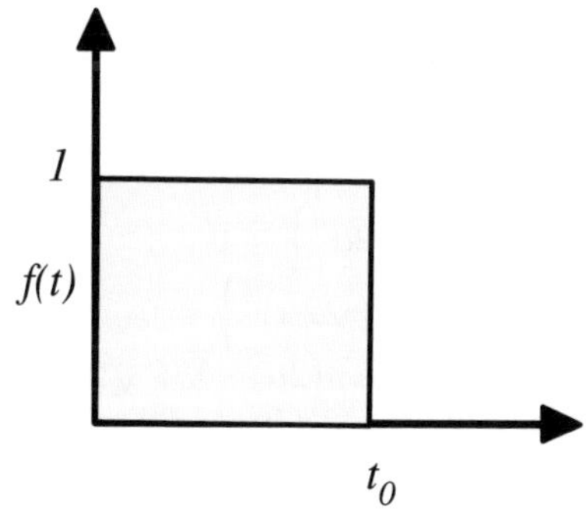

Figure 2.5. *Unit step forcing,* $t_0 = \frac{2\pi}{\omega_1}$

This problem can be solved by using the following steps:

1) *Obtain the system matrices:* The mass, stiffness and the damping matrices are given by

$$\mathbf{M}=\begin{bmatrix} m & 0 & 0\\ 0 & m & 0\\ 0 & 0 & m\end{bmatrix}, \quad \mathbf{K}=\begin{bmatrix} 2k & -k & 0\\ -k & 2k & -k\\ 0 & -k & 2k\end{bmatrix} \quad \text{and} \quad \mathbf{C}=\begin{bmatrix} c & 0 & 0\\ 0 & c & 0\\ 0 & 0 & c\end{bmatrix}. \qquad [2.146]$$

Note that the damping matrix is mass proportional, so that the system is proportionally damped.

2) *Obtain the undamped natural frequencies:* For notational convenience assume that the eigenvalues $\lambda_j = \omega_j^2$. The three DOF system has three eigenvalues and they are the roots of the following characteristic equation

$$\det\left[\mathbf{K}-\lambda\mathbf{M}\right]=0. \qquad [2.147]$$

Using the mass and the stiffness matrices from equation [2.146], this can be simplified as

$$\begin{aligned} &\det\begin{bmatrix} 2k-\lambda m & -k & 0\\ -k & 2k-\lambda m & -k\\ 0 & -k & 2k-\lambda m\end{bmatrix}=0\\ \text{or}\quad &m\det\begin{bmatrix} 2\alpha-\lambda & -\alpha & 0\\ -\alpha & 2\alpha-\lambda & -\alpha\\ 0 & -\alpha & 2\alpha-\lambda\end{bmatrix}=0 \quad \text{where} \quad \alpha=\frac{k}{m}. \end{aligned} \qquad [2.148]$$

Expanding the determinant in [2.148], we have

$$\begin{aligned} &(2\alpha-\lambda)\left\{(2\alpha-\lambda)^2-\alpha^2\right\}-\alpha\alpha\,(2\alpha-\lambda)=0\\ \text{or}\quad &(2\alpha-\lambda)\left\{(2\alpha-\lambda)^2-2\alpha^2\right\}=0\\ \text{or}\quad &(2\alpha-\lambda)\left\{(2\alpha-\lambda)^2-\left(\sqrt{2}\alpha\right)^2\right\}=0\\ \text{or}\quad &(2\alpha-\lambda)\left(2\alpha-\lambda-\sqrt{2}\alpha\right)\left(2\alpha-\lambda+\sqrt{2}\alpha\right)=0\\ \text{or}\quad &(2\alpha-\lambda)\left((2-\sqrt{2})\,\alpha-\lambda\right)\left((2+\sqrt{2})\,\alpha-\lambda\right)=0. \end{aligned} \qquad [2.149]$$

It implies that the three roots (in the increasing order) are

$$\lambda_1=\left(2-\sqrt{2}\right)\alpha, \quad \lambda_2=2\alpha, \quad \text{and} \quad \lambda_3=\left(2+\sqrt{2}\right)\alpha. \qquad [2.150]$$

Since $\lambda_j=\omega_j^2$, the natural frequencies are

$$\omega_1=\sqrt{\left(2-\sqrt{2}\right)\alpha}, \quad \omega_2=\sqrt{2\alpha}, \quad \text{and} \quad \omega_3=\sqrt{\left(2+\sqrt{2}\right)\alpha}. \qquad [2.151]$$

3) *Obtain the undamped mode shapes:* From equation [2.55] the eigenvalue equation can be written as

$$[\mathbf{K} - \lambda_j \mathbf{M}]\,\mathbf{x}_j = 0. \qquad [2.152]$$

For this problem, substituting **K** and **M** from equation [2.146] and dividing by m, we have

$$\begin{bmatrix} 2\alpha - \lambda & -\alpha & 0 \\ -\alpha & 2\alpha - \lambda & -\alpha \\ 0 & -\alpha & 2\alpha - \lambda \end{bmatrix} \begin{Bmatrix} x_{1j} \\ x_{2j} \\ x_{3j} \end{Bmatrix} = 0. \qquad [2.153]$$

Here, x_{1j}, x_{2j} and x_{3j} are the three components of jth eigenvector corresponding to the three masses. To obtain $\mathbf{x}_j$ we need to substitute λ_j from equation [2.150] in the above equation and solve for each component of $\mathbf{x}_j$ for every j.

The first eigenvector, $j = 1$:

Substituting $\lambda = \lambda_1 = \left(2 - \sqrt{2}\right)\alpha$ in [2.153], we have

$$\begin{bmatrix} 2\alpha - \left(2 - \sqrt{2}\right)\alpha & -\alpha & 0 \\ -\alpha & 2\alpha - \left(2 - \sqrt{2}\right)\alpha & -\alpha \\ 0 & -\alpha & 2\alpha - \left(2 - \sqrt{2}\right)\alpha \end{bmatrix} \begin{Bmatrix} x_{11} \\ x_{21} \\ x_{31} \end{Bmatrix} = 0 \qquad [2.154]$$

or

$$\begin{bmatrix} \sqrt{2}\alpha & -\alpha & 0 \\ -\alpha & \sqrt{2}\alpha & -\alpha \\ 0 & -\alpha & \sqrt{2}\alpha \end{bmatrix} \begin{Bmatrix} x_{11} \\ x_{21} \\ x_{31} \end{Bmatrix} = 0 \quad \text{or} \quad \begin{bmatrix} \sqrt{2} & -1 & 0 \\ -1 & \sqrt{2} & -1 \\ 0 & -1 & \sqrt{2} \end{bmatrix} \begin{Bmatrix} x_{11} \\ x_{21} \\ x_{31} \end{Bmatrix} = 0. \qquad [2.155]$$

This can be separated into three equations as

$$\sqrt{2}x_{11} - x_{21} = 0, \quad -x_{11} + \sqrt{2}x_{21} - x_{31} = 0 \quad \text{and} \quad -x_{21} + \sqrt{2}x_{31} = 0. \qquad [2.156]$$

These three equations cannot be solved uniquely but once we fix one element, the other two elements can be expressed in terms of it. This implies that the ratios between the modal amplitudes are unique. Solving the system of linear equations [2.156], we have $x_{21} = \sqrt{2}x_{11}$ and $x_{21} = \sqrt{2}x_{31}$, that is $x_{11} = x_{31} = \gamma_1$ (say). Therefore, the first eigenvector is given by

$$\mathbf{x}_1 = \begin{Bmatrix} x_{11} \\ x_{21} \\ x_{31} \end{Bmatrix} = \gamma_1 \begin{Bmatrix} 1 \\ \sqrt{2} \\ 1 \end{Bmatrix}. \qquad [2.157]$$

The constant γ_1 can be obtained from the mass normalization condition in [2.63], that is

$$\mathbf{x}_1^T \mathbf{M} \mathbf{x}_1 = 1 \quad \text{or} \quad \gamma_1 \begin{Bmatrix} 1 \\ \sqrt{2} \\ 1 \end{Bmatrix}^T \begin{bmatrix} m & 0 & 0 \\ 0 & m & 0 \\ 0 & 0 & m \end{bmatrix} \gamma_1 \begin{Bmatrix} 1 \\ \sqrt{2} \\ 1 \end{Bmatrix} = 1 \qquad [2.158]$$

$$\text{or} \quad \gamma_1^2 m \begin{Bmatrix} 1 \\ \sqrt{2} \\ 1 \end{Bmatrix}^T \begin{bmatrix} 1 & 0 & 0 \\ 0 & 1 & 0 \\ 0 & 0 & 1 \end{bmatrix} \begin{Bmatrix} 1 \\ \sqrt{2} \\ 1 \end{Bmatrix} = 1 \qquad [2.159]$$

$$\text{or} \quad \gamma_1^2 m \left(1 + \sqrt{2}\sqrt{2} + 1\right) = 1 \quad \text{that is} \quad \gamma_1 = \frac{1}{2\sqrt{m}}. \qquad [2.160]$$

Thus, the mass normalized first eigenvector is given by

$$\mathbf{x}_1 = \frac{1}{2\sqrt{m}} \begin{Bmatrix} 1 \\ \sqrt{2} \\ 1 \end{Bmatrix}. \qquad [2.161]$$

The second eigenvector, $j = 2$:

Substituting $\lambda = \lambda_2 = 2\alpha$ in [2.153], we have

$$\begin{bmatrix} 2\alpha - 2\alpha & -\alpha & 0 \\ -\alpha & 2\alpha - 2\alpha & -\alpha \\ 0 & -\alpha & 2\alpha - 2\alpha \end{bmatrix} \begin{Bmatrix} x_{12} \\ x_{22} \\ x_{32} \end{Bmatrix} = 0 \qquad [2.162]$$

$$\text{or} \quad \begin{bmatrix} 0 & -\alpha & 0 \\ -\alpha & 0 & -\alpha \\ 0 & -\alpha & 0 \end{bmatrix} \begin{Bmatrix} x_{12} \\ x_{22} \\ x_{32} \end{Bmatrix} = 0 \quad \text{or} \quad \begin{bmatrix} 0 & 1 & 0 \\ 1 & 0 & 1 \\ 0 & 1 & 0 \end{bmatrix} \begin{Bmatrix} x_{12} \\ x_{22} \\ x_{32} \end{Bmatrix} = 0. \qquad [2.163]$$

This implies that $x_{22} = 0$ and $x_{12} = -x_{32} = \gamma_2$ (say). Therefore, the second eigenvector is given by

$$\mathbf{x}_2 = \begin{Bmatrix} x_{12} \\ x_{22} \\ x_{32} \end{Bmatrix} = \gamma_2 \begin{Bmatrix} 1 \\ 0 \\ -1 \end{Bmatrix}. \qquad [2.164]$$

Using the mass normalization condition

$$\mathbf{x}_2^T \mathbf{M} \mathbf{x}_2 = 1 \quad \text{or} \quad \gamma_2 \begin{Bmatrix} 1 \\ 0 \\ -1 \end{Bmatrix}^T \begin{bmatrix} m & 0 & 0 \\ 0 & m & 0 \\ 0 & 0 & m \end{bmatrix} \gamma_2 \begin{Bmatrix} 1 \\ 0 \\ -1 \end{Bmatrix} = 1 \qquad [2.165]$$

$$\text{or} \quad \gamma_2^2 m \left(1 + 1\right) = 1 \quad \text{that is} \quad \gamma_2 = \frac{1}{\sqrt{2m}} = \frac{\sqrt{2}}{2\sqrt{m}}. \qquad [2.166]$$

Thus, the mass normalized second eigenvector is given by

$$\mathbf{x}_2 = \frac{1}{2\sqrt{m}} \begin{Bmatrix} \sqrt{2} \\ 0 \\ -\sqrt{2} \end{Bmatrix}. \qquad [2.167]$$

The third eigenvector, $j = 3$:

Substituting $\lambda = \lambda_3 = \left(2 + \sqrt{2}\right)\alpha$ in [2.153], we have

$$\begin{bmatrix} 2\alpha - \left(2+\sqrt{2}\right)\alpha & -\alpha & 0 \\ -\alpha & 2\alpha - \left(2+\sqrt{2}\right)\alpha & -\alpha \\ 0 & -\alpha & 2\alpha - \left(2+\sqrt{2}\right)\alpha \end{bmatrix} \begin{Bmatrix} x_{13} \\ x_{23} \\ x_{33} \end{Bmatrix} = 0 \qquad [2.168]$$

$$\text{or} \quad \begin{bmatrix} -\sqrt{2}\alpha & -\alpha & 0 \\ -\alpha & -\sqrt{2}\alpha & -\alpha \\ 0 & -\alpha & -\sqrt{2}\alpha \end{bmatrix} \begin{Bmatrix} x_{13} \\ x_{23} \\ x_{33} \end{Bmatrix} = 0 \quad \text{or} \quad \begin{bmatrix} \sqrt{2} & 1 & 0 \\ 1 & \sqrt{2} & 1 \\ 0 & 1 & \sqrt{2} \end{bmatrix} \begin{Bmatrix} x_{13} \\ x_{23} \\ x_{33} \end{Bmatrix} = 0. \qquad [2.169]$$

This implies that

$$\begin{aligned} \sqrt{2}x_{13} + x_{23} &= 0 \Rightarrow x_{23} = -\sqrt{2}x_{13} \\ x_{13} + \sqrt{2}x_{23} + x_{33} &= 0 \\ \text{and} \quad x_{23} + \sqrt{2}x_{33} &= 0 \Rightarrow x_{23} = -\sqrt{2}x_{33} \end{aligned} \qquad [2.170]$$

that is $x_{13} = x_{33} = \gamma_3$ (say). Therefore, the third eigenvector is given by

$$\mathbf{x}_3 = \begin{Bmatrix} x_{13} \\ x_{23} \\ x_{33} \end{Bmatrix} = \gamma_3 \begin{Bmatrix} 1 \\ -\sqrt{2} \\ 1 \end{Bmatrix}. \qquad [2.171]$$

Using the mass normalization condition

$$\mathbf{x}_3^T \mathbf{M} \mathbf{x}_3 = 1 \quad \text{or} \quad \gamma_3 \begin{Bmatrix} 1 \\ -\sqrt{2} \\ 1 \end{Bmatrix}^T \begin{bmatrix} m & 0 & 0 \\ 0 & m & 0 \\ 0 & 0 & m \end{bmatrix} \gamma_3 \begin{Bmatrix} 1 \\ -\sqrt{2} \\ 1 \end{Bmatrix} = 1 \qquad [2.172]$$

$$\text{or} \quad \gamma_3^2 m \begin{Bmatrix} 1 \\ -\sqrt{2} \\ 1 \end{Bmatrix}^T \begin{bmatrix} 1 & 0 & 0 \\ 0 & 1 & 0 \\ 0 & 0 & 1 \end{bmatrix} \begin{Bmatrix} 1 \\ -\sqrt{2} \\ 1 \end{Bmatrix} = 1 \qquad [2.173]$$

$$\text{or} \quad \gamma_3^2 m \left(1 + \sqrt{2}\sqrt{2} + 1\right) = 1, \quad \text{that is,} \quad \gamma_3 = \frac{1}{2\sqrt{m}}. \qquad [2.174]$$

Thus, the mass normalized third eigenvector is given by

$$\mathbf{x}_3 = \frac{1}{2\sqrt{m}} \begin{Bmatrix} 1 \\ -\sqrt{2} \\ 1 \end{Bmatrix}. \tag{2.175}$$

Combining the three eigenvectors, the mass normalized undamped modal matrix is now given by

$$\mathbf{X} = [\mathbf{x}_1, \mathbf{x}_2, \mathbf{x}_3] = \frac{1}{2\sqrt{m}} \begin{bmatrix} 1 & \sqrt{2} & 1 \\ \sqrt{2} & 0 & -\sqrt{2} \\ 1 & -\sqrt{2} & 1 \end{bmatrix}. \tag{2.176}$$

Here, the modal matrix turns out to be symmetric. But in general this is not the case.

4) *Obtain the modal damping factors:* The damping matrix in the modal coordinate can be obtained from [2.97] as

$$\mathbf{C}' = \mathbf{X}^T\mathbf{C}\mathbf{X} = \frac{1}{2\sqrt{m}} \begin{bmatrix} 1 & \sqrt{2} & 1 \\ \sqrt{2} & 0 & -\sqrt{2} \\ 1 & -\sqrt{2} & 1 \end{bmatrix}^T \begin{bmatrix} c & 0 & 0 \\ 0 & c & 0 \\ 0 & 0 & c \end{bmatrix} \frac{1}{2\sqrt{m}} \begin{bmatrix} 1 & \sqrt{2} & 1 \\ \sqrt{2} & 0 & -\sqrt{2} \\ 1 & -\sqrt{2} & 1 \end{bmatrix} = \frac{c}{m}\mathbf{I}. \tag{2.177}$$

Therefore,

$$2\zeta_j\omega_j = \frac{c}{m} \quad \text{or} \quad \zeta_j = \frac{c}{2m\omega_j}. \tag{2.178}$$

Since ω_j becomes bigger for higher modes, modal damping gets smaller, i.e. higher modes are less damped.

5) *Response due to applied loading:* The applied loading $\mathbf{f}(t) = \{f(t), 0, 0\}^T$ where $f(t) = 1 - U(t - t_0)$ with $t_0 = \frac{2\pi}{\omega_1}$. In the Laplace domain

$$\bar{f}(s) = \mathcal{L}\left[1 - U(t - t_0)\right] = 1 - \frac{e^{-st_0}}{s}. \tag{2.179}$$

Thus, the term $\mathbf{x}_j^T\bar{\mathbf{f}}(s)$ can be obtained as

$$\mathbf{x}_j^T\bar{\mathbf{f}}(s) = \begin{Bmatrix} x_{1j} \\ x_{2j} \\ x_{3j} \end{Bmatrix}^T \begin{Bmatrix} \bar{f}(s) \\ 0 \\ 0 \end{Bmatrix} = x_{1j}\left(1 - \frac{e^{-st_0}}{s}\right) \quad \forall j. \tag{2.180}$$

Since the initial conditions are zero, the dynamic response in the Laplace domain can be obtained from equation [2.134] as

$$\bar{\mathbf{q}}(s) = \sum_{j=1}^{3} \left\{ \frac{x_{1j}\left(1 - \frac{e^{-st_0}}{s}\right)}{s^2 + 2s\zeta_j\omega_j + \omega_j^2} \right\} \mathbf{x}_j = \sum_{j=1}^{3} \left\{ \frac{x_{1j}\left(s - e^{-st_0}\right)}{s\left(s^2 + 2s\zeta_j\omega_j + \omega_j^2\right)} \right\} \mathbf{x}_j. \qquad [2.181]$$

In the frequency domain, the response is given by

$$\bar{\mathbf{q}}(\mathrm{i}\omega) = \sum_{j=1}^{3} \left\{ \frac{x_{1j}\left(\mathrm{i}\omega - e^{-\mathrm{i}\omega t_0}\right)}{\mathrm{i}\omega\left(-\omega^2 + 2\mathrm{i}\omega\zeta_j\omega_j + \omega_j^2\right)} \right\} \mathbf{x}_j. \qquad [2.182]$$

For the numerical calculations we assume $m = 1$, $k = 1$ and $c = 0.2$. Using these values, from [2.182], the absolute value of the frequency-domain response of the three masses are plotted in Figure 2.6. The three peaks in the diagram correspond to the three natural frequencies of the system. The time-domain response can be obtained by evaluating the convolution integral in [2.142] and substituting $a_j(t)$ in equation [2.137]. In practice, usually numerical integration methods are used to evaluate this integral. For this problem, however, a closed-form expression can be obtained. We have

$$\mathbf{x}_j^T \mathbf{f}(\tau) = x_{1j} f(\tau). \qquad [2.183]$$

From Figure 2.5, it can be noted that

$$f(\tau) = \begin{cases} 1 & \text{if} \quad \tau < t_0, \\ 0 & \text{if} \quad \tau > t_0. \end{cases} \qquad [2.184]$$

Because of this, the limit of the integral in [2.142] can be changed as

$$\begin{aligned} a_j(t) &= \int_0^t \frac{1}{\omega_{d_j}} \mathbf{x}_j^T \mathbf{f}(\tau) e^{-\zeta_j\omega_j(t-\tau)} \sin\left(\omega_{d_j}(t-\tau)\right) d\tau \\ &= \int_0^{t_0} \frac{1}{\omega_{d_j}} x_{1j} e^{-\zeta_j\omega_j(t-\tau)} \sin\left(\omega_{d_j}(t-\tau)\right) d\tau \\ &= \frac{x_{1j}}{\omega_{d_j}} \int_0^{t_0} e^{-\zeta_j\omega_j(t-\tau)} \sin\left(\omega_{d_j}(t-\tau)\right) d\tau. \end{aligned} \qquad [2.185]$$

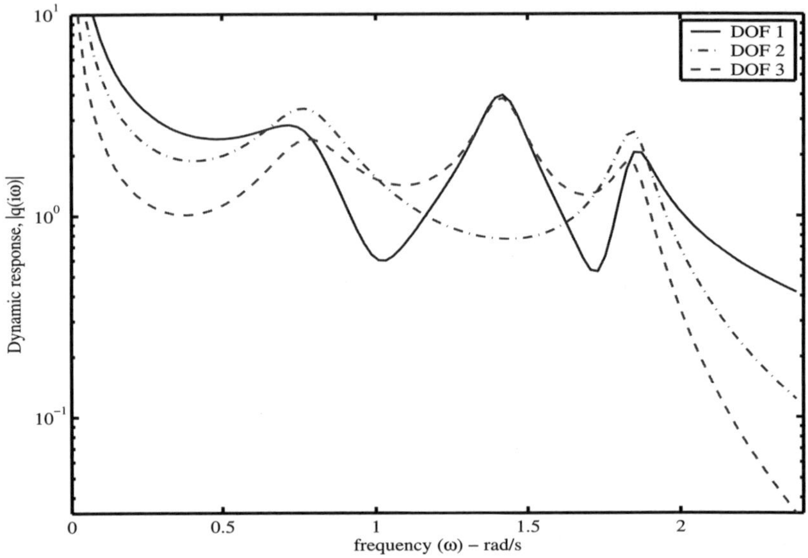

Figure 2.6. *Absolute value of the frequency-domain response of the three masses due to applied step loading at first DOF*

By making a substitution $\tau' = t - \tau$, this integral can be evaluated as

$$a_j(t) = \frac{x_{1j}}{\omega_{d_j}} \frac{e^{-\zeta_j \omega_j t}}{\omega_j^2} \left\{ \alpha_j \sin\left(\omega_{d_j} t\right) + \beta_j \cos\left(\omega_{d_j} t\right) \right\} \qquad [2.186]$$

where $\alpha_j = \left\{ \omega_{d_j} \sin\left(\omega_{d_j} t_0\right) + \zeta_j \omega_j \cos\left(\omega_{d_j} t_0\right) \right\} e^{\zeta_j \omega_j t_0} - \zeta_j \omega_j \qquad [2.187]$

and $\beta_j = \left\{ \omega_{d_j} \cos\left(\omega_{d_j} t_0\right) - \zeta_j \omega_j \sin\left(\omega_{d_j} t_0\right) \right\} e^{\zeta_j \omega_j t_0} - \omega_{d_j}. \qquad [2.188]$

The time-domain responses of the three masses obtained using equation [2.137] are shown in Figure 2.7. From the figure it is observed that, because the forcing is applied to the first mass (DOF 1), it moves earlier and comparatively more than the other two masses.

6) *Response due to initial displacement:* When $\mathbf{q}_0 = \{0, 1, 0\}^T$, we have

$$\mathbf{x}_j^T \mathbf{C} \mathbf{q}_0 = \begin{Bmatrix} x_{1j} \\ x_{2j} \\ x_{3j} \end{Bmatrix}^T \begin{bmatrix} c & 0 & 0 \\ 0 & c & 0 \\ 0 & 0 & c \end{bmatrix} \cdot \begin{Bmatrix} 0 \\ 1 \\ 0 \end{Bmatrix} = x_{2j} c \quad \forall j. \qquad [2.189]$$

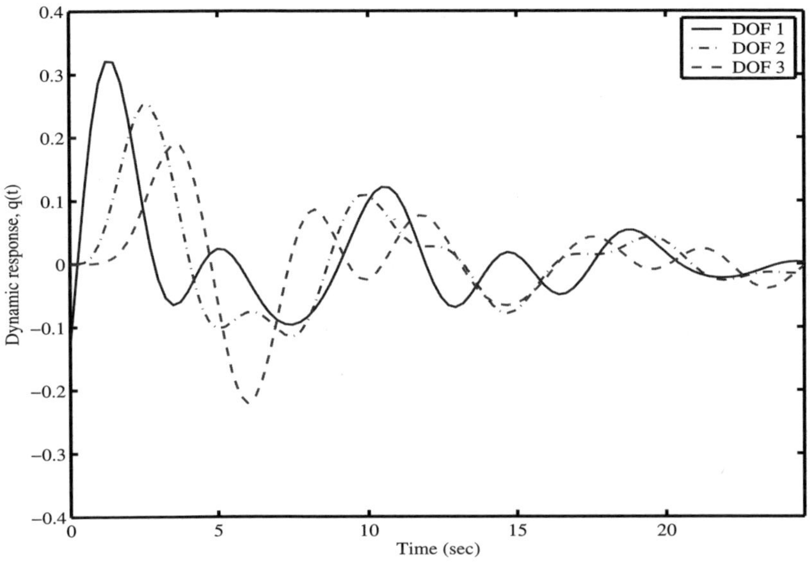

Figure 2.7. *Time-domain response of the three masses due to applied step loading at first DOF*

Similarly,

$$\mathbf{x}_j^T \mathbf{M} \mathbf{q}_0 = x_{2j} m \quad \forall j. \tag{2.190}$$

The dynamic response in the Laplace domain can be obtained from equation [2.134] as

$$\begin{aligned} \bar{\mathbf{q}}(s) &= \sum_{j=1}^{3} \left\{ \frac{x_{2j} c}{s^2 + 2s\zeta_j \omega_j + \omega_j^2} + \frac{x_{2j} m s}{s^2 + 2s\zeta_j \omega_j + \omega_j^2} \right\} \mathbf{x}_j \\ &= \sum_{j=1}^{3} x_{2j} \left\{ \frac{c + ms}{s^2 + 2s\zeta_j \omega_j + \omega_j^2} \right\} \mathbf{x}_j. \end{aligned} \tag{2.191}$$

From equation [2.178] note that $c = 2\zeta_j \omega_j m$. Substituting this in the above equation, we have

$$\bar{\mathbf{q}}(s) = \sum_{j=1}^{3} x_{2j} m \left\{ \frac{2\zeta_j \omega_j + s}{s^2 + 2s\zeta_j \omega_j + \omega_j^2} \right\} \mathbf{x}_j. \tag{2.192}$$

In the frequency domain, the response is given by

$$\bar{\mathbf{q}}(\mathrm{i}\omega) = \sum_{j=1}^{3} x_{2j} m \left\{ \frac{2\zeta_j \omega_j + \mathrm{i}\omega}{-\omega^2 + 2\mathrm{i}\omega\zeta_j\omega_j + \omega_j^2} \right\} \mathbf{x}_j. \qquad [2.193]$$

Figure 2.8 shows the responses of the three masses. Note that the second peak is "missing" and the responses of the first and the third masses are exactly the same. This is due to the fact that, in the second mode of vibration, the middle mass remains stationary while the two other masses move an equal amount in the opposite directions (see the second mode in equation [2.167]). The time-domain response can be obtained by directly taking the inverse Laplace transform of [2.192] as

$$\begin{aligned} \mathbf{q}(t) &= \mathcal{L}^{-1}\left[\bar{\mathbf{q}}(s)\right] = \sum_{j=1}^{3} x_{2j} m \mathcal{L}^{-1}\left[\frac{2\zeta_j\omega_j + s}{s^2 + 2s\zeta_j\omega_j + \omega_j^2}\right]\mathbf{x}_j \\ &= \sum_{j=1}^{3} x_{2j} m e^{-\zeta_j\omega_j t}\cos\left(\omega_{d_j} t\right)\mathbf{x}_j. \end{aligned} \qquad [2.194]$$

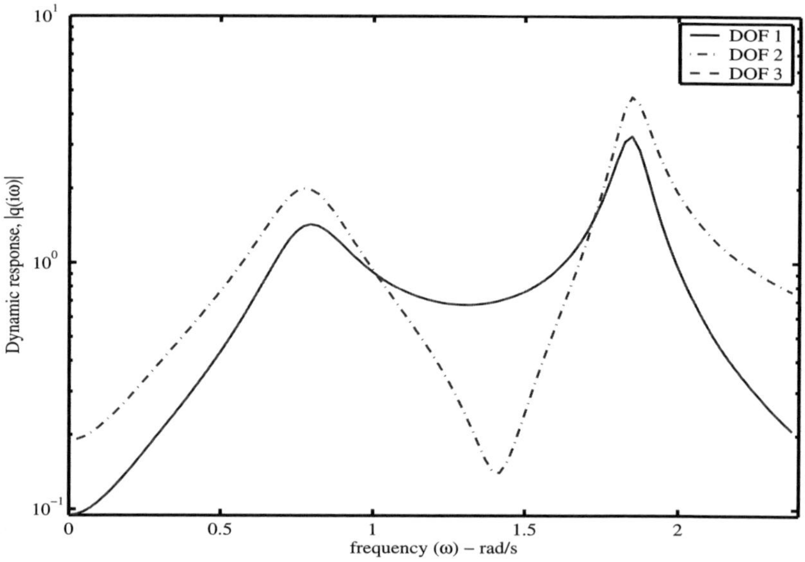

Figure 2.8. *Absolute value of the frequency-domain response of the three masses due to unit initial displacement at second DOF*

This expression is plotted in Figure 2.9. Observe that initial displacement of the second mass is unity, which verifies that the initial condition has been applied correctly. Because of the symmetry of the system the displacements of the two other masses are identical.

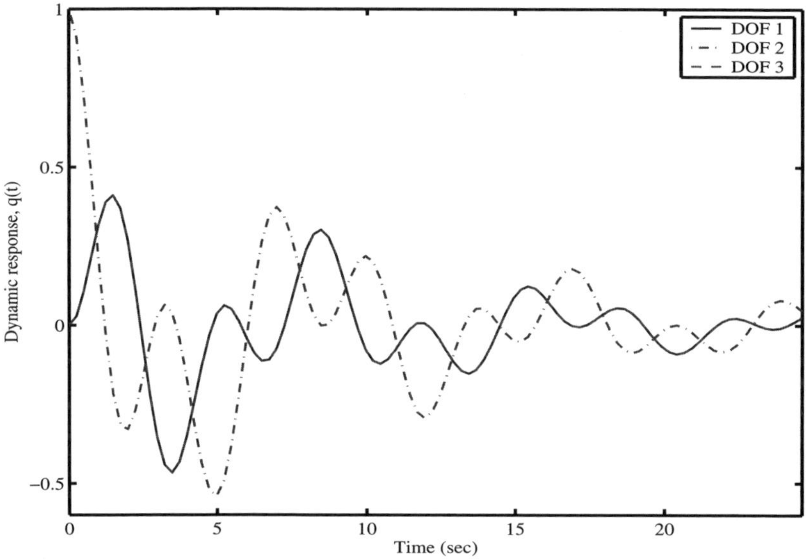

Figure 2.9. *Time-domain response of the three masses due to unit initial displacement at second DOF*

7) *Response due to initial velocity:* When $\dot{\mathbf{q}}_0 = \{0, 1, 1\}^T$, we have

$$\mathbf{x}_j^T \mathbf{M} \dot{\mathbf{q}}_0 = \begin{Bmatrix} x_{1j} \\ x_{2j} \\ x_{3j} \end{Bmatrix}^T \begin{bmatrix} m & 0 & 0 \\ 0 & m & 0 \\ 0 & 0 & m \end{bmatrix} \begin{Bmatrix} 0 \\ 1 \\ 1 \end{Bmatrix} = (x_{2j} + x_{3j})\, m \quad \forall j. \qquad [2.195]$$

The dynamic response in the Laplace domain can be obtained from equation [2.134] as

$$\bar{\mathbf{q}}(s) = \sum_{j=1}^{3} \left\{ \frac{(x_{2j} + x_{3j})\, m}{s^2 + 2s\zeta_j\omega_j + \omega_j^2} \right\} \mathbf{x}_j. \qquad [2.196]$$

The time-domain response can be obtained using the inverse Laplace transform as

$$\mathbf{q}(t) = \sum_{j=1}^{3} (x_{2j} + x_{3j}) \frac{m}{\omega_{d_j}} e^{-\zeta_j \omega_j t} \sin\left(\omega_{d_j} t\right) \mathbf{x}_j. \qquad [2.197]$$

The responses of the three masses in the frequency domain and in the time domain are respectively shown in Figures 2.10 and 2.11. In this case, all the modes of the system can be observed. Because the initial conditions of the second and the third masses are the same, their initial displacements are close to each other. However, as the time passes the displacements of these two masses start differing from each other.

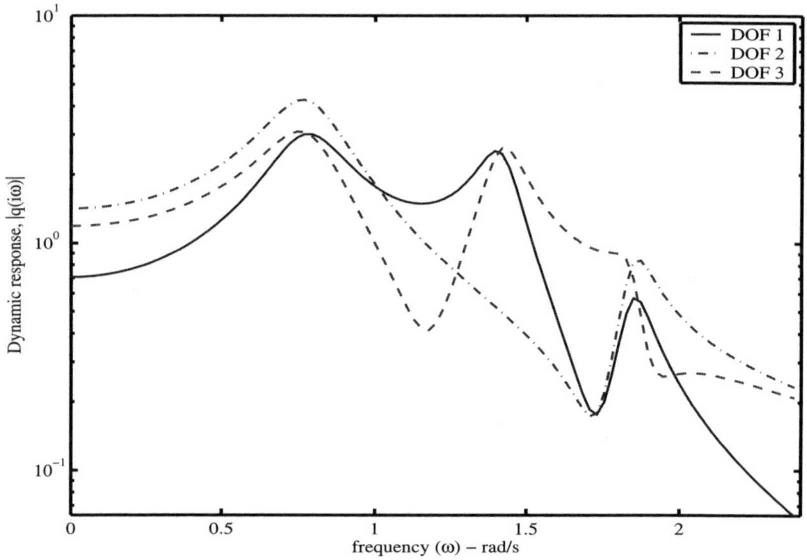

Figure 2.10. *Absolute value of the frequency-domain response of the three masses due to unit initial velocity at the second and third DOF*

2.5. Non-proportionally damped systems

Modes of proportionally damped systems preserve the simplicity of the real normal modes as in the undamped case. Unfortunately, there is no physical reason why a general system should behave like this. In fact, practical experience in modal testing shows that most real-life structures do not do so, as they possess complex modes instead of real normal modes. This implies that, in general, linear systems are non-classically damped. When the system is non-classically damped, some or all of the N differential equations in the modal coordinate [2.96] are coupled through the $\mathbf{X}^T\mathbf{C}\mathbf{X}$ term and cannot be reduced to N second-order uncoupled equations. This coupling brings several complications to the system dynamics – the eigenvalues and the eigenvectors no longer remain real and also the eigenvectors do not satisfy the classical orthogonality relationships as given by equations [2.63] and [2.64].

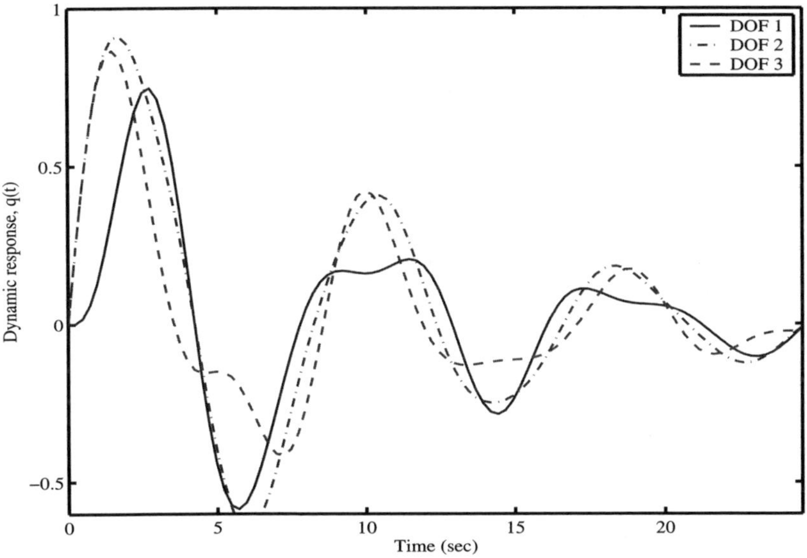

Figure 2.11. *Time-domain response of the three masses due to unit initial velocity at the second and third DOF*

2.5.1. *Free vibration and complex modes*

The complex eigenvalue problem associated with the equation of motion [2.95] can be represented by

$$s_j^2\mathbf{M}\mathbf{z}_j + s_j\mathbf{C}\mathbf{z}_j + \mathbf{K}\mathbf{z}_j = \mathbf{0} \qquad [2.198]$$

where $s_j \in \mathbb{C}$ is the jth eigenvalue and $\mathbf{z}_j \in \mathbb{C}^N$ is the jth eigenvector. The eigenvalues, s_j, are the roots of the characteristic polynomial

$$\det\left[s^2\mathbf{M} + s\mathbf{C} + \mathbf{K}\right] = 0. \qquad [2.199]$$

The order of the polynomial is $2N$ and if the roots are complex they appear in complex conjugate pairs. The methods for solving this kind of complex problem follow mainly two routes: the state-space method and the methods in configuration space or "N-space". Brief discussions on these two approaches are taken up in the following sections.

2.5.1.1. *The state-space method*

The state-space method is based on transforming the N second-order coupled equations into a set of $2N$ first-order coupled equations by augmenting the displacement response vectors with the velocities of the corresponding coordinates. We can write equation [2.95] together with a trivial equation $\mathbf{M}\dot{\mathbf{q}}(t) - \mathbf{M}\dot{\mathbf{q}}(t) = 0$ in a matrix form as

$$\begin{bmatrix} \mathbf{C} & \mathbf{M} \\ \mathbf{M} & \mathbf{O} \end{bmatrix} \begin{Bmatrix} \dot{\mathbf{q}}(t) \\ \ddot{\mathbf{q}}(t) \end{Bmatrix} + \begin{bmatrix} \mathbf{K} & \mathbf{O} \\ \mathbf{O} & -\mathbf{M} \end{bmatrix} \begin{Bmatrix} \mathbf{q}(t) \\ \dot{\mathbf{q}}(t) \end{Bmatrix} = \begin{Bmatrix} \mathbf{f}(t) \\ \mathbf{0} \end{Bmatrix} \quad [2.200]$$

$$\text{or} \quad \mathbf{A}\,\dot{\mathbf{u}}(t) + \mathbf{B}\,\mathbf{u}(t) = \mathbf{r}(t) \quad [2.201]$$

where

$$\mathbf{A} = \begin{bmatrix} \mathbf{C} & \mathbf{M} \\ \mathbf{M} & \mathbf{O} \end{bmatrix} \in \mathbb{R}^{2N\times 2N}, \quad \mathbf{B} = \begin{bmatrix} \mathbf{K} & \mathbf{O} \\ \mathbf{O} & -\mathbf{M} \end{bmatrix} \in \mathbb{R}^{2N\times 2N},$$
$$\mathbf{u}(t) = \begin{Bmatrix} \mathbf{q}(t) \\ \dot{\mathbf{q}}(t) \end{Bmatrix} \in \mathbb{R}^{2N}, \quad \text{and} \quad \mathbf{r}(t) = \begin{Bmatrix} \mathbf{f}(t) \\ \mathbf{0} \end{Bmatrix} \in \mathbb{R}^{2N}. \quad [2.202]$$

In the above equation, $\mathbf{O}$ is the $N \times N$ null matrix. This form of equation of motion is also known as the "Duncan form". The eigenvalue problem associated with equation [2.201] can be expressed as

$$s_j \mathbf{A}\boldsymbol{\phi}_j + \mathbf{B}\boldsymbol{\phi}_j = \mathbf{0} \quad [2.203]$$

where $s_j \in \mathbb{C}$ is the jth eigenvalue and $\boldsymbol{\phi}_j \in \mathbb{C}^{2N}$ is the jth eigenvector. This eigenvalue problem is similar to the undamped eigenvalue problem [2.55] except (1) the dimension of the matrices are $2N$ as opposed to N, and (2) the matrices are not positive definite. Because of (1) the computational cost to obtain the eigensolutions of [2.203] is much higher compared to the undamped eigensolutions and due to (2) the eigensolutions, in general, becoming complex valued. From a phenomenological point of view, this implies that the modes are not synchronous, i.e. there is a "phase lag" so that different degrees of freedom do not simultaneously reach their corresponding "peaks" and "troughs". Thus, both from computational and conceptual point of view, complex modes significantly complicate the problem and in practice they are often avoided. From the expression of $\mathbf{u}(t)$ in equation [2.202], the state-space complex eigenvectors $\boldsymbol{\phi}_j$ can be related to the jth eigenvector of the second-order system as

$$\boldsymbol{\phi}_j = \left\{ \begin{array}{c} \mathbf{z}_j \\ s_j \mathbf{z}_j \end{array} \right\}. \qquad [2.204]$$

Since $\mathbf{A}$ and $\mathbf{B}$ are real matrices, taking a complex conjugate ($(\bullet)^*$ denotes complex conjugation) of the eigenvalue equation [2.203], it is trivial to see that

$$s_j^* \mathbf{A} \boldsymbol{\phi}_j^* + \mathbf{B} \boldsymbol{\phi}_j^* = \mathbf{0}. \qquad [2.205]$$

This implies that the eigensolutions must appear in complex conjugate pairs. For convenience, arrange the eigenvalues and the eigenvectors so that

$$s_{j+N} = s_j^* \qquad [2.206]$$

$$\boldsymbol{\phi}_{j+N} = \boldsymbol{\phi}_j^*, \quad j = 1, 2, \cdots, N \qquad [2.207]$$

Like the real normal modes, complex modes in the state-space also satisfy orthogonal relationships over the $\mathbf{A}$ and $\mathbf{B}$ matrices. For distinct eigenvalues it is easy to show that

$$\boldsymbol{\phi}_j^T \mathbf{A} \boldsymbol{\phi}_k = 0 \quad \text{and} \quad \boldsymbol{\phi}_j^T \mathbf{B} \boldsymbol{\phi}_k = 0; \quad \forall j \neq k. \qquad [2.208]$$

Premultiplying equation [2.203] by $\mathbf{y}_j^T$, we obtain

$$\boldsymbol{\phi}_j^T \mathbf{B} \boldsymbol{\phi}_j = -s_j \boldsymbol{\phi}_j^T \mathbf{A} \boldsymbol{\phi}_j. \qquad [2.209]$$

The eigenvectors may be normalized so that

$$\boldsymbol{\phi}_j^T \mathbf{A} \boldsymbol{\phi}_j = \gamma_j \qquad [2.210]$$

where $\gamma_j \in \mathbb{C}$ is the normalization constant. In view of the expressions of $\boldsymbol{\phi}_j$ in equation [2.204], the above relationship can be expressed in terms of the eigensolutions of the second-order system as

$$\mathbf{z}_j^T \left[2 s_j \mathbf{M} + \mathbf{C} \right] \mathbf{z}_j = \gamma_j. \qquad [2.211]$$

There are several ways in which the normalization constants can be selected. The one that is most consistent with traditional modal analysis practice is to choose $\gamma_j = 2 s_j$. Observe that this degenerates to the familiar mass normalization relationship $\mathbf{z}_j^T \mathbf{M} \mathbf{z}_j = 1$ when the damping is zero.

2.5.1.2. *Approximate methods in the configuration space*

It has been pointed out that the state-space approach toward the solution of equation of motion in the context of linear structural dynamics is not only computationally expensive but also fails to provide the physical insight which modal analysis in the configuration space or N-space offers.

2.5.1.2.1. Light damping assumption

Assuming that the damping is light, a simple *first-order perturbation method* is used to obtain complex modes and frequencies in terms of undamped modes and frequencies.

The undamped modes form a *complete* set of vectors so that each complex mode $\mathbf{z}_j$ can be expressed as a linear combination of $\mathbf{x}_j$. Assume that

$$\mathbf{z}_j = \sum_{k=1}^{N} \alpha_k^{(j)} \mathbf{x}_k \qquad [2.212]$$

where $\alpha_k^{(j)}$ are complex constants which we want to determine. Since damping is assumed light, $\alpha_k^{(j)} \ll 1, \forall j \neq k$ and $\alpha_j^{(j)} = 1, \forall j$. Suppose the complex natural frequencies are denoted by λ_j, which are related to complex eigenvalues s_j through

$$s_j = \mathrm{i}\lambda_j. \qquad [2.213]$$

Substituting s_j and $\mathbf{z}_j$ in the eigenvalue equation [2.198], we have

$$\left[-\lambda_j^2 \mathbf{M} + \mathrm{i}\lambda_j \mathbf{C} + \mathbf{K}\right] \sum_{k=1}^{N} \alpha_k^{(j)} \mathbf{x}_k = \mathbf{0}. \qquad [2.214]$$

Premultiplying by $\mathbf{x}_j^T$ and using the orthogonality conditions [2.63] and [2.64], we have

$$-\lambda_j^2 + \mathrm{i}\lambda_j \sum_{k=1}^{N} \alpha_k^{(j)} C'_{jk} + \omega_j^2 = 0 \qquad [2.215]$$

where $C'_{jk} = \mathbf{x}_j^T \mathbf{C} \mathbf{x}_k$ is the jkth element of the modal damping matrix $\mathbf{C}'$. Because of small damping assumption we can neglect the product $\alpha_k^{(j)} C'_{jk}, \forall j \neq k$ since they are small compared to $\alpha_j^{(j)} C'_{jj}$. Thus, equation [2.215] can be approximated as

$$-\lambda_j^2 + \mathrm{i}\lambda_j \alpha_j^{(j)} C'_{jj} + \omega_j^2 \approx 0. \qquad [2.216]$$

By solving the quadratic equation, we have

$$\lambda_j \approx \pm\omega_j + \mathrm{i}C'_{jj}/2. \qquad [2.217]$$

This is the expression of approximate complex natural frequencies. Premultiplying equation [2.214] by $\mathbf{x}_k^T$ and using the orthogonality conditions [2.63] and [2.64] and light damping assumption, it can be shown that

$$\alpha_k^{(j)} \approx \frac{\mathrm{i}\omega_j C'_{kj}}{\omega_j^2 - \omega_k^2}, \quad k \neq j. \qquad [2.218]$$

Substituting this in [2.212], approximate complex modes can be given by

$$\mathbf{z}_j \approx \mathbf{x}_j + \sum_{k\neq j}^{N} \frac{\mathrm{i}\omega_j C'_{kj}\mathbf{x}_k}{\omega_j^2 - \omega_k^2}. \qquad [2.219]$$

This expression shows that (1) the imaginary parts of complex modes are approximately orthogonal to the real parts, and (2) the "complexity" of the modes will be more if ω_j and ω_k are close, i.e. modes will be significantly complex when the natural frequencies of a system are closely spaced. It should be recalled that the first-order perturbation expressions are only valid when damping is small. For moderate to large damping values, more general analysis, discussed in Chapter 6, should be used.

2.5.1.2.2. Light non-proportional damping

Here, we assume that the system has light non-proportional damping. Therefore, a higher amount of damping can be used provided it not significantly non-proportional. Equation [2.214] can be rewritten as

$$-\lambda_j^2\alpha_k^{(j)} + \mathrm{i}\lambda_j\left(\alpha_j^{(j)}C'_{ki} + \alpha_k^{(j)}C'_{kk} + \sum_{l\neq k\neq i}^{N}\alpha_j^{(j)}C'_{kl}\right) + \omega_k^2\alpha_k^{(j)} = 0 \qquad [2.220]$$

which leads to

$$\alpha_k^{(j)} \approx -\frac{\mathrm{i}\lambda_j C'_{kj}}{\omega_k^2 - \lambda_j^2 + \mathrm{i}\lambda_j C'_{kk}}. \qquad [2.221]$$

By replacing ω_k from equation [2.217] for the k-th set, the denominator appearing in the above expression can be factorized as

$$\omega_k^2 - \lambda_j^2 + \mathrm{i}\lambda_j C'_{kk} \approx -(\lambda_j - \lambda_k)(\lambda_j - \lambda_k^*) \qquad [2.222]$$

where $(\bullet)^*$ denotes complex conjugation. With the help of this factorization, equation [2.221] can be expressed as

$$\alpha_k^{(j)} \approx \frac{\mathrm{i}\lambda_j C'_{kj}}{(\lambda_j - \lambda_k)(\lambda_j - \lambda_k^*)} \qquad [2.223]$$

and consequently from the series sum of equation [2.212] the approximate expression for the complex modes reads

$$\mathbf{z}_j \approx \mathbf{x}_j + \sum_{k=1}^{N} \frac{\mathrm{i}\lambda_j C'_{kj} \mathbf{x}_k}{(\lambda_j - \lambda_k)(\lambda_j - \lambda_k^*)}. \qquad [2.224]$$

The approach taken here is very similar to the classical perturbation analysis described before, but the $\alpha_k^{(j)}$ expressed by equation [2.223] appear to be slightly different. The classical expression for this is given by [2.218], which is equivalent to replacing the complex natural frequencies by undamped natural frequencies in equation [2.223]. Numerical calculations using the $\alpha_k^{(j)}$ described by equation [2.223] yield more accurate results than those of the classical analysis. Further discussion on approximate methods in the configuration space is given in Chapter 6.

EXAMPLE 2.4.– To illustrate the accuracy of different approximation, consider a two degree-of-freedom system shown in Figure 2.12. The system matrices for the example taken can be expressed by

$$\mathbf{M} = \begin{bmatrix} 1 & 0 \\ 0 & 1 \end{bmatrix}; \quad \mathbf{C} = \begin{bmatrix} 4 & -4 \\ -4 & 4 \end{bmatrix} \quad \text{and} \quad \mathbf{K} = \begin{bmatrix} 1100 & -100 \\ -100 & 1200 \end{bmatrix} \qquad [2.225]$$

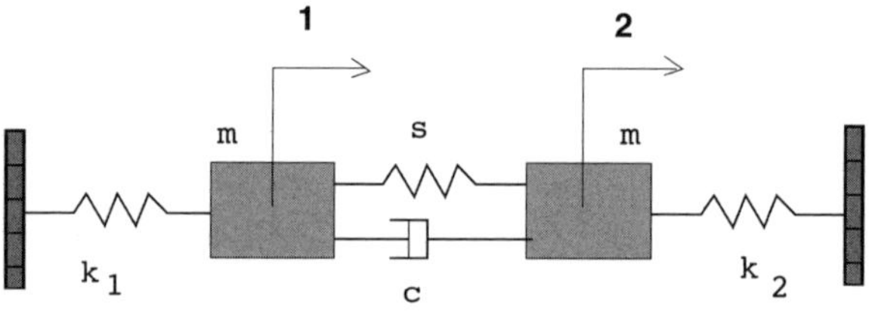

Figure 2.12. *A two degree-of-system, $m = 1$ kg, $k_1 = 1{,}000$ N/m, $s = 100$ N/m, $k_2 = 1.1k_1$, $c = 4.0$ Ns/m*

Using equation [2.224], the approximate complex mode shapes can be calculated as

$$\mathbf{z}_1 = \begin{bmatrix} 0.7870 - 0.0661\mathrm{i} \\ 0.6287 + 0.1070\mathrm{i} \end{bmatrix}; \quad \mathbf{z}_2 = \begin{bmatrix} -0.6406 - 0.1179\mathrm{i} \\ 0.7797 - 0.0729\mathrm{i} \end{bmatrix}. \qquad [2.226]$$

The corresponding exact mode shapes from the state-space formulation can be obtained as

$$\mathbf{u}_1^e = \begin{bmatrix} 0.7870 - 0.0661\mathrm{i} \\ 0.6443 + 0.1018\mathrm{i} \end{bmatrix}; \quad \mathbf{u}_2^e = \begin{bmatrix} -0.6406 - 0.1179\mathrm{i} \\ 0.7631 - 0.0603\mathrm{i} \end{bmatrix} \qquad [2.227]$$

while the classical perturbation method gives

$$\mathbf{u}_1^c = \begin{bmatrix} 0.7870 - 0.0661\mathrm{i} \\ 0.4658 + 0.2362\mathrm{i} \end{bmatrix}; \quad \mathbf{u}_2^c = \begin{bmatrix} -0.6406 - 0.1179\mathrm{i} \\ 0.8866 - 0.3989\mathrm{i} \end{bmatrix}. \qquad [2.228]$$

For comparison, the modes are normalized to have the same numerical value in the first element, so that only the second element differs. It is clear that the results obtained from the expression suggested here are closer to the exact values than the classical analysis. Besides this, the expression in equation [2.224] also indicates a conceptual difference from the classical analysis: the "correction terms" no longer remain purely imaginary. This in turn means that the real part of the complex mode shapes are not the undamped mode shapes. This fact is indeed true and can be verified from the exact analysis.

2.5.2. *Dynamic response*

Once the complex mode shapes and natural frequencies are obtained (either from the state-space method or from an approximate method), the dynamic response can be obtained using the orthogonality properties of the complex eigenvectors in the state-space. We will derive the expressions for the general dynamic response in the frequency and time domain.

2.5.2.1. *Frequency-domain analysis*

Taking the Laplace transform of the state-space equation of motion [2.201], we have

$$s\mathbf{A}\bar{\mathbf{u}}(s) - \mathbf{A}\mathbf{u}_0 + \mathbf{B}\bar{\mathbf{u}}(s) = \bar{\mathbf{r}}(s) \qquad [2.229]$$

where $\bar{\mathbf{u}}(s)$ is the Laplace transform of $\mathbf{u}(t)$, $\mathbf{u}_0$ is the vector of initial conditions in the state-space and $\bar{\mathbf{r}}(s)$ is the Laplace transform of $\mathbf{r}(t)$. From the expressions of $\mathbf{u}(t)$ and $\mathbf{r}(t)$ in equation [2.202], it is obvious that

$$\bar{\mathbf{u}}(s) = \begin{Bmatrix} \bar{\mathbf{q}}(s) \\ s\bar{\mathbf{q}}(s) \end{Bmatrix} \in \mathbb{C}^{2N}, \quad \bar{\mathbf{r}}(s) = \begin{Bmatrix} \bar{\mathbf{f}}(s) \\ \mathbf{0} \end{Bmatrix} \in \mathbb{C}^{2N} \quad \text{and} \quad \mathbf{u}_0 = \begin{Bmatrix} \mathbf{q}_0 \\ \dot{\mathbf{q}}_0 \end{Bmatrix} \in \mathbb{R}^{2N} \qquad [2.230]$$

For distinct eigenvalues the mode shapes $\boldsymbol{\phi}_k, k = 1, 2, \cdots, 2N$, form a complete set of vectors. Therefore, the solution of equation [2.229] can be expressed in terms of a linear combination of $\boldsymbol{\phi}_k$ as

$$\bar{\mathbf{u}}(s) = \sum_{k=1}^{2N} \beta_k(s)\boldsymbol{\phi}_k. \qquad [2.231]$$

We only need to determine the constants $\beta_k(s)$ to obtain the complete solution. Substituting $\bar{\mathbf{u}}(s)$ from [2.231] into equation [2.229], we have

$$[s\mathbf{A} + \mathbf{B}] \sum_{k=1}^{2N} \beta_k(s)\boldsymbol{\phi}_k = \bar{\mathbf{r}}(s) + \mathbf{A}\mathbf{u}_0. \qquad [2.232]$$

Premultiplying by $\boldsymbol{\phi}_j^T$ and using the orthogonality and normalization relationships [2.208]–[2.210], we have

$$\gamma_j (s - s_j) \beta_j(s) = \boldsymbol{\phi}_j^T \{\bar{\mathbf{r}}(s) + \mathbf{A}\mathbf{u}_0\}$$

$$\text{or} \quad \beta_j(s) = \frac{1}{\gamma_j} \frac{\boldsymbol{\phi}_j^T \bar{\mathbf{r}}(s) + \boldsymbol{\phi}_j^T \mathbf{A}\mathbf{u}_0}{s - s_j}. \qquad [2.233]$$

Using the expressions of $\mathbf{A}$, $\boldsymbol{\phi}_j$ and $\bar{\mathbf{r}}(s)$ from equations [2.202], [2.204] and [2.230], the term $\boldsymbol{\phi}_j^T \bar{\mathbf{r}}(s) + \boldsymbol{\phi}_j^T \mathbf{A}\mathbf{u}_0$ can be simplified and $\beta_k(s)$ can be related with the mode shapes of the second-order system as

$$\beta_j(s) = \frac{1}{\gamma_j} \frac{\mathbf{z}_j^T \left\{\bar{\mathbf{f}}(s) + \mathbf{M}\dot{\mathbf{q}}_0 + \mathbf{C}\mathbf{q}_0 + s\mathbf{M}\mathbf{q}_0\right\}}{s - s_j}. \qquad [2.234]$$

Since we are only interested in the displacement response, we only need to determine the first N rows of equation [2.231]. Using the partition of $\mathbf{u}(s)$ and $\boldsymbol{\phi}_j$, we have

$$\bar{\mathbf{q}}(s) = \sum_{j=1}^{2N} \beta_j(s)\mathbf{z}_j. \qquad [2.235]$$

Substituting $\beta_j(s)$ from equation [2.234] into the preceding equation, we have

$$\bar{\mathbf{q}}(s) = \sum_{j=1}^{2N} \frac{1}{\gamma_j} \mathbf{z}_j \frac{\mathbf{z}_j^T \left\{\bar{\mathbf{f}}(s) + \mathbf{M}\dot{\mathbf{q}}_0 + \mathbf{C}\mathbf{q}_0 + s\mathbf{M}\mathbf{q}_0\right\}}{s - s_j} \qquad [2.236]$$

$$\text{or} \quad \bar{\mathbf{q}}(s) = \mathbf{H}(s) \left\{\bar{\mathbf{f}}(s) + \mathbf{M}\dot{\mathbf{q}}_0 + \mathbf{C}\mathbf{q}_0 + s\mathbf{M}\mathbf{q}_0\right\} \qquad [2.237]$$

where

$$\mathbf{H}(s) = \sum_{j=1}^{2N} \frac{1}{\gamma_j} \frac{\mathbf{z}_j \mathbf{z}_j^T}{s - s_j} \qquad [2.238]$$

is the transfer function matrix. Recalling that the eigensolutions appear in complex conjugate pairs, using [2.206], the transfer function matrix [2.238] can be expanded as

$$\mathbf{H}(s) = \sum_{j=1}^{2N} \frac{\mathbf{z}_j \mathbf{z}_j^T}{\gamma_j (s - s_j)} = \sum_{j=1}^{N} \left[\frac{\mathbf{z}_j \mathbf{z}_j^T}{\gamma_j (s - s_j)} + \frac{\mathbf{z}_j^* \mathbf{z}_j^{*T}}{\gamma_j^* (s - s_j^*)} \right]. \qquad [2.239]$$

The transfer function matrix is often expressed in terms of the complex natural frequencies λ_j. Substituting $s_j = \mathrm{i}\lambda_j$ in the preceding expression, we have

$$\mathbf{H}(s) = \sum_{j=1}^{N} \left[\frac{\mathbf{z}_j \mathbf{z}_j^T}{\gamma_j (s - \mathrm{i}\lambda_j)} + \frac{\mathbf{z}_j^* \mathbf{z}_j^{*T}}{\gamma_j^* (s + \mathrm{i}\lambda_j^*)} \right] \quad \text{and} \quad \gamma_j = \mathbf{z}_j^T \left[2\mathrm{i}\lambda_j \mathbf{M} + \mathbf{C} \right] \mathbf{z}_j. \qquad [2.240]$$

It can be shown that the transfer function matrix $\mathbf{H}(s)$ in [2.240] reduces to its equivalent expression for the undamped case in [2.85]. In the undamped limit $\mathbf{C} = 0$. This results in $\lambda_j = \omega_j = \lambda_j^*$ and $\mathbf{z}_j = \mathbf{x}_j = \mathbf{z}_j^*$. In view of the mass normalization relationship, we also have $\gamma_j = 2\mathrm{i}\omega_j$. Consider a typical term in [2.240]

$$\begin{aligned} &\left[\frac{\mathbf{z}_j \mathbf{z}_j^T}{\gamma_j (s - \mathrm{i}\lambda_j)} + \frac{\mathbf{z}_j^* \mathbf{z}_j^{*T}}{\gamma_j^* (s + \mathrm{i}\lambda_j^*)} \right] = \left[\frac{1}{2\mathrm{i}\omega_j} \frac{1}{\mathrm{i}\omega - \mathrm{i}\omega_j} + \frac{1}{-2\mathrm{i}\omega_j} \frac{1}{\mathrm{i}\omega + \mathrm{i}\omega_j} \right] \mathbf{x}_j \mathbf{x}_j^T \\ &= \frac{1}{2\mathrm{i}^2 \omega_j} \left[\frac{1}{\omega - \omega_j} - \frac{1}{\omega + \omega_j} \right] \mathbf{x}_j \mathbf{x}_j^T \qquad [2.241] \\ &= -\frac{1}{2\omega_j} \left[\frac{\omega + \omega_j - \omega + \omega_j}{(\omega - \omega_j)(\omega + \omega_j)} \right] \mathbf{x}_j \mathbf{x}_j^T = -\frac{1}{2\omega_j} \left[\frac{2\omega_j}{\omega^2 - \omega_j^2} \right] \mathbf{x}_j \mathbf{x}_j^T = \frac{\mathbf{x}_j \mathbf{x}_j^T}{\omega_j^2 - \omega^2}. \end{aligned}$$

Note that this term was derived before for the transfer function matrix of the undamped system in equation [2.85]. Therefore, equation [2.237] is the most general expression of the dynamic response of damped linear dynamic systems. In a similar manner, it can also be verified that when the system is proportionally damped, equation [2.237] reduces to [2.133] as expected.

2.5.2.2. *Time-domain analysis*

Combining equations [2.237] and [2.240], we have

$$\bar{\mathbf{q}}(s) = \sum_{j=1}^{N} \left\{ \frac{\mathbf{z}_j^T \bar{\mathbf{f}}(s) + \mathbf{z}_j^T \mathbf{M}\dot{\mathbf{q}}_0 + \mathbf{z}_j^T \mathbf{C}\mathbf{q}_0 + s\mathbf{z}_j^T \mathbf{M}\mathbf{q}_0}{\gamma_j (s - \mathrm{i}\lambda_j)} \mathbf{z}_j + \frac{{\mathbf{z}_j^*}^T \bar{\mathbf{f}}(s) + {\mathbf{z}_j^*}^T \mathbf{M}\dot{\mathbf{q}}_0 + {\mathbf{z}_j^*}^T \mathbf{C}\mathbf{q}_0 + s{\mathbf{z}_j^*}^T \mathbf{M}\mathbf{q}_0}{\gamma_j^* (s + \mathrm{i}\lambda_j^*)} \mathbf{z}_j^* \right\}. \quad [2.242]$$

From the table of Laplace transforms, we know that

$$\mathcal{L}^{-1}\left[\frac{1}{s-a}\right] = e^{at}$$

and $$\mathcal{L}^{-1}\left[\frac{s}{s-a}\right] = ae^{at}, \quad t > 0. \quad [2.243]$$

Taking the inverse Laplace transform of [2.242], dynamic response in the time domain can be obtained as

$$\mathbf{q}(t) = \mathcal{L}^{-1}\left[\bar{\mathbf{q}}(s)\right] = \sum_{j=1}^{N} \frac{1}{\gamma_j} a_{1_j}(t)\mathbf{z}_j + \frac{1}{\gamma_j^*} a_{2_j}(t)\mathbf{z}_j^* \quad [2.244]$$

where

$$\begin{aligned} a_{1_j}(t) &= \mathcal{L}^{-1}\left[\frac{\mathbf{z}_j^T \bar{\mathbf{f}}(s)}{s - \mathrm{i}\lambda_j}\right] + \mathcal{L}^{-1}\left[\frac{1}{s - \mathrm{i}\lambda_j}\right] \left(\mathbf{z}_j^T \mathbf{M}\dot{\mathbf{q}}_0 + \mathbf{z}_j^T \mathbf{C}\mathbf{q}_0\right) \\ &\quad + \mathcal{L}^{-1}\left[\frac{s}{s - \mathrm{i}\lambda_j}\right] \mathbf{z}_j^T \mathbf{M}\mathbf{q}_0 \\ &= \int_0^t e^{\mathrm{i}\lambda_j (t-\tau)} \mathbf{z}_j^T \mathbf{f}(\tau) d\tau + e^{\mathrm{i}\lambda_j t} \left(\mathbf{z}_j^T \mathbf{M}\dot{\mathbf{q}}_0 + \mathbf{z}_j^T \mathbf{C}\mathbf{q}_0 + \mathrm{i}\lambda_j \mathbf{z}_j^T \mathbf{M}\mathbf{q}_0\right), \\ &\quad \text{for} \quad t > 0 \end{aligned} \quad [2.245]$$

and similarly

$$\begin{aligned} a_{2_j}(t) &= \mathcal{L}^{-1}\left[\frac{\mathbf{z}_j^{*^T}\bar{\mathbf{f}}(s)}{s+\mathrm{i}\lambda_j}\right] + \mathcal{L}^{-1}\left[\frac{1}{s+\mathrm{i}\lambda_j}\right]\left(\mathbf{z}_j^{*^T}\mathbf{M}\dot{\mathbf{q}}_0 + \mathbf{z}_j^{*^T}\mathbf{C}\mathbf{q}_0\right) \\ &\quad + \mathcal{L}^{-1}\left[\frac{s}{s+\mathrm{i}\lambda_j}\right]\mathbf{z}_j{*}^T\mathbf{M}\mathbf{q}_0 \\ &= \int_0^t e^{-\mathrm{i}\lambda_j(t-\tau)}\mathbf{z}_j^{*^T}\mathbf{f}(\tau)d\tau + e^{-\mathrm{i}\lambda_j t}\left(\mathbf{z}_j^{*^T}\mathbf{M}\dot{\mathbf{q}}_0 + \mathbf{z}_j^{*^T}\mathbf{C}\mathbf{q}_0 - \mathrm{i}\lambda_j\mathbf{z}_j^{*^T}\mathbf{M}\mathbf{q}_0\right), \\ &\quad \text{for} \quad t > 0. \end{aligned} \qquad [2.246]$$

EXAMPLE 2.5.– Figure 2.13 shows a three DOF spring-mass system. This system is identical to the system used in example 2.3, except that the damper attached with the middle block is now disconnected.

1) Show that in general the system possesses complex modes.

2) Obtain approximate expressions for complex natural frequencies (using the first-order perturbation method).

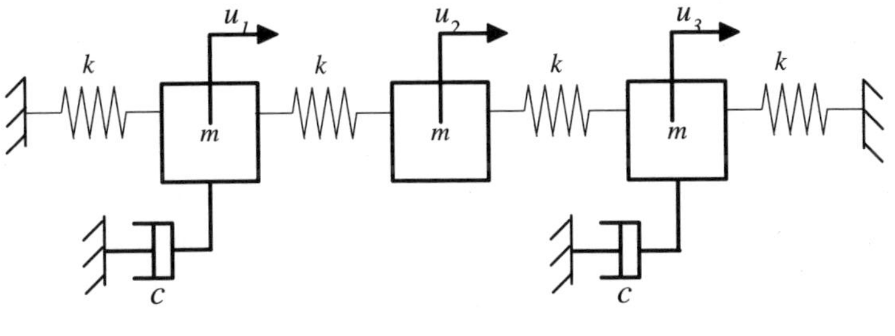

Figure 2.13. *There DOF damped spring-mass system*

SOLUTION–. The mass and the stiffness matrices of the system are same as in example 2.3 (given in equation [2.146]). The damping matrix is clearly given by

$$\mathbf{C} = \begin{bmatrix} c & 0 & 0 \\ 0 & 0 & 0 \\ 0 & 0 & c \end{bmatrix}. \qquad [2.247]$$

From this (and also from observation), it clear that the damping matrix is neither proportional to the mass matrix nor is it proportional to the stiffness matrix. Therefore, it is *likely* that the system will not have classical normal mode. In order to be sure, we need to check whether Caughey and O'Kelly's [CAU 65] criterion, i.e. $\mathbf{CM}^{-1}\mathbf{K} = \mathbf{KM}^{-1}\mathbf{C}$, is satisfied or not. Since $\mathbf{M} = m\mathbf{I}$ is a diagonal matrix, $\mathbf{M}^{-1} = \frac{1}{m}\mathbf{I}$. Recall that for any matrix $\mathbf{A}$, $\mathbf{IA} = \mathbf{AI} = \mathbf{A}$. Using the system matrices, we have

$$\begin{aligned}\mathbf{CM}^{-1}\mathbf{K} = \mathbf{C}\frac{1}{m}\mathbf{IK} &= \frac{1}{m}\mathbf{CK} = \frac{ck}{m}\begin{bmatrix}1&0&0\\0&0&0\\0&0&1\end{bmatrix}\begin{bmatrix}2&-1&0\\-1&2&-1\\0&-1&2\end{bmatrix}\\ &= \frac{ck}{m}\begin{bmatrix}2&-1&0\\0&0&0\\0&-1&2\end{bmatrix}\end{aligned} \qquad [2.248]$$

and
$$\begin{aligned}\mathbf{KM}^{-1}\mathbf{C} = \mathbf{K}\frac{1}{m}\mathbf{IC} &= \frac{1}{m}\mathbf{KC} = \frac{ck}{m}\begin{bmatrix}2&-1&0\\-1&2&-1\\0&-1&2\end{bmatrix}\begin{bmatrix}1&0&0\\0&0&0\\0&0&1\end{bmatrix}\\ &= \frac{ck}{m}\begin{bmatrix}2&0&0\\-1&0&-1\\0&0&2\end{bmatrix}.\end{aligned} \qquad [2.249]$$

It is clear that $\mathbf{CM}^{-1}\mathbf{K} \neq \mathbf{KM}^{-1}\mathbf{C}$, i.e. Caughey and O'Kelly's [CAU 65] condition is not satisfied by the system matrices. This confirms that the system does not possess classical normal modes but has complex modes.

Exact complex modes of the system can be obtained using the state-space approach outlined in section 2.5.1.1. For a three DOF system we need to solve an eigenvalue problem of the order six. Here, we will obtain approximate natural frequencies using the first-order perturbation method described in section 2.5.1.2. Using the undamped modal matrix in equation [2.176], the damping matrix in the modal coordinated can be obtained as

$$\begin{aligned}\mathbf{C}' &= \mathbf{X}^T\mathbf{CX} = \frac{1}{2\sqrt{m}}\begin{bmatrix}1&\sqrt{2}&1\\\sqrt{2}&0&-\sqrt{2}\\1&-\sqrt{2}&1\end{bmatrix}^T\begin{bmatrix}c&0&0\\0&0&0\\0&0&c\end{bmatrix}\frac{1}{2\sqrt{m}}\begin{bmatrix}1&\sqrt{2}&1\\\sqrt{2}&0&-\sqrt{2}\\1&-\sqrt{2}&1\end{bmatrix}\\ &= \frac{c}{4m}\begin{bmatrix}1&\sqrt{2}&1\\\sqrt{2}&0&-\sqrt{2}\\1&-\sqrt{2}&1\end{bmatrix}\begin{bmatrix}1&0&0\\0&0&0\\0&0&1\end{bmatrix}\begin{bmatrix}1&\sqrt{2}&1\\\sqrt{2}&0&-\sqrt{2}\\1&-\sqrt{2}&1\end{bmatrix}\\ &= \frac{c}{4m}\begin{bmatrix}1&\sqrt{2}&1\\\sqrt{2}&0&-\sqrt{2}\\1&-\sqrt{2}&1\end{bmatrix}\begin{bmatrix}1&\sqrt{2}&1\\0&0&0\\1&-\sqrt{2}&1\end{bmatrix} = \frac{c}{4m}\begin{bmatrix}2&0&2\\0&4&0\\2&0&2\end{bmatrix} = \frac{c}{m}\begin{bmatrix}1/2&0&1/2\\0&1&0\\1/2&0&1/2\end{bmatrix}.\end{aligned} \qquad [2.250]$$

Note that unlike example 2.3, $\mathbf{C}'$ is not a diagonal matrix, i.e. the equation of motion in the modal coordinates is coupled through the off-diagonal terms of the $\mathbf{C}'$ matrix. Approximate complex natural frequencies can be obtained from equation [2.217] as

$$\lambda_1 \approx \pm\omega_1 + \mathrm{i}C'_{11}/2 = \pm\sqrt{\left(2-\sqrt{2}\right)\alpha} + \mathrm{i}\frac{c}{4m}, \quad [2.251]$$

$$\lambda_2 \approx \pm\omega_2 + \mathrm{i}C'_{22}/2 = \pm\sqrt{2\alpha} + \mathrm{i}\frac{c}{2m} \quad [2.252]$$

and $$\lambda_3 \approx \pm\omega_3 + \mathrm{i}C'_{33}/2 = \pm\sqrt{\left(2+\sqrt{2}\right)\alpha} + \mathrm{i}\frac{c}{4m}. \quad [2.253]$$

In the above equations, undamped eigenvalues obtained in example 2.3 (equation [2.151]) have been used. The second complex mode is the most heavily damped (as the imaginary part is twice compared to the other two modes). This is because in the second mode the middle mass is stationary (look at the second mode shape in [2.164]), while the two "damped" masses move a maximum distance away from it. This causes maximum "stretch" to both the dampers and results in maximum damping in this mode.

2.6. Rayleigh quotient for damped systems

In this section, we aim to develop qualitative understanding of the natural frequencies of damped systems. In his classical treatise on the theory of sound [RAY 77], Lord Rayleigh introduced the notion of a quotient of two quadratics representing the potential and kinetic energies of a vibrating system. Since then, the Rayleigh quotient has been widely applied to the analysis of many vibrating systems and their associated linear algebraic eigenvalue problems. The Rayleigh quotient provides a variational approach to estimate the eigenvalues of an algebraic, generalized eigenvalue problem, as in the case of determining the natural frequencies of a vibrating system. Numerical methods to solve eigenvalue problems [BAT 76] such as the shifted inverse power method rely on the properties of Rayleigh quotients for speedier convergence [STR 88]. Thus, the practical utility of the Rayleigh quotient is wide-ranging. This section is devoted to the Rayleigh quotient for damped systems.

Traditionally, and in many text books on vibration [MEI 97, NEW 87] and linear algebra [STR 88, WIL 88], the Rayleigh quotient is defined as a ratio of two quadratics. In the case of a generalized eigenvalue problem involving two real and symmetric matrices $\mathbf{A}$ and $\mathbf{B}$, the Rayleigh quotient is defined as follows

$$R(\mathbf{x}) = \frac{\mathbf{x}^T\mathbf{A}\mathbf{x}}{\mathbf{x}^T\mathbf{B}\mathbf{x}}, \quad \mathbf{A}\mathbf{x} = \lambda\mathbf{B}\mathbf{x} \quad [2.254]$$

where the eigenvalue problem is abbreviated as EVP. The stationarity properties of this "classical" Rayleigh quotient are well established [STR 88]. The objective of the present investigation is to explore whether similar Rayleigh-like quotients with stationary properties exist for a vibrating system with dissipation. Discrete vibrating systems are chosen here for the purpose of illustration; generalization of the results for continuous systems is straightforward.

Rayleigh quotients for discrete systems are defined in section 2.6.1. Three quotients are introduced in the case of a viscously damped system and their stationary properties are investigated in section 2.6.2 and section 2.6.3. An application of Rayleigh quotients is discussed in section 2.6.4. The main conclusions emerging from this study are summarized in section 2.6.5. Throughout this study the terms modes and eigenvectors are used interchangeably.

2.6.1. *Rayleigh quotients for discrete systems*

In the context of vibration analysis of undamped systems, the two quadratic functions in the Rayleigh quotient assume the physical meaning of the kinetic and potential energies. Thus, associated with any admissible deformation vector (ϕ), we can define the following quantities:

$$\begin{aligned} \mathcal{U} &= \phi^T \mathbf{K} \phi, \quad \mathcal{T} = \phi^T \mathbf{M} \phi \\ R(\phi) &= \tfrac{\mathcal{U}}{\mathcal{T}} = \frac{\phi^T \mathbf{K} \phi}{\phi^T \mathbf{M} \phi} \end{aligned} \qquad [2.255]$$

where $\mathcal{T}$ and $\mathcal{U}$ are the kinetic and potential energies of the system and R is the *classical* Rayleigh quotient.

However, when systems with dissipation are considered, we are faced with three quadratics. In this situation, we can define *three* independent quotients as follows

$$\begin{aligned} \mathcal{U} &= \phi^T \mathbf{K} \phi, \quad \mathcal{T} = \phi^T \mathbf{M} \phi, \quad \mathcal{D} = \phi^T \mathbf{C} \phi \\ R_1(\phi) &= \frac{\mathcal{U}}{\mathcal{T}} = \frac{\phi^T \mathbf{K} \phi}{\phi^T \mathbf{M} \phi} \\ R_2(\phi) &= \frac{\mathcal{D}}{\mathcal{T}} = \frac{\phi^T \mathbf{C} \phi}{\phi^T \mathbf{M} \phi} \\ R_3(\phi) &= \frac{\mathcal{D}}{\mathcal{U}} = \frac{\phi^T \mathbf{C} \phi}{\phi^T \mathbf{K} \phi}. \end{aligned} \qquad [2.256]$$

Note that for the Rayleigh quotient to be finite, the denominator terms in the above equation should not be equal to zero. This requires that $\mathbf{M}$ be positive definite for R_1 and R_2 to be finite, and $\mathbf{K}$ be positive definite for R_3 to be finite. For the majority of vibrating systems, $\mathbf{M}$ is positive definite while $\mathbf{K}$ need not be. Thus, the existence of R_3 is case specific.

It is the objective of this work to investigate the stationarity properties of the quotients defined in equations [2.255] and [2.256]. The proof of stationary property of the quotient defined in equation [2.255] is well known [STR 88, WIL 88, COU 89]. However, it is repeated here for the sake of completeness and also since the proof of stationarity for other quotients closely follows a similar procedure.

2.6.1.1. *Stationarity of* $R(\phi)$

Let a vector ϕ be chosen such that it is close to one of the eigenvectors (modes) $\mathbf{x}_r$ of the system so that we can express ϕ as

$$\phi = \sum_i c_i \mathbf{x}_i = \mathbf{x}_r + \sum_{i \neq r} \epsilon_i \mathbf{x}_i, \quad \epsilon_i = \frac{c_i}{c_r} << 1 \qquad [2.257]$$

where ϵ_i is a small real quantity. Now the Rayleigh quotient reads

$$R(\phi) = \frac{\phi^T \mathbf{K} \phi}{\phi^T \mathbf{M} \phi} = \frac{\mathbf{x}_r^T \mathbf{K} \mathbf{x}_r + 2 \sum_{i \neq r} \epsilon_i \mathbf{x}_i^T \mathbf{K} \mathbf{x}_r + O(\epsilon^2)}{\mathbf{x}_r^T \mathbf{M} \mathbf{x}_r + 2 \sum_{i \neq r} \epsilon_i \mathbf{x}_i^T \mathbf{M} \mathbf{x}_r + O(\epsilon^2)} \qquad [2.258]$$

here the symmetry of $\mathbf{M}$ and $\mathbf{K}$ is assumed. Because of the orthogonality properties of the eigenvectors [WIL 88, STR 88]

$$\mathbf{x}_i^T \mathbf{M} \mathbf{x}_j = \delta_{ij}, \quad \mathbf{x}_i^T \mathbf{K} \mathbf{x}_j = 0, \quad \mathbf{x}_i^T \mathbf{K} \mathbf{x}_i = \lambda_i. \qquad [2.259]$$

Equation [2.258] simplifies to

$$R(\phi) = \frac{\lambda_r + O(\epsilon^2)}{1 + O(\epsilon^2)} = \lambda_r (1 + O(\epsilon^2)). \qquad [2.260]$$

The above result proves the stationarity of the Rayleigh quotient, i.e. first-order changes in ϕ lead to second-order changes in $R(\phi)$. When ϕ is close to one of the eigenvectors, the corresponding value of the quotient is stationary. Further choosing the first eigenvector as the trial vector ϕ leads to a minimum value of $R(\phi)$. $R(\phi)$ is maximum when the trial vector is close to the eigenvector corresponding to the highest eigenvalue. For intermediate eigenvectors $R(\phi)$ is neither a minimum nor a maximum, that is $R(\phi)$ is at a saddle point. A mini-max (or inf-sup) principle due to Courant and Fischer applies in this case [COU 89, STR 88].

2.6.2. *Proportional damping*

Consider first the case of proportional damping, as discussed in detail in section 2.4. Here, proportional damping is defined in the sense that the same vector ϕ simultaneously diagonalizes the three quadratics $\mathcal{T}$, $\mathcal{U}$ and $\mathcal{D}$. In other words, the three matrices $\mathbf{M}$, $\mathbf{K}$ and $\mathbf{C}$ can be simultaneously diagonalized. In the case of proportionally damped systems, the eigenvectors are real but the eigenvalues are not, i.e. the undamped modes are also the modes of the proportionally damped system. Thus, the proof of stationarity of the first Rayleigh quotient $R_1(\phi)$ is the same as that given in section 2.6.1.1.

Consider the second Rayleigh quotient associated with any admissible deformation vector ϕ as defined in equation [2.257]

$$R_2(\phi) = \frac{\phi^T\mathbf{C}\phi}{\phi^T\mathbf{M}\phi} = \frac{\mathbf{x}_r^T\mathbf{C}\mathbf{x}_r + 2\sum_{i\neq r}\epsilon_i\mathbf{x}_i^T\mathbf{C}\mathbf{x}_r + O(\epsilon^2)}{\mathbf{x}_r^T\mathbf{M}\mathbf{x}_r + 2\sum_{i\neq r}\epsilon_i\mathbf{x}_i^T\mathbf{M}\mathbf{x}_r + O(\epsilon^2)} \qquad [2.261]$$

here the symmetry of $\mathbf{M}$ and $\mathbf{C}$ is assumed. Because of the orthogonality properties of the eigenvectors

$$\mathbf{x}_i^T\mathbf{M}\mathbf{x}_j = \delta_{ij}. \qquad [2.262]$$

Define

$$\mathbf{x}_i^T\mathbf{C}\mathbf{x}_j = C'_{ij}, \quad \mathbf{x}_i^T\mathbf{C}\mathbf{x}_i = C'_{ii}. \qquad [2.263]$$

With the above definition, equation [2.261] can be expressed as

$$R_2(\phi) = \frac{C'_{rr} + 2\sum_{i\neq r}\epsilon_i C'_{ir} + O(\epsilon^2)}{1 + O(\epsilon^2)}. \qquad [2.264]$$

When damping is proportional, the matrix $\mathbf{C}'_{ir}$ is diagonal, i.e. $\mathbf{C}'_{ir} = 0$ for $i \neq r$. In this case, the above equation simplifies to

$$R_2(\phi) = \frac{C'_{rr} + O(\epsilon^2)}{1 + O(\epsilon^2)} = C'_{rr}(1 + O(\epsilon^2)) \qquad [2.265]$$

which proves the stationarity of Rayleigh quotient in the case of a proportionally damped system.

Similar proof can be constructed for $R_3(\phi)$. The equation corresponding to equation [2.265] in this case will read as

$$R_3(\phi) = \frac{C'_{rr} + O(\epsilon^2)}{\omega_r^2 + O(\epsilon^2)} = \frac{C'_{rr}}{\omega_r^2}(1 + O(\epsilon^2)). \qquad [2.266]$$

2.6.3. *Non-proportional damping*

Consider the case of non-proportional damping wherein the damping matrix $\mathbf{C}$ cannot be diagonalized simultaneously with $\mathbf{M}$ and $\mathbf{K}$ matrices. As a result, the vector ϕ is not necessarily real. Vibrating systems with non-proportional damping are known to possess complex modes in general. Physically, the complex modes represent *nearly* standing waves. For systems with small dissipation, a perturbation theory originally put forward by Rayleigh [RAY 77] can be used to represent the complex modes in terms of the real modes of the undamped system.

According to the first-order perturbation theory [WOO 98], the complex modes of a viscously damped system are related to the corresponding undamped modes by

$$\mathbf{z}_n \approx \mathbf{x}_n + \mathrm{i}\sum_{k \neq n} \alpha_{kn}\mathbf{x}_k, \quad \text{where} \quad \alpha_{kn} = \frac{\omega_n C'_{kn}}{\omega_n^2 - \omega_k^2} < 1. \qquad [2.267]$$

The undamped modes are mass normalized, i.e. $\mathbf{x}_n^T\mathbf{M}\mathbf{x}_n = 1$. In the above equation, $\mathbf{C}'$ is the damping matrix in modal coordinates, i.e. $C'_{kn} = \mathbf{x}_k^T\mathbf{C}\mathbf{x}_n$. The assumption in the perturbation theory is that the terms of the order α_{kn}^2 are very small and hence negligible.

When k and n refer to two adjacent modes the coefficient α_{kn} can be related to the modal overlap factor defined as $\mu_{kn} \equiv \zeta_n\omega_n/(\omega_k - \omega_n)$ and the ratio $\gamma_{kn} = C'_{kn}/C'_{nn}$ by: $\alpha_{kn} \approx (1/2)\,\mu_{kn}\gamma_{kn}$. Note that γ_{kn} is a measure of the diagonal dominance of the $\mathbf{C}'$ matrix. μ_{kn} is a measure of the spacing of adjacent modes normalized with respect to the half power band width of each mode. Thus, significantly complex modes are to be expected when the modal damping matrix is not diagonally dominant and the modal overlap is not small. Unless the modal overlap factor is of the order of unity the second and higher order powers of the α_{kn} can be safely ignored. If not, then the perturbation expansion has to be extended suitably until the imaginary part of the complex mode converges. Adequacy of the first-order theory for systems with small damping has been shown in [WOO 98].

Since the eigenvectors $\mathbf{z}_i, i = 1, \ldots, n$, form the complete basis in an n-dimensional complex vector space, any arbitrary complex vector ψ can be written as

$$\psi = \sum_i c_i \mathbf{z}_i. \qquad [2.268]$$

Select a vector close to $\mathbf{z}_r$ which can be written as

$$\psi = \mathbf{z}_r + \sum_{i \neq r} \epsilon_i \mathbf{z}_i, \quad | \epsilon_i |=| \frac{c_i}{c_r} |<< 1. \qquad [2.269]$$

Consider the first real-valued Rayleigh quotient associated with the above trial vector:

$$R_1(\psi) = \frac{\psi^H \mathbf{K} \psi}{\psi^H \mathbf{M} \psi} = \frac{\mathbf{z}_r^H \mathbf{K} \mathbf{z}_r + 2 \sum_{i \neq r} \Re(\epsilon_i) \mathbf{z}_i^H \mathbf{K} \mathbf{z}_r + O(| \epsilon |^2)}{\mathbf{z}_r^H \mathbf{M} \mathbf{z}_r + 2 \sum_{i \neq r} \Re(\epsilon_i) \mathbf{z}_i^H \mathbf{M} \mathbf{z}_r + O(| \epsilon |^2)}. \qquad [2.270]$$

Noting the orthogonality properties given in equation [2.259], we can deduce the following equations

$$\mathbf{z}_r^H \mathbf{M} \mathbf{z}_r = 1 + O(\alpha^2) \qquad [2.271]$$

and

$$\mathbf{z}_i^H \mathbf{M} \mathbf{z}_r = O(\alpha^2). \qquad [2.272]$$

Similarly, with $\mathbf{K}$ we can show that

$$\mathbf{z}_r^H \mathbf{K} \mathbf{z}_r = \omega_r^2 + O(\alpha^2) \qquad [2.273]$$

and

$$\mathbf{z}_i^H \mathbf{K} \mathbf{z}_r = O(\alpha^2). \qquad [2.274]$$

Substituting [2.271]–[2.274] in equation [2.270], we obtain

$$R_1(\psi) = \frac{\omega_r^2 + O(\alpha^2) + O(\epsilon^2)}{1 + O(\alpha^2) + O(\epsilon^2)} \approx \omega_r^2 (1 - O(\epsilon^2)) \qquad [2.275]$$

which proves the stationarity of $R_1(\psi)$.

Consider the second real-valued Rayleigh quotient defined as follows

$$R_2(\psi) = \frac{\mathbf{z}^H \mathbf{C} \mathbf{z}}{\mathbf{z}^H \mathbf{M} \mathbf{z}}. \qquad [2.276]$$

Substituting equation [2.267] in the above equation leads to

$$R_2(\psi) = \frac{\psi^H \mathbf{C} \psi}{\psi^H \mathbf{M} \psi} = \frac{\mathbf{z}_r^H \mathbf{C} \mathbf{z}_r + 2\sum_{i\neq r} \Re(\epsilon_i) \mathbf{z}_i^H \mathbf{C} \mathbf{z}_r + O(|\, \epsilon\, |^2)}{\mathbf{z}_r^H \mathbf{M} \mathbf{z}_r + 2\sum_{i\neq r} \Re(\epsilon_i) \mathbf{z}_i^H \mathbf{M} \mathbf{z}_r + O(|\, \epsilon\, |^2)}. \qquad [2.277]$$

Now $\mathbf{z}_r^H \mathbf{C} \mathbf{z}_r$ can be expanded as

$$\begin{aligned} \mathbf{z}_r^H \mathbf{C} \mathbf{z}_r &= [\mathbf{x}_r^T - \mathrm{i}\sum_{k\neq r} \alpha_{kr} \mathbf{x}_k^T] \mathbf{C} [\mathbf{x}_r + \mathrm{i}\sum_{k\neq r} \alpha_{kr} \mathbf{x}_k] \\ &= C'_{rr} - \mathrm{i}\sum_{k\neq r} \alpha_{kr} \left[\mathbf{x}_k^T \mathbf{C} \mathbf{x}_r - \mathbf{x}_r^T \mathbf{C} \mathbf{x}_k\right] + O(\alpha^2) = C'_{rr} + O(\alpha^2). \end{aligned} \qquad [2.278]$$

Note that $\mathbf{C}$ is assumed to be symmetric in simplifying the above equation. Similarly, we can write

$$\begin{aligned} \mathbf{z}_i^H \mathbf{C} \mathbf{z}_r &= [\mathbf{x}_i^T - \mathrm{i}\sum_{k\neq i} \alpha_{ki} \mathbf{x}_k^T] \mathbf{C} [\mathbf{x}_r + \mathrm{i}\sum_{k\neq r} \alpha_{kr} \mathbf{x}_k] \\ &= C'_{ir} - \mathrm{i}\sum_{k\neq i} \alpha_{ki} C'_{kr} + \mathrm{i}\sum_{k\neq r} \alpha_{kr} C'_{ik} + O(\alpha^2) = C'_{ir} + O(\alpha) + O(\alpha^2). \end{aligned}$$

$$[2.279]$$

Substituting equations [2.271] and [2.272] and equations [2.278] and [2.279] in equation [2.277], we obtain

$$R_2(\psi) = \frac{C'_{rr} + 2\sum_{i\neq r} \Re(\epsilon_i) C'_{ir} + O(\epsilon) O(\alpha) + O(\epsilon^2) + O(\alpha^2)}{1 + O(\epsilon^2) + O(\alpha^2)}. \qquad [2.280]$$

It can be seen that first-order changes in ψ lead to first-order changes in $R_2(\psi)$. However, if the modal damping matrix is diagonally dominant, i.e.

$$\frac{C'_{ir}}{C'_{rr}} << 1 \qquad [2.281]$$

then first-order changes in ψ lead to second-order changes in $R_2(\psi)$. In this case, stationarity of the Rayleigh quotient is obtained.

Returning to the third quotient $R_3(\psi)$

$$R_3(\psi) = \frac{\psi^H \mathbf{C} \psi}{\psi^H \mathbf{K} \psi} = \frac{C'_{rr} + 2\sum_{i \neq r} \Re(\epsilon_i) C'_{ir} + O(\epsilon)O(\alpha) + O(\epsilon^2) + O(\alpha^2)}{\omega_r^2 + O(\epsilon^2) + O(\alpha^2)}. \qquad [2.282]$$

The above quotient is not stationary in general. However, when the modal damping matrix is diagonally dominant in accordance with equation [2.281], stationarity of $R_3(\psi)$ can be shown as earlier (see section 2.6.2).

In the case of a complex vector ψ, we are also tempted to define complex-valued Rayleigh quotients by replacing the Hermitian transpose (complex conjugate transpose) with the ordinary transpose operator. The stationarity property of these complex-valued quotients, however, cannot be shown. Hence, the discussion of these quotients will not be pursued any further.

2.6.4. *Application of Rayleigh quotients*

In the case of an SDOF system with viscous damping, the three quotients simplify to $R_1 = \omega^2$, $R_2 = 2\zeta\omega$ and $R_3 = 2\zeta/\omega$, where ω and ζ denote the natural frequency and the critical damping factor, respectively. The response of the system in the time domain is described by $\exp\left(-\zeta\omega t - \mathrm{i}\omega\sqrt{1-\zeta^2 t}\right)$. We note that R_2 governs the decay rate (or real part of the complex eigenvalue) of vibration in the time domain. The same will be true for a multi-degree-of-freedom system, provided that response can be decomposed into SDOF systems using modal summation, i.e. damping is proportional [MEI 97].

The Rayleigh quotient R_1 and its usefulness in solving the eigenvalue problem associated with the undamped system is well documented [MEI 97, WIL 88, STR 88]. Consequent to the stationary property of R_1, a theorem originally put forward by Rayleigh, known as Rayleigh's principle or interlacing theorem, gives the influence of constraints. It states that the eigenvalues of the constrained system (ω') interlace with the eigenvalues of the un-constrained system (ω) such that $\omega_n \leq \omega'_n \leq \omega_{n+1}$.

Similar results follow from the stationarity of R_2 and R_3. In this context, we refer to Rayleigh's original statement in §88 of [RAY 77] "... theorems, of importance in other branches of science, may be stated for systems such that only T and F, or only

V and F, are sensible." We note that $T \equiv \mathcal{T}$, $V \equiv \mathcal{U}$ and $F \equiv \mathcal{D}$. Thus, stationarity of R_2 implies that the decay rates of each normal mode is stationary. The interlacing theorem would suggest that the decay rates of each normal mode also interlace when a constraint is applied. The interlacing property was discussed in §88 of [RAY 77] and a lesser-known work of Rayleigh [RAY 85]. The present study extends these ideas to the general case of damped systems with viscous or non-viscous damped processes.

In a viscoelastic system, we deal with elastic potentials and damped potentials. Stationarity of R_3 has important consequences for such problems, especially in conjunction with the interlacing theorem. A noteworthy work on applying the Rayleigh quotients to determine the elastic and material loss constants of orthotropic sheet materials was undertaken in [MCI 78, MCI 88]. Our primary aim in this work has been to show the range of applicability of stationarity principles in damped viscous systems. Further application of these results to non-viscously damped systems will be explored in section 5.7.

2.6.5. *Synopses*

Rayleigh quotients are revisited in the context of damped systems. The study of their stationarity properties leads to the following conclusions:

1) In the case of a proportionally damped viscous system, the three Rayleigh quotients associated with the damped system are stationary.

2) In the case of a non-proportionally damped system, the Rayleigh quotient involving a mass and stiffness matrix is stationary while the remaining two involving the damping matrix are not. Stationarity in this case is subject to the diagonal dominance of the modal damping matrix. For an arbitrarily chosen viscous damping matrix, the stationarity property does not hold true. However, this negative conclusion is to be balanced by the wide variety of practical engineering structures where the modal damping is diagonally dominant, consequently, Rayleigh quotients are stationary.

2.7. Summary

Dynamics of SDOF and MDOF undamped and viscously damped systems were discussed. The basic concepts of the natural frequency and the damping factor of a damped SDOF system were introduced. Frequency response and impulse response functions of SDOF undamped and viscously damped systems were derived. The nature of the dynamic response of viscously damped SDOF system is governed by the value of the damping factor. The expressions of general dynamic response due to arbitrary forcing function and initial conditions were derived for SDOF undamped and damped systems. These expressions were then generalized to MDOF undamped and proportionally damped systems using the eigenvalues, eigenvectors and modal damping factors. It was shown that the classical modal analysis can be used to obtain the dynamic response of undamped and proportionally damped systems in a similar manner.

The concept of proportional damping was investigated in detail. Proportional damping is the most common approach for modeling dissipative forces in complex engineering structures and it has been used in various dynamic problems for more than 10 decades. One of the main limitations of the mass and stiffness proportional damping approximation comes from the fact that the arbitrary variation of damping factors with respect to the vibration frequency cannot be modeled accurately by using this approach. Experimental results, however, suggest that damping factors can vary with frequency. In this chapter, a new generalized proportional damping model is discussed in order to capture the frequency-variation of the damping factors accurately. The generalized proportional damping expresses the damping matrix in terms of smooth continuous functions involving specially arranged mass and stiffness matrices so that the system still possesses classical normal modes. This enables us to model variations in the modal damping factors with respect to the frequency in a simplified manner.

Dynamic analysis of general viscously damped systems requires the calculation of complex natural frequencies and complex modes. This can be achieved in a straightforward, but computationally expensive, manner by using the state-space approach. A symmetric state-space formulation is outlined in this chapter. The complex eigenvectors in the state-space satisfy orthogonality relationships, similar to that of the classical normal modes in the configuration space. By exploiting the orthogonality relationships, it is possible to extend the modal analysis to obtain the dynamic response using the modal superposition like undamped or proportionally damped systems. The modes, however, are complex modes and require the solution of a general quadratic eigenvalue problem. A simple perturbation-based method based on the light damping assumption is given. More details on the computational methods for complex modes are given in Chapter 6.

The concept of the Rayleigh quotient for damped systems was discussed. Stationarity properties were examined based on the perturbation theory. It was shown that Rayleigh quotients with stationary properties exist for systems with proportional damping. It was also shown that the stationarity property of Rayleigh quotients in the case of non-proportional damping is conditional upon the diagonal dominance of the modal damping matrix. The studies undertaken in this chapter give the necessary platform to discuss more advanced damping models. In the next chapter, we begin the discussion on dynamics of SDOF systems with non-viscous damping.

Chapter 3

Non-Viscously Damped Single-Degree-of-Freedom Systems

Increasing use of modern composite materials and active control mechanisms in aerospace and automotive industries demands sophisticated treatment of dissipative forces for proper analysis and design. Viscous damping is the most common model for modeling of vibration damping in linear systems. This model, first introduced by Lord Rayleigh [RAY 77], assumes that the instantaneous generalized velocities are the only relevant variables that determine damping. In the previous chapter, we discussed the dynamics of viscously damped systems in some detail. Viscous damping models are used widely for their simplicity and mathematical convenience even though the energy dissipation behavior of real structural materials may not be accurately represented by simple viscous models. It is well recognized that in general, a physically realistic model of damping will not be viscous. Damping models in which the dissipative forces depend on any quantity other than the instantaneous generalized velocities are non-viscous damping models. Mathematically, any causal model that makes the energy dissipation functional non-negative is a possible candidate for a non-viscous damping model. Clearly, a wide range of choice is possible (see Table 1.1), either based on the physics of the problem, or by *a priori* selecting a model and fitting its parameters from experiments. Among various damping models, the "exponential damping model" is particularly promising and has been used by many authors. With this model, the damping force is expressed as

$$f_d(t) = \int_0^t c\, \mu e^{-\mu(t-\tau)} \dot{u}(\tau)\, \mathrm{d}\tau \qquad [3.1]$$

where c is the viscous damping constant, μ is the relaxation parameter and $q(t)$ is the displacement as a function of time. In the context of viscoelastic materials, the physical basis for exponential models has been well established (see, for example, [CRE 73]). A selected literature review including the justifications for considering exponential damping model may be found in [WAG 03]. In [ADH 14b], Chapter 3, we have proposed a few methods by which the damping parameters in equation [3.1] can be obtained from experimental measurements. In [ADH 14b], Chapter 4, we have proposed four non-viscosity indices in order to quantify non-viscous damping in multiple-degree-of-freedom dynamic systems. Studies reported in [ADH 01c] show that the use of viscous damping model when the "true" damping model is non-viscous can lead to modeling errors.

Methods for the analysis of linear multiple-degree-of-freedom systems with damping of the form [3.1] have been considered by many authors, for example [MCT 93, WOO 98, ADH 02b, WAG 03, MUR 97b, MUR 98b]. See section 1.3 for a brief review of existing works. Although these publications provide excellent analytical and numerical tools for the analysis of non-viscously damped systems, most of the physical understandings are still from the point of view of a viscously damped dynamic system. In this chapter, we address the issue of how far we can extend the classical concepts borrowed from viscously damped systems to non-viscously damped systems. Historically, the majority of vibration engineers are trained from a viscous damping point of view, and most vibration analysis textbooks, finite element packages and modal analysis softwares only allow viscous damping models. For these reasons, wherever possible, we have attempted to relate the newly developed results to a viscously damped system and have pointed out whether there are any conceptual differences. The study in this chapter is limited to a single-degree-of-freedom (SDOF) dynamic system with energy dissipation characteristics given by equation [3.1]. The outline of the chapter is as follows. In section 3.1, the equation of motion is introduced and the exact analytical solutions of the eigenvalues are derived. The conditions for sustainable oscillatory motion are discussed in section 3.2. The critical damping factors of a non-viscously damped dynamic system are discussed in section 3.3. The nature of the eigenvalues of the dynamic system is considered in section 3.4. The frequency response function of the system is derived in section 3.5. The characteristics of the response amplitude are discussed in section 3.6. In section 3.7, a simplified analysis of dynamic response is proposed. Finally, our main findings are summarized in section 3.8.

3.1. The equation of motion

An SDOF system with non-viscous damping is shown in Figure 3.1. The equation of motion of the system with damping characteristics given by equation [3.1] can be expressed as

$$m\,\ddot{q}(t) + \int_0^t c\,\mu e^{-\mu(t-\tau)}\dot{q}(\tau)\,\mathrm{d}\tau + k\,q(t) = f(t) \qquad [3.2]$$

together with the initial conditions

$$q(0) = q_0 \quad \text{and} \quad \dot{q}(0) = \dot{q}_0 \tag{3.3}$$

where m is the mass of the dynamic system, k is the spring stiffness, $f(t)$ is the applied forcing and $\dot{(\bullet)}$ represents derivative with respect to time. Transforming equation [3.2] into the Laplace domain, we obtain

$$s^2 m\,\overline{q}(s) + s\,c\left(\frac{\mu}{s+\mu}\right)\overline{q}(s) + k\,\overline{q}(s) = \overline{f}(s) + m\,\dot{q}_0 + \left(s\,m + c\frac{\mu}{s+\mu}\right)q_0 \tag{3.4}$$

where s is the complex Laplace domain parameter and $\overline{(\bullet)}$ is the Laplace transform of $(\bullet)$. For convenience, we introduce the constants ω_n, ζ and β as follows

$$\omega_n = \sqrt{\frac{k}{m}}, \quad \zeta = \frac{c}{2\sqrt{k\,m}} \quad \text{and} \quad \beta = \frac{\omega_n}{\mu} \tag{3.5}$$

where ω_n is the undamped natural frequency, ζ is the viscous damping factor and β is the non-viscous damping factor. When $\zeta \to 0$, the dynamic system is effectively undamped. When $\beta \to 0$, then $\mu \to \infty$ and the dynamic system is effectively viscously damped. We will use these limiting cases frequently to develop our physical understandings of the results to be derived in this chapter.

Using the constants in [3.5], equation [3.4] can be rewritten as

$$\overline{d}(s)\,\overline{q}(s) = \overline{p}(s) \tag{3.6}$$

where the dynamic stiffness coefficient $\overline{d}(s)$ and the equivalent forcing function $\overline{p}(s)$ are given by

$$\overline{d}(s) = s^2 + s\,2\zeta\omega_n\left(\frac{\omega_n}{s\beta + \omega_n}\right) + \omega_n^2 \tag{3.7}$$

$$\text{and} \quad \overline{p}(s) = \frac{\overline{f}(s)}{m} + \dot{q}_0 + \left(s + 2\zeta\omega_n\frac{\omega_n}{s\beta + \omega_n}\right)q_0. \tag{3.8}$$

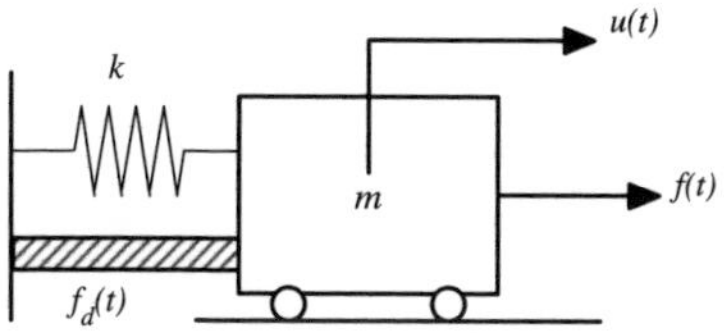

Figure 3.1. *Single-degree-of-freedom non-viscously damped dynamic system, damping force* $f_d(t) = \int_0^t c\,\mu e^{-\mu(t-\tau)}\dot{u}(\tau)\,\mathrm{d}\tau$

The aim of a dynamic analysis is often to obtain the dynamic response, either in the time domain or in the frequency domain. For an SODF dynamic system, it is a relatively simple task, we can either directly integrate equation [3.2] with the initial conditions [3.3], or alternatively can invert the coefficient associated with $\overline{q}(s)$ in equation [3.6]. Such an approach is not suitable for multiple-degree-of-freedom systems with non-proportional damping and may not provide much physical insight. We pursue an approach that involves eigensolutions of the dynamic system. The eigenvalues are the zeros of the dynamic stiffness coefficient and can be obtained by setting $\overline{d}(s) = 0$. Therefore, using equation [3.7], the eigenvalues are the solutions of the characteristics equation

$$\beta\, s^3 + \omega_n s^2 + (\beta\omega_n^2 + 2\zeta\omega_n^2)s + \omega_n^3 = 0. \qquad [3.9]$$

In contrast to a viscously damped dynamic system where we obtain a quadratic equation, this is a cubic equation.

The three roots of equation [3.9] can appear in two distinct forms: (1) one root is real and the other two roots are in a complex conjugate pair, or (2) all roots are real. Case (1) represents an *underdamped system* that usually arises when the "small damping" assumption is made. The complex conjugate pair of roots corresponds to the "vibration" of the dynamic system while the third root corresponds to a purely dissipative motion. Case (2) represents an *overdamped system* in which the system cannot sustain any oscillatory motion. For simplicity, we introduce a (in general complex valued) non-dimensional frequency parameter

$$r = \frac{s}{\omega_n} \qquad [3.10]$$

and transform the characteristics equation [3.9] to

$$\beta r^3 + r^2 + (\beta + 2\zeta)r + 1 = 0 \qquad [3.11]$$

$$\text{or} \quad r^3 + \sum_{j=0}^{2} a_j\, r^j = 0. \qquad [3.12]$$

The constants associated with the powers of r are given by

$$a_0 = \frac{1}{\beta},\; a_1 = 1 + 2\,\frac{\zeta}{\beta} \quad \text{and} \quad a_2 = \frac{1}{\beta}. \qquad [3.13]$$

The cubic equation [3.12] can be solved exactly in closed form (see, for example, [ABR 65] (section 3.8)). Define the following constants

$$Q = \frac{3a_1 - a_2^2}{9} = \frac{\left(3\,\beta^2 + 6\,\beta\,\zeta - 1\right)}{9\beta^2} \qquad [3.14]$$

$$\text{and} \quad R = \frac{9a_2a_1 - 27a_0 - 2a_2^3}{54} = -\frac{\left(9\,\beta^2 - 9\,\beta\,\zeta + 1\right)}{27\beta^3}. \qquad [3.15]$$

From these, calculate the negative of the discriminant

$$D = Q^3 + R^2 = \frac{1}{27\beta^4}\left(\beta^4 + 6\,\beta^3\zeta + 2\,\beta^2 + 12\,\beta^2\zeta^2 - 10\,\beta\,\zeta + 1 + 8\,\beta\,\zeta^3 - \zeta^2\right) \qquad [3.16]$$

and define two new constants

$$S = \sqrt[3]{R + \sqrt{D}} \quad \text{and} \quad T = \sqrt[3]{R - \sqrt{D}}. \qquad [3.17]$$

Using these constants, the three roots of equation [3.12] can be expressed by Cardano's formula as

$$r_1 = -\frac{a_2}{3} - \frac{1}{2}(S+T) + \mathrm{i}\frac{\sqrt{3}}{2}(S-T) \qquad [3.18]$$

$$r_2 = -\frac{a_2}{3} - \frac{1}{2}(S+T) - \mathrm{i}\frac{\sqrt{3}}{2}(S-T) \qquad [3.19]$$

$$\text{and} \quad r_3 = -\frac{a_2}{3} + (S+T). \qquad [3.20]$$

These are the normalized eigenvalues of the system. The actual eigenvalues, that is the solutions of equation [3.9], can be obtained as $s_j = \omega_n r_j$, $j = 1, 2, 3$. Throughout the chapter, we will use these normalized eigenvalues to investigate the dynamic properties of the non-viscously damped system. If the non-viscous damping factor β is zero, equation [3.11] reduces to the quadratic equation

$$r^2 + 2\zeta r + 1 = 0. \qquad [3.21]$$

As expected, this is the characteristic equation for a viscously damped system. For this special case, the two solutions of equation [3.21] are given by

$$r_1 = -\zeta + \mathrm{i}\sqrt{1-\zeta^2} \quad \text{and} \quad r_2 = -\zeta - \mathrm{i}\sqrt{1-\zeta^2}. \qquad [3.22]$$

Since the nature of these solutions is very well understood, throughout the chapter we will compare the new results with them. Particularly, we will point out in what ways the normalized eigensolutions of the non-viscously damped dynamic system given by equations [3.18]–[3.20] differ from the normalized eigensolutions of the corresponding viscously damped dynamic system given by equation [3.22].

3.2. Conditions for oscillatory motion

We consider the following questions of fundamental interest:

– Under what conditions can a non-viscously damped dynamic system sustain oscillatory motions?

– Is there any "critical damping factor" for a non-viscously damped dynamic system so that beyond this value the system becomes overdamped?

For a viscously damped dynamic system, the answers to the above questions are well known. From equation [3.22], it is clear that if the viscous damping factor ζ is more than 1, then the dynamic system becomes overdamped and consequently it will not be able to sustain any oscillatory motions. This simple fact is no longer true for a non-viscously damped system. We will show that there is a precisely defined parameter space where the dynamic system becomes overdamped. We will also derive an expression of critical viscous damping factor ζ as a function of non-viscous damping factor β in section 3.3.

Roots r_1 and r_2 in equations [3.18] and [3.19] will be in a complex conjugate pair provided $S - T \neq 0$. The motion corresponding to the complex conjugate roots r_1 and r_2 is oscillatory (and decaying) in nature, while the motion corresponding to the real root r_3 is a pure non-oscillatory decay. Considering the expressions of S and T in equation [3.17], it is easy to observe that the system will oscillate provided $D > 0$. Therefore, the critical condition is given by

$$D = 0. \qquad [3.23]$$

From the expression of D in [3.16], this condition can be rewritten as

$$27\beta^4 D = 8\,\beta\zeta^3 + \left(12\,\beta^2 - 1\right)\zeta^2 + \left(6\,\beta^3 - 10\,\beta\right)\zeta + \left(1 + 2\,\beta^2 + \beta^4\right) = 0. \qquad [3.24]$$

In Figure 3.2, $D(\zeta, \beta) = 0$ is plotted for $0 \leq \zeta \leq 6$ and $0 \leq \beta \leq 0.5$. This plot shows the parameter domain where the system can have oscillatory motion. For a viscously damped dynamic system, $\beta = 0$, which is represented by the x-axis of Figure 3.2. Along the x-axis when $\zeta > 1$, the oscillatory motion is not possible, which is well known. But the scenario changes in an interesting way for non-zero β (i.e.

for a non-viscously damped system). For example, if $\beta \approx 0.1$, the system can have oscillatory motion even when $\zeta > 2$, which is more than twice the critical viscous damping factor. Conversely, there are also regions where the system may not have oscillatory motion even when $\zeta < 1$. Perhaps the most interesting observation from Figure 3.2 is that if β is more than about 0.2, then the dynamic system will *always* have oscillatory motion no matter what the value of the viscous damping factor is. We explain these observations by solving a constrained optimization problem.

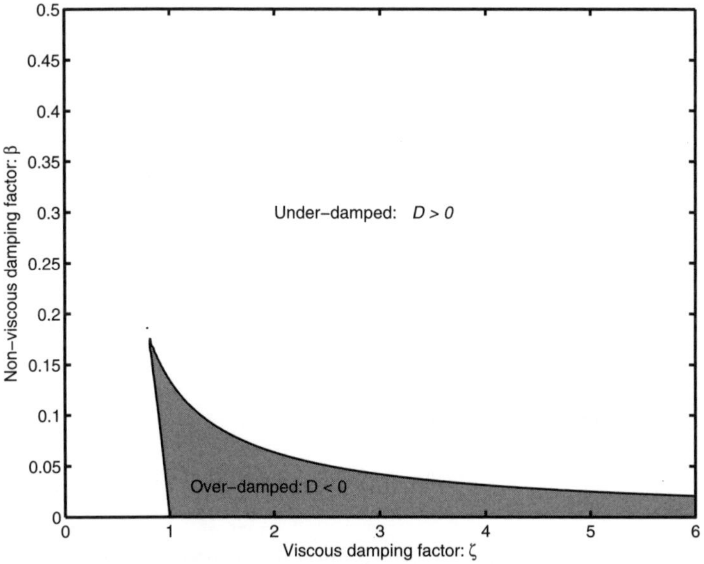

Figure 3.2. *The boundary between oscillatory and non-oscillatory motion*

In view of Figure 3.2, we want to find the maximum value of β for which $D(\zeta, \beta) = 0$. To solve this constrained optimization problem, construct the Lagrangian

$$\mathcal{L}(\zeta, \beta) = \beta + \gamma D(\zeta, \beta) \tag{3.25}$$

where γ is the Lagrange multiplier. Differentiating the Lagrangian with respect to ζ and β and setting them to zero, the optimality conditions can be expressed by

$$\frac{\partial \mathcal{L}(\zeta, \beta)}{\partial \zeta} = \gamma \frac{\partial D(\zeta, \beta)}{\partial \zeta} = 0 \tag{3.26}$$

$$\text{and} \quad \frac{\partial \mathcal{L}(\zeta, \beta)}{\partial \beta} = 1 + \gamma \frac{\partial D(\zeta, \beta)}{\partial \beta} = 0. \tag{3.27}$$

Since the Lagrange multiplier cannot be zero, from equation [3.26], we have $\frac{\partial D}{\partial \zeta} = 0$. Differentiating equation [3.24], this condition can be expressed as

$$12\,\beta\zeta^2 + \left(12\,\beta^2 - 1\right)\zeta + 3\,\beta^3 - 5\,\beta = 0. \tag{3.28}$$

Solving this quadratic equation, the positive value of ζ is given by

$$\zeta = \frac{\left(1 - 12\,\beta^2\right) + \sqrt{1 + 216\,\beta^2}}{24\beta}. \tag{3.29}$$

We need to satisfy the constrain $D(\zeta, \beta) = 0$. Substituting ζ from [3.29] in the expression of D in equation [3.24] and simplifying, we obtain

$$5832\,\beta^4 + 540\,\beta^2 - 1 - \left(1 + 216\,\beta^2\right)^{3/2} = 0. \tag{3.30}$$

Letting

$$z = 27\beta^2 \tag{3.31}$$

equation [3.30] can be expressed as

$$8z^2 + 20z - 1 - (1 + 8z)^{3/2} = 0 \quad \text{or} \quad 8z^2 + 20z - 1 = (1 + 8z)^{3/2}. \tag{3.32}$$

This equation has only one positive solution that is precisely at $z = 1$. Therefore, using equation [3.31], the maximum value of β so that $D = 0$ is obtained as

$$\beta_c = \frac{1}{3\sqrt{3}} = 0.1925. \tag{3.33}$$

Substituting this value of β in the expression of ζ in equation [3.29], the critical value of ζ can be obtained as

$$\zeta_c = \frac{4}{3\sqrt{3}} = 0.7698. \tag{3.34}$$

In Figure 3.3, we have (again) plotted the surface $D = 0$ concentrating around the derived critical values of ζ and β. The shaded region corresponds to the parameter combinations for which oscillatory motion is not possible. Based on this analysis, it can be said that a non-viscously damped system will *always* have oscillatory motions if $\zeta < \zeta_c$ and/or $\beta > \beta_c$ (parameter regions C_1 and A in the figure). If $\beta < \beta_c$, then it is possible to have overdamped motion even if $\zeta < 1$ as in the parameter region B shown in Figure 3.3. When $\beta < \beta_c$, there are two distinct parameter regions (shown as C_1 and C_2 in the figure) in which oscillatory motion is possible. Therefore, we can think of two critical damping factors for a non-viscously damped dynamic system. We will analyze this issue in the next section. From this analysis, we have the following fundamental result:

THEOREM 3.1.– An exponential non-viscously damped system will have oscillatory motions if $\zeta < \frac{4}{3\sqrt{3}}$ or $\beta > \frac{1}{3\sqrt{3}}$.

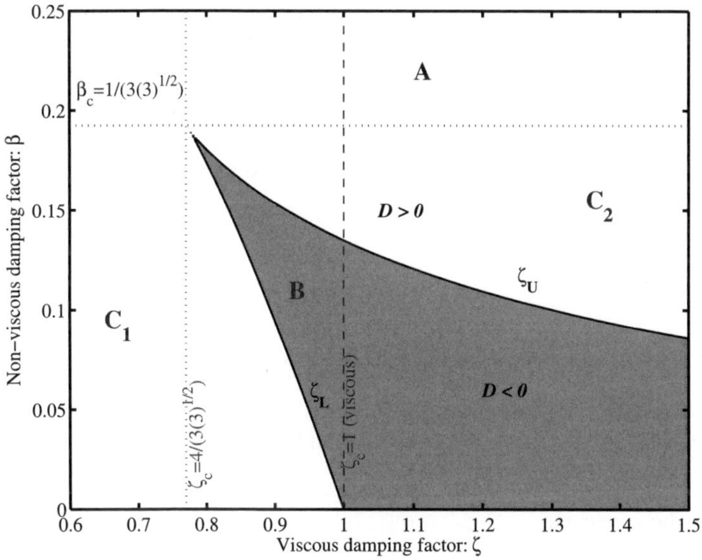

Figure 3.3. *Critical values of ζ and β for oscillatory motion*

It is often useful to express the critical value of β in equation [3.33] in terms of the relaxation time and the natural time period of the corresponding undamped dynamic system. Using the definition of β in [3.5], it follows that a non-viscously damped system will always have oscillatory motion if

$$\frac{2\pi/T_n}{\mu} > \frac{1}{3\sqrt{3}} \quad \text{or} \quad \frac{1}{\mu} > \frac{1}{3\sqrt{3}}\frac{T_n}{2\pi} \qquad [3.35]$$

that is if the relaxation time is more than $\frac{1}{3\sqrt{3}}$ times the natural time period of the corresponding undamped dynamic system. An equivalent result for a viscoelastic system has been obtained by Muravyov and Hutton [MUR 98b]. In their study, the hereditary term was associated with the stiffness coefficient as opposed to the damping coefficient. Using the properties of the roots of the characteristic equation, they have shown that oscillatory motion is always possible when the relaxation time $1/\alpha$ is greater than $\frac{1}{\sqrt{3}}\frac{T_n}{2\pi}$. This implies that Muravyov and Hutton's limiting value of the relaxation time is three times more than what has been derived in this chapter.

3.3. Critical damping factors

From Figure 3.3, it can be noted that when $\beta < \beta_c$, there are two distinct parameter regions for which oscillatory motion is possible (regions C_1 and C_2 in the figure). Therefore, for each value of β, two critical damping factors can be defined. Using the notations ζ_L and ζ_U, the dynamic system will have overdamped motion when $\zeta_L < \zeta < \zeta_U$. We call ζ_L the lower critical damping factor and ζ_U the upper critical damping factor. To obtain the critical damping factors, it is required to solve $D = 0$ for ζ. From equation [3.24], note that $D = 0$ is a cubic equation and we will again use Cardano's formula to obtain the explicit solutions. Rewrite equation [3.24] as

$$\zeta^3 + \sum_{j=0}^{2} b_j \zeta^j = 0 \qquad [3.36]$$

where the coefficients associated with the powers of ζ are given by

$$b_0 = \frac{1 + 2\beta^2 + \beta^4}{8\beta}, \; b_1 = \frac{3\beta^2 - 5}{4} \quad \text{and} \quad b_2 = \frac{12\beta^2 - 1}{8\beta}. \qquad [3.37]$$

Obtain the constants

$$Q_\zeta = \frac{3b_1 - b_2^2}{9} = -\frac{216\,\beta^2 + 1}{576\beta^2} \qquad [3.38]$$

$$\text{and} \quad R_\zeta = \frac{9b_2 b_1 - 27b_0 - 2b_2^3}{54} = -\frac{5832\,\beta^4 + 540\,\beta^2 - 1}{13824\beta^3} \qquad [3.39]$$

and calculate

$$D_\zeta = Q_\zeta^3 + R_\zeta^2 = \frac{-2187\,\beta^4 + 81\,\beta^2 - 1 + 19683\,\beta^6}{110592\beta^4}. \qquad [3.40]$$

It can be easily verified that when $\beta = \beta_c$, $D_\zeta = 0$ and when $\beta < \beta_c$, $D_\zeta < 0$. This implies that all the three solutions of equation [3.24] will be real when $\beta < \beta_c$. Defining

$$\theta = \arccos\left(R_\zeta/\sqrt{-Q_\zeta^3}\right) = \arccos\left(-\frac{5832\,\beta^4 + 540\,\beta^2 - 1}{(216\,\beta^2 + 1)^{3/2}}\right) \qquad [3.41]$$

the three solutions of equation [3.36] can be given by

$$\zeta_1 = 2\sqrt{-Q_\zeta}\cos\left(\frac{\theta}{3}\right) - b_2/3, \qquad [3.42]$$

$$\zeta_2 = 2\sqrt{-Q_\zeta}\cos\left(\frac{2\pi+\theta}{3}\right) - b_2/3 \qquad [3.43]$$

and $$\zeta_3 = 2\sqrt{-Q_\zeta}\cos\left(\frac{4\pi+\theta}{3}\right) - b_2/3. \qquad [3.44]$$

It turns out that $\zeta_1 > \zeta_3$ and ζ_2 is always negative when $\beta < \beta_c$. Therefore, we discard ζ_2 and assign $\zeta_L = \zeta_3$ and $\zeta_U = \zeta_1$. After some simplifications, the lower and the upper critical damping factors for a non-viscously damped system can be expressed as

$$\zeta_L = \frac{1}{24\beta}\left(1 - 12\beta^2 + 2\sqrt{1+216\beta^2} + \cos((4\pi+\theta)/3)\right) \qquad [3.45]$$

and $$\zeta_U = \frac{1}{24\beta}\left(1 - 12\beta^2 + 2\sqrt{1+216\beta^2} + \cos(\theta/3)\right). \qquad [3.46]$$

The above two equations are plotted in Figure 3.3. When $\beta \rightarrow \beta_c$, the critical damping factors approach each other and eventually when $\beta = \beta_c$, both critical damping factors become the same and equal to ζ_c. When $\beta < \beta_c$, the absolute value of the argument of arccos in equation [3.41] will always be less than 1 so that the computation of θ is always possible. The existence of two critical damping factors is a significantly different property from a viscously damped system. In the limiting case when $\beta \rightarrow 0$, it can be verified that $\zeta_L \rightarrow 1$ and $\zeta_U \rightarrow \infty$. This indeed implies that a viscously damped system has only one critical damping factor and that is $\zeta = 1$. The results obtained here can be summarized in the following theorem:

THEOREM 3.2.– When $\beta < \frac{1}{3\sqrt{3}}$, an exponential non-viscously damped system will have oscillatory motions if and only if $\zeta \notin [\zeta_L, \zeta_U]$.

3.4. Characteristics of the eigenvalues

Dynamic properties of linear systems are characterized by their eigensolutions. The normalized eigenvalues of the non-viscously damped system are expressed in terms of the non-dimensional parameters ζ and β in equations [3.18]–[3.20]. We are interested in the natural frequency and the decay rate corresponding to the oscillating eigenvalues r_1 and r_2 given by equations [3.18] and [3.19]. The decay rate of the non-oscillating eigenvalue r_3 given by equation [3.20] is also a quantity of interest in this study.

3.4.1. *Characteristics of the natural frequency*

The normalized natural frequency is given by the imaginary part of r_1 (or r_2) in equation [3.18]. The contours of $\Im(r_1(\zeta, \beta))$ when $0 \leq \zeta \leq 3$ and $0 \leq \beta \leq 0.6$ are shown in Figure 3.4. The range of these parameters is selected on the basis of numerical studies. Majority of the interesting features exist within these parameter ranges. There are three distinct regions (marked by the letters A, B and C) in Figure 3.4, which are consistent with our findings in the previous sections. In region A, where $\zeta < \zeta_c$, the system always has non-zero natural frequency. For increasing ζ, while $\beta < \beta_c$, the natural frequency decreases and eventually in region B the natural frequency becomes zero (i.e. the system cannot oscillate). In region C, where $\beta > \beta_c$, the normalized natural frequency is always greater than 1. For a fixed value of ζ, the natural frequency decreases for increasing β. An interesting fact emerges from Figure 3.4 that for a fixed value of β in region C, the natural frequency increases for increasing ζ. This is exactly the opposite behavior compared to a viscously damped dynamic system, where the damped natural frequency decreases for increasing ζ. A physical explanation of these facts can be given.

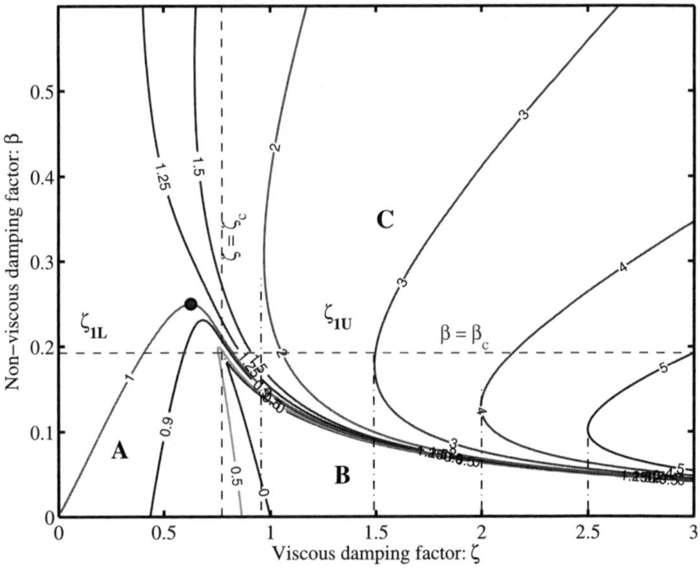

Figure 3.4. *Contours of the normalized natural frequency* ($\Im(r_1)$)

For convenience, express the damping force in equation [3.1] as

$$f_d(t) = \int_0^t g(t-\tau)\dot{q}(\tau)\,\mathrm{d}\tau, \quad \text{where} \quad g(t) = c\,\mu e^{-\mu\,t}. \qquad [3.47]$$

Integrating by parts, we have

$$\begin{aligned} f_d(t) &= |q(\tau)g(t-\tau)|_0^t + \int_0^t \dot{g}(t-\tau)q(\tau)\mathrm{d}\tau = q(t)g(0) - q(0)g(t) \\ &\quad + \int_0^t \dot{g}(t-\tau)q(\tau)\mathrm{d}\tau. \end{aligned}$$

Assuming zero initial conditions and utilizing the expression of $g(t)$, this equation can be expressed as

$$\begin{aligned} f_d(t) &= c\mu\,q(t) - \mu\int_0^t c\,\mu e^{-\mu(t-\tau)}q(\tau)\,\mathrm{d}\tau = \frac{1}{m}\left(\frac{\zeta}{\beta}\right)q(t) \\ &\quad - \mu\int_0^t c\,\mu e^{-\mu(t-\tau)}q(\tau)\,\mathrm{d}\tau. \end{aligned}$$

It is evident that the term $\frac{1}{m}\left(\frac{\zeta}{\beta}\right)$ would associate with the stiffness term in the equation of motion [3.2]. This implies that for higher values of ζ, the effective stiffness (hence the natural frequency) of the dynamic system would increase and it would decrease for higher values of β. This analysis explains the observations in Figure 3.4.

The contour line 1 in Figure 3.4 is particularly interesting. For these parameter combinations, the damped natural frequency will be the same as the natural frequency of the equivalent undamped system. Viscously damped systems do not show this behavior and it is unique to non-viscously damped systems. In order to obtain these parameter combinations analytically, we substitute

$$r = -\sigma + \mathrm{i} \qquad [3.48]$$

in the characteristic equation [3.12]. Here, σ is an arbitrary positive real number. This substitution ensures that the normalized natural frequency is equal to 1. Carrying out the substitution [3.48] and separating the real and imaginary parts of the resulting equation, we obtain

$$\beta\sigma^2 - 2\beta - \sigma + 2\zeta = 0 \qquad [3.49]$$

$$\text{and} \quad 3\beta\sigma^2 - 2\sigma + 2\zeta = 0 \qquad [3.50]$$

when $\zeta \neq 0$ and $\beta \neq 0$. Eliminating σ from these equations, we have

$$8\beta\zeta^2 - \left(1 + 24\beta^2\right)\zeta + 2\beta\left(1 + 9\beta^2\right) = 0. \qquad [3.51]$$

The system parameters must satisfy this equation for the damped natural frequency to be the same as the undamped natural frequency. The two solutions for ζ can be obtained from equation [3.51] as

$$\zeta_{1L} = \left(1 + 24\beta^2 - \sqrt{1 - 16\beta^2}\right)/16\beta \qquad [3.52]$$

$$\text{and} \quad \zeta_{1U} = \left(1 + 24\beta^2 + \sqrt{1 - 16\beta^2}\right)/16\beta. \qquad [3.53]$$

These equations are shown in Figure 3.4. Note that in general $\zeta_{1L} < \zeta_{1U}$. A real value of ζ can only be obtained from equations [3.52] and [3.53] if

$$1 - 16\beta^2 \geq 0 \quad \text{or} \quad \beta \leq \frac{1}{4}. \qquad [3.54]$$

This implies that for a non-viscously damped system, the damped natural frequency cannot be the same as the undamped natural frequency if $\beta > \frac{1}{4}$. When $\beta < \frac{1}{4}$, such situation may arise for two values of ζ given by equations [3.52] and [3.53]. When $\beta = \frac{1}{4}$, from equations [3.52] and [3.53], we have $\zeta_{1L} = \zeta_{1U} = \frac{5}{8}$. This point is shown by a dot in Figure 3.4. For a given value of β, if $\zeta < \zeta_{1L}$ or $\zeta > \zeta_{1U}$, then the damped natural frequency will be more than the undamped natural frequency and vice versa. Based on these discussions, we have the following results.

THEOREM 3.3.– If $\beta > 1/4$, the natural frequency of an exponential non-viscously damped system will be more than that of an equivalent undamped system.

THEOREM 3.4.– When $\beta < 1/4$, the natural frequency of an exponential non-viscously damped system will be more than that of an equivalent undamped system if and only if $\zeta \notin [\zeta_{1L}, \zeta_{1U}]$.

In order to develop further understandings of the normalized natural frequency, we substitute

$$r = -\sigma + \mathrm{i}\lambda \qquad [3.55]$$

in the characteristic equation [3.12]. Here, λ is an arbitrary positive real number. Carrying out the substitution [3.55] and separating the real and imaginary parts of the resulting equation, we obtain

$$(3\beta\sigma - 1)\,\lambda^2 - \beta\sigma^3 - 2\zeta\sigma + \sigma^2 - \beta\sigma + 1 = 0 \qquad [3.56]$$

$$\text{and} \quad 3\beta\sigma^2 - \beta\lambda^2 - 2\sigma + \beta + 2\zeta = 0. \qquad [3.57]$$

Eliminating σ from these equations, we have

$$h_\lambda(\zeta, \beta) = 0 \qquad [3.58]$$

where

$$\begin{aligned} h_\lambda \; = \; & 8\beta\zeta^3 + \left(12\beta^2 - 36\lambda^2\beta^2 - 1\right)\zeta^2 + \left(12\beta\lambda^2 + 6\beta^3 - 36\beta^3\lambda^2 - 10\beta\right. \\ & \left. +48\beta^3\lambda^4\right) + (1 - \lambda^2 - 8\beta^2\lambda^4 + 2\beta^2 - 9\beta^4\lambda^2 + \beta^4 - 16\beta^4\lambda^6\zeta \\ & +6\beta^2\lambda^2 + 24\beta^4\lambda^4)\,. \end{aligned} \qquad [3.59]$$

For the normalized natural frequency to be λ, the constants ζ and β must satisfy equation [3.58]. Every contour line in Figure 3.4 is a plot of equation [3.58] for different values of λ. When $\lambda > 2$, note that there is a minimum value of ζ for the natural frequency to be λ. To obtain the minimum values of ζ for every λ, we need to solve the following constrained optimization problem

$$\min \quad \zeta \quad \text{subject to} \quad h_\lambda(\zeta, \beta) = 0. \qquad [3.60]$$

Constructing the Lagrangian

$$\mathcal{L}_\lambda(\zeta, \beta) = \zeta + \gamma_\lambda h_\lambda(\zeta, \beta) \qquad [3.61]$$

and setting $\frac{\partial \mathcal{L}_\lambda(\zeta,\beta)}{\partial \zeta} = \frac{\partial \mathcal{L}_\lambda(\zeta,\beta)}{\partial \beta} = 0$, it is possible to solve this problem in closed form. Due to the complexity of $h_\lambda(\zeta, \beta)$ in equation [3.59], the resulting expressions are complicated. It can shown that the optimization problem [3.60] has real solutions only if $\lambda > 1$. This can also be observed from Figure 3.4. We will pursue a simplified approach based on the second derivative of the Lagrangian. The optimal conditions will be satisfied if $\frac{\partial^2 \mathcal{L}_\lambda(\zeta,\beta)}{\partial \zeta^2} > 0$ and $\frac{\partial^2 \mathcal{L}_\lambda(\zeta,\beta)}{\partial \beta^2} > 0$. Since the Lagrange multiplier $\gamma_\lambda \neq 0$, the critical condition can be expressed by $\frac{\partial^2 h_\lambda(\zeta,\beta)}{\partial \zeta^2} = 0$, that is

$$12(3\lambda^2 - 1)\beta^2 + 24\zeta\beta - 1 = 0. \qquad [3.62]$$

Solving for β, we have

$$\beta = \frac{\zeta \pm \sqrt{36\zeta^2 - 9\lambda^2 + 3}}{3\lambda^2 - 1}. \qquad [3.63]$$

To obtain a real solution from this equation, we need $36\zeta^2 - 9\lambda^2 + 3 > 0$, that is

$$\zeta > \sqrt{(\lambda/2)^2 - 1/12}. \qquad [3.64]$$

In the above formulation, we have not solved the true optimization problem. Due to this, equation [3.64] does not produce accurate results for all $\lambda > 1$. Numerical calculations show that equation [3.64] works well when $\lambda \geq 2$. For large λ, the term $\sqrt{(\lambda/2)^2 - 1/12}$ can be approximated as $(\lambda/2)$ without significant loss of accuracy. From these discussions, for all practical purposes, we have the following general result:

THEOREM 3.5.– The normalized natural frequency of an exponential non-viscously damped system will be more than λ if and only if $\zeta > \sqrt{(\lambda/2)^2 - 1/12} \approx \lambda/2$ for all $\lambda \geq 2$.

The minimum values of ζ required for different values of λ using this theorem are shown by a dashed-dotted line in Figure 3.4. From the numerical results, the applicability of theorem 3.5 can be confirmed.

3.4.2. *Characteristics of the decay rate corresponding to the oscillating mode*

The normalized decay rate corresponding to the oscillating mode is given by the real part of r_1 or r_2 in equations [3.18] and [3.19]. Figure 3.5 shows the contours of $-\Re(r_1(\zeta,\beta))$ when $0 \leq \zeta \leq 3$ and $0 \leq \beta \leq 0.6$. To understand the characteristics of the decay rate, it is helpful to divide Figure 3.5 into three regions as done earlier. In region A, where $\zeta < \zeta_c$, the decay rate is less than 1 for the values of β considered. In region B, where $\zeta > \zeta_c$ and $\beta < \beta_c$, the system does not oscillate. In region C, where $\beta > \beta_c$ and $\zeta > 1$, the decay rate decreases for increasing β. Interestingly, if β is close to 0.5 or more, the decay rate can be less than 1 even when $\zeta \geq 3$. This is in sharp contrast with a viscously damped system where the normalized decay rate is always ζ. For a better understanding of these observations, we again substitute $r = -\sigma + \mathrm{i}\lambda$ in the characteristic equation [3.12]. Eliminating λ from equations [3.56] and [3.57], we have

$$\left(4\sigma^3 + \sigma\right)\beta^2 + \left(2\zeta\sigma - 4\sigma^2\right)\beta + \sigma - \zeta = 0. \qquad [3.65]$$

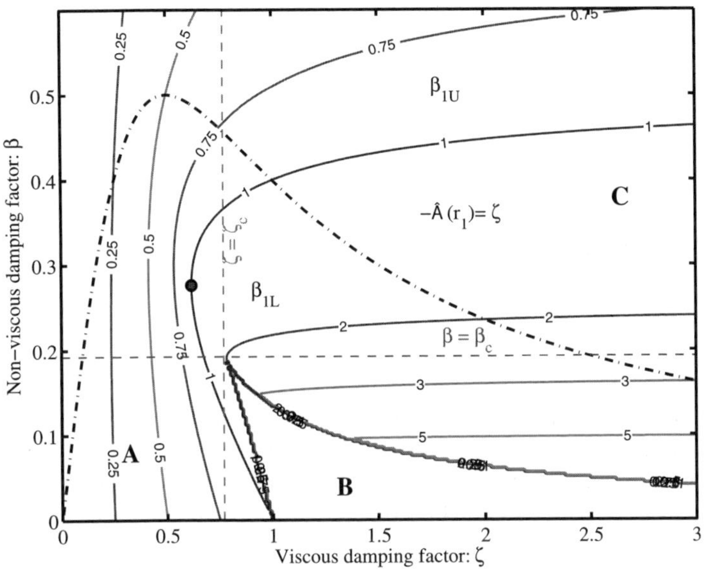

Figure 3.5. *Contours of the normalized decay rate (-$\Re(r_1)$) corresponding to the oscillating mode*

For the normalized decay rate to be σ, the constants ζ and β must satisfy equation [3.65]. For a given value of ζ, equation [3.65] can be solved for β as

$$\beta_{\sigma L} = \frac{2\sigma - \zeta - \sqrt{\zeta^2 - 1 + \zeta/\sigma}}{1 + 4\sigma^2} \qquad [3.66]$$

and $$\beta_{\sigma U} = \frac{2\sigma - \zeta + \sqrt{\zeta^2 - 1 + \zeta/\sigma}}{1 + 4\sigma^2}. \qquad [3.67]$$

It follows that for a known value of ζ, a particular decay rate corresponding to the oscillating mode can occur for two values of β given by equations [3.66] and [3.67]. This fact can be verified from Figure 3.5 by considering any contour level. Considering $\sigma = 1$, the plots of β_{1L} and β_{1U} are shown in Figure 3.5 for illustration.

When β is known, by solving equation [3.65] for ζ, we have

$$\zeta_\sigma = \frac{\sigma\left(4\beta^2\sigma^2 + \beta^2 - 4\sigma\beta + 1\right)}{1 - 2\sigma\beta}, \quad \beta \neq 1/2\sigma. \qquad [3.68]$$

This implies that for a known value of β, a particular decay rate corresponding to the oscillating mode can occur only for a unique value of ζ. This can also be verified from figure 3.5. From the contour levels in Figure 3.5, it can be observed that there is a minimum value of ζ for a particular decay rate. To obtain the minimum value of ζ, we consider the expressions of $\beta_{\sigma L}$ and $\beta_{\sigma U}$. Real values of $\beta_{\sigma L}$ and $\beta_{\sigma U}$ can be obtained from equations [3.66] and [3.67] if and only if

$$\zeta^2 - 1 + \zeta/\sigma \geq 0. \qquad [3.69]$$

Considering the positive solution, the condition in equation [3.69] can be expressed as

$$\zeta \geq \left(\sqrt{1+4\sigma^2} - 1\right)/2\sigma. \qquad [3.70]$$

It follows that no matter what the value of β is, the decay rate will always be less than σ if ζ is less than $\left(\sqrt{1+4\sigma^2} - 1\right)/2\sigma$. To illustrate this point, consider the $\sigma = 1$ curve in Figure 3.5. From [3.70], the minimum value of ζ for which σ can be 1 is obtained as $\zeta = (\sqrt{5} - 1)/2$. This point is shown by a dot in Figure 3.5.

Another interesting observation from Figure 3.5 is that when $\zeta > 1$, the decay rate, in general, decreases for increasing values of β and there is a critical value of β beyond which a particular decay rate cannot occur. Considering asymptotic expansion of the upper value $\beta_{\sigma U}$ given in equation [3.67], we have

$$\beta_{\sigma U} = \frac{2\sigma - \zeta + \sqrt{\zeta^2 - 1 + \zeta/\sigma}}{1+4\sigma^2} \approx \frac{1}{2\sigma} - \frac{1}{8\sigma^2}\frac{1}{\zeta} + \frac{1}{16\sigma^3}\frac{1}{\zeta^2} - \frac{5+4\sigma^2}{128\sigma^4}\frac{1}{\zeta^3} + \cdots. \qquad [3.71]$$

Therefore, when $\zeta \to \infty$, we have

$$(\beta_\sigma)_{\max} = \lim_{\zeta\to\infty} \frac{2\sigma - \zeta + \sqrt{\zeta^2 - 1 + \zeta/\sigma}}{1+4\sigma^2} = \frac{1}{2\sigma}. \qquad [3.72]$$

This analysis indicates that no matter what the value of ζ is, the decay rate will always be less than σ if β is more than $1/2\sigma$. To illustrate this, again consider the $\sigma = 1$ curve in Figure 3.5 and observe that the decay rate is always less than 1 if $\beta > 1/2$. From these discussions, we have the following universal results:

THEOREM 3.6.– The normalized decay rate corresponding to the oscillating mode of an exponential non-viscously damped system will be less than σ if $\beta > 1/2\sigma$ or $\zeta < \left(\sqrt{1+4\sigma^2} - 1\right)/2\sigma$ for any $\sigma \in \mathbb{R}^+$.

THEOREM 3.7.– When $\zeta > \left(\sqrt{1+4\sigma^2}-1\right)/2\sigma$, the normalized decay rate corresponding to the oscillating mode of an exponential non-viscously damped system will be less than σ if and only if $\beta \notin [\beta_{\sigma L}, \beta_{\sigma U}]$ for any $\sigma \in \mathbb{R}^+$.

It is useful to compare the decay rate of a non-viscously damped dynamic system to that of an equivalent viscously damped system. From equation [3.22], recall that the decay rate of a viscously damped system is ζ. Therefore, substituting $\sigma = \zeta$ in equation [3.65] and considering $\zeta \neq 0$ and $\beta \neq 0$, we have

$$4\zeta^2\beta + \beta - 2\zeta = 0. \qquad [3.73]$$

Solving this equation for β, we have

$$\beta_\zeta = \frac{2\zeta}{1+4\zeta^2}. \qquad [3.74]$$

If this condition is satisfied, then the decay rate of a non-viscously damped system will be the same as that of an equivalent viscously damped system. The parameter combinations that satisfy equation [3.74] are plotted by a dashed-dotted line in Figure 3.5. If $\beta > \frac{2\zeta}{1+4\zeta^2}$, then the decay rate will be less than ζ and vice versa. Observe that there is a maximum value of β beyond which the decay rate will always be less than ζ. To obtain this critical value of β, we solve equation [3.73] for ζ and obtain

$$\zeta_{\zeta L} = \left(1 - \sqrt{1-4\beta^2}\right)/4\beta \qquad [3.75]$$

$$\text{and} \quad \zeta_{\zeta U} = \left(1 + \sqrt{1-4\beta^2}\right)/4\beta. \qquad [3.76]$$

A real solution can be obtained only if $\left(1-4\beta^2\right) \geq 0$, that is when

$$\beta \leq \frac{1}{2}. \qquad [3.77]$$

This result implies that if $\beta > 1/2$, the decay rate will always be less than ζ. From equations [3.75] and [3.76], it is also clear that for a given value of $\beta < 1/2$, the normalized decay rate will be more than ζ if $\zeta_{\zeta L} < \zeta < \zeta_{\zeta U}$ and vice versa. Based on these discussions, we have the following theorems:

THEOREM 3.8.– If $\beta > 1/2$, the decay rate corresponding to the oscillating mode of an exponential non-viscously damped system will be less than that of an equivalent viscously damped system.

THEOREM 3.9.– When $\beta < 1/2$, the decay rate corresponding to the oscillating mode of an exponential non-viscously damped system will be less than that of an equivalent viscously damped system if and only if $\zeta \notin [\zeta_{\zeta L}, \zeta_{\zeta U}]$ or $\beta > 2\zeta/(1+4\zeta^2)$.

3.4.3. *Characteristics of the decay rate corresponding to the non-oscillating mode*

The overdamped or non-oscillating eigenvalue r_3 is given by equation [3.20]. The contours of $-r_3(\zeta, \beta)$ when $0 \leq \zeta \leq 3$ and $0 \leq \beta \leq 0.6$ are shown in Figure 3.6. There are two distinct regions denoted by A and B in the figure. In region A, where $\zeta < 1$ and β is small, the decay rate is very high. This is expected because this eigenvalue would not have existed for a viscously damped system, that is when $\beta = 0$. In region B, where $\zeta > 1$, interestingly the decay rate is not high, even for small values of β. Moreover, for increasing values of ζ, the decay rate decreases, which appears to be counter-intuitive. Also note the very sharp change in the values of r_3 around $\zeta = 1$ for small values of β ($\beta < 0.05$). The reasons behind these behaviors are not entirely clear to us and further research is required for proper explanations.

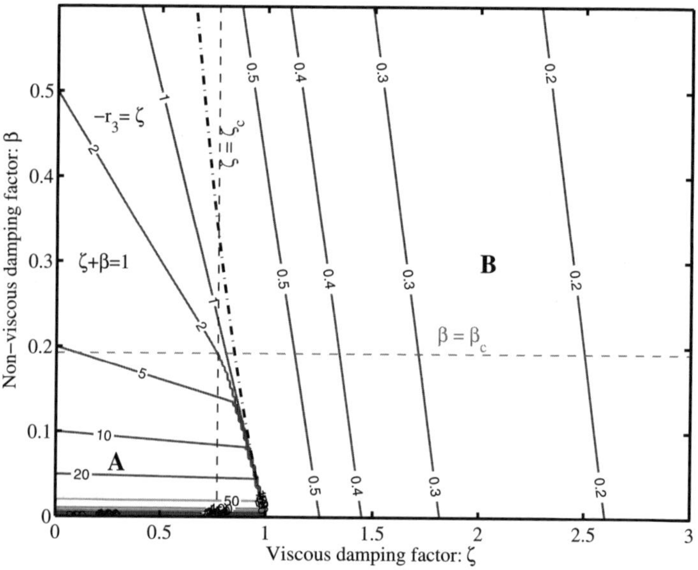

Figure 3.6. *Contours of the normalized decay rate corresponding to the overdamped eigenvalue ($-r_3$) as a function of ζ and β*

To compare the decay rate of the overdamped mode with the decay rate of a viscously damped system, it is required to know under what condition the decay rate will be ζ. Substituting $r = -\zeta$ in the characteristic equation [3.12] and solving for β, we have

$$\beta = \frac{1 - \zeta^2}{\zeta(1 + \zeta^2)}. \qquad [3.78]$$

This equation is plotted in Figure 3.6 by a dashed-dotted line. Clearly, to obtain a positive value of β from equation [3.78], we need $\zeta < 1$. From this analysis, we have the following result:

THEOREM 3.10.– The decay rate corresponding to the non-oscillating mode of an exponential non-viscously damped system will be less than that of an equivalent viscously damped system if $\beta > \frac{1-\zeta^2}{\zeta(1+\zeta^2)}$.

For a general value of the decay rate, we substitute $r = -\sigma$ in the characteristic equation [3.12]. After some simplifications, the required condition can be expressed as

$$\frac{\beta}{1/\sigma^2} + \frac{\zeta}{1/2(\sigma + 1/\sigma)} = 1. \qquad [3.79]$$

This is equation of a straight line. To illustrate this result, consider $\sigma = 1$. According to equation [3.79], this can happen if $\zeta + \beta = 1$. This line is shown in Figure 3.6. From the figure it is evident that if $\zeta + \beta > 1$, the decay rate will be less than 1. From this discussion, we have the following generalization:

THEOREM 3.11.– The normalized decay rate corresponding to the non-oscillating mode of an exponential non-viscously damped system will be less than σ if $\frac{\beta}{1/\sigma^2} + \frac{\zeta}{1/2(\sigma+1/\sigma)} > 1$ for any $\sigma \in \mathbb{R}^+$.

Another useful result, which follows from this theorem, is:

COROLLARY 3.1.– If $\zeta + \beta > 1$, the normalized decay rate corresponding to the non-oscillating mode of an exponential non-viscously damped system will be less than 1.

3.5. The frequency response function

The results given in the previous section define the conditions under which a non-viscously damped system can sustain oscillatory motions. The remaining chapter is aimed at gaining insights into the nature of the dynamic response. The frequency response function of linear systems contains complete information regarding the dynamic response. The direct computation of the frequency response function of an SDOF system is a trivial task. To gain further insight into the dynamic response characteristics, it is often useful to express the frequency response function in terms of the eigenvalues of the system. The aim of this section is to establish a connection to the results given in the previous section, which gives the expression of the eigenvalues as a function of ζ and β.

We begin with the normalized frequency response function $\overline{h}(s)$, which is defined as the solution of equation [3.6] with the forcing function $\overline{p}(s) = 1$. Therefore, from equation [3.6], we obtain

$$\overline{h}(s) = \frac{1}{\overline{d}(s)}, \quad \text{where} \quad \overline{d}(s) = s^2 + s\,2\zeta\omega_n\left(\frac{\omega_n}{s\beta + \omega_n}\right) + \omega_n^2. \qquad [3.80]$$

Noting that $\overline{d}(s)$ has zeros at $s = \lambda_j, j = 1, 2, 3$, where the eigenvalues $\lambda_j = \omega_n r_j$, the frequency response function can be conveniently expressed by the pole-residue form as

$$\overline{h}(s) = \sum_{j=1}^{3} \frac{R_j}{s - \lambda_j}. \qquad [3.81]$$

Here, the residues are

$$\begin{aligned} R_j &= \lim_{s\to\lambda_j} \frac{s-\lambda_j}{\overline{d}(s)} = \frac{1}{\frac{\partial \overline{d}(s)}{\partial s}\big|_{s=\lambda_j}} \\ &= \frac{1}{2}\frac{1}{\lambda_j + \zeta\omega_n(\omega_n/(\beta\lambda_j+\omega_n))^2}. \end{aligned} \qquad [3.82]$$

Because λ_1 and λ_2 appear in a complex conjugate pair, it is convenient to write $\lambda_1 = \lambda$ and $\lambda_2 = \lambda^*$, where $(\bullet)^*$ denotes the complex conjugation. We denote the real eigenvalue by $\lambda_3 = \nu$. Using these notations and substituting $s = \mathrm{i}\omega$, the frequency response function in equation [3.81] can be expressed as

$$\overline{h}(\mathrm{i}\omega) = \frac{R_\lambda}{\mathrm{i}\omega - \lambda} + \frac{R_\lambda^*}{\mathrm{i}\omega - \lambda^*} + \frac{R_\nu}{\mathrm{i}\omega - \nu} \qquad [3.83]$$

where

$$R_\lambda = \frac{1}{2}\frac{1}{\lambda + \zeta\omega_n\left(1 + (\beta\lambda/\omega_n)\right)^{-2}} \quad \text{and} \quad R_\nu = \frac{1}{2}\frac{1}{\nu + \zeta\omega_n\left(1 + (\beta\nu/\omega_n)\right)^{-2}}. \qquad [3.84]$$

For the special cases, when the system is undamped ($\zeta = 0$) or viscously damped ($\beta = 0$), equation [3.83] reduces to its corresponding familiar forms as follows:

– For undamped systems, $\zeta = 0$ and ν does not exist. The eigenvalue λ is purely imaginary so that $\lambda = \mathrm{i}\omega_n$. From equation [3.84], we obtain $R_\lambda = 1/(2\mathrm{i}\omega_n)$. Substitution of these values in equation [3.83] results

$$\begin{aligned}\overline{h}(\mathrm{i}\omega) &= \frac{1}{2\mathrm{i}\omega_n}\frac{1}{\mathrm{i}\omega - \mathrm{i}\omega_n} - \frac{1}{2\mathrm{i}\omega_n}\frac{1}{\mathrm{i}\omega + \mathrm{i}\omega_n} = \frac{1}{2\mathrm{i}\omega_n}\left[\frac{1}{\mathrm{i}\omega - \mathrm{i}\omega_n} - \frac{1}{\mathrm{i}\omega + \mathrm{i}\omega_n}\right] \\ &= \frac{1}{\omega_n^2 - \omega^2}.\end{aligned} \quad [3.85]$$

– For viscously damped systems, $\beta = 0$ and ν does not exist. The eigenvalue λ can be expressed as

$$\lambda = -\zeta\omega_n + \mathrm{i}\omega_d \quad \text{where} \quad \omega_d = \omega_n\sqrt{1-\zeta^2}. \quad [3.86]$$

From equation [3.84], we obtain $R_\lambda = 1/(2(-\zeta\omega_n + \mathrm{i}\omega_d + \zeta\omega_n)) = 1/(2\mathrm{i}\omega_d)$. Substituting these in equation [3.83], we obtain

$$\begin{aligned}\overline{h}(\mathrm{i}\omega) &= \frac{1}{2\mathrm{i}\omega_d}\frac{1}{\mathrm{i}\omega - (-\zeta\omega_n + \mathrm{i}\omega_d)} - \frac{1}{2\mathrm{i}\omega_d}\frac{1}{\mathrm{i}\omega - (-\zeta\omega_n - \mathrm{i}\omega_d)} \\ &= \frac{1}{2\mathrm{i}\omega_d}\left[\frac{2\mathrm{i}\omega_d}{(\zeta\omega_n + \mathrm{i}\omega)^2 - (\mathrm{i}\omega_d)^2}\right] \\ &= \frac{1}{\omega_n^2 + 2\mathrm{i}\omega\zeta\omega_n - \omega^2}.\end{aligned} \quad [3.87]$$

In the time domain, the impulse response function can be obtained by taking the inverse Laplace transform of $\overline{h}(s)$ as

$$h(t) = \Re\left\{\frac{e^{\lambda t}}{\lambda + \zeta\omega_n\left(1 + (\beta\lambda/\omega_n)\right)^{-2}}\right\} + \frac{1}{2}\frac{e^{\nu t}}{\nu + \zeta\omega_n\left(1 + (\beta\nu/\omega_n)\right)^{-2}}. \quad [3.88]$$

The first term in equation [3.88] is oscillating in nature because λ is complex, while the second term is purely decaying in nature because ν is real and negative.

It is convenient to define a non-dimensional driving frequency parameter as

$$\tilde{\omega} = \frac{\omega}{\omega_n}. \quad [3.89]$$

Substituting $s = \mathrm{i}\omega = \mathrm{i}\tilde{\omega}\omega_n$ in equation [3.80], we have

$$\overline{h}(\mathrm{i}\omega) = \frac{1}{\omega_n^2}\left\{\frac{1}{-\tilde{\omega}^2 + 2\mathrm{i}\zeta\tilde{\omega}\left(\frac{1}{\mathrm{i}\beta\tilde{\omega}+1}\right) + 1}\right\}. \quad [3.90]$$

Separating the real and imaginary parts, the non-dimensional frequency response function can be expressed as

$$G(\mathrm{i}\omega) = \omega_n^2\, \overline{h}(\mathrm{i}\omega) = \frac{1 + \mathrm{i}\beta\tilde{\omega}}{(1 - \tilde{\omega}^2) + \mathrm{i}\tilde{\omega}\,(2\zeta + \beta - \beta\tilde{\omega}^2)}. \qquad [3.91]$$

From equation [3.91], the amplitude of vibration can be obtained as

$$|G(\mathrm{i}\omega)| = \sqrt{G(\mathrm{i}\omega)\, G^*(\mathrm{i}\omega)} = \sqrt{\frac{1 + \beta^2\tilde{\omega}^2}{(1 - \tilde{\omega}^2)^2 + \tilde{\omega}^2\,(2\zeta + \beta - \beta\tilde{\omega}^2)^2}}. \qquad [3.92]$$

Figure 3.7 shows the amplitude of the non-dimensional frequency response $|G(\mathrm{i}\omega)|$ as a function of the normalized frequency ω/ω_n. The numerical values of β and ζ are selected such that Figure 3.7 shows the general overall behavior. The static case, that is when $\omega/\omega_n = 0$, the amplitude of vibration is 1. Therefore, as the frequency changes, the values of $|G(\mathrm{i}\omega)|$ in equation [3.92] can be regarded as the amplification factors. When $\beta < \beta_c = 1/(3\sqrt{3})$, the frequency response function is similar to that of the viscously damped system. This is expected because the value of β is relatively small. The amplitude of the peak response of the non-viscously damped system is more than that of the viscously damped system. In general, the higher the values of β, the higher the values of the amplitudes of the peak response. Another interesting fact that can be seen from Figure 3.7 is that the dynamic response amplitude has a peak even when $\zeta > 1/\sqrt{2}$. For example, in Figure 3.7(d), the viscously damped system does not have any response peak as $\zeta = 1$ (critical viscous damping). However, for the non-viscously damped system, the response amplitude has a peak when $\beta = 1$ or $\beta = 0.75$, but not if $\beta < 0.5$. These interesting response behaviors are explored further in the next section.

3.6. Characteristics of the response amplitude

The maximum vibration amplitude of a linear system near the resonance is of fundamental engineering interest because it can lead to damage or even failure of a structure. For viscously damped system, it is known from section 2.2.2.1 that if $\zeta < 1/\sqrt{2}$, then the frequency response function has a peak when $\omega/\omega_n = \sqrt{1 - 2\zeta^2}$. At this frequency, the amplitude of the maximum dynamic response is given by

$$|G|_{\max} = \frac{1}{2\zeta\sqrt{1 - \zeta^2}}. \qquad [3.93]$$

Recently, Vinokur [VIN 03] derived a closed-form expression of the frequency point where the vibration amplitude of a hysteretically damped system reaches its maximum value. We are interested in the equivalent results for non-viscously damped systems. Specifically, we ask the following questions of fundamental engineering interest:

– Under what conditions can the amplitude of the frequency response function reach a maximum value?

– At what frequency will it occur?

– What will be the value of the maximum amplitude of the frequency response function?

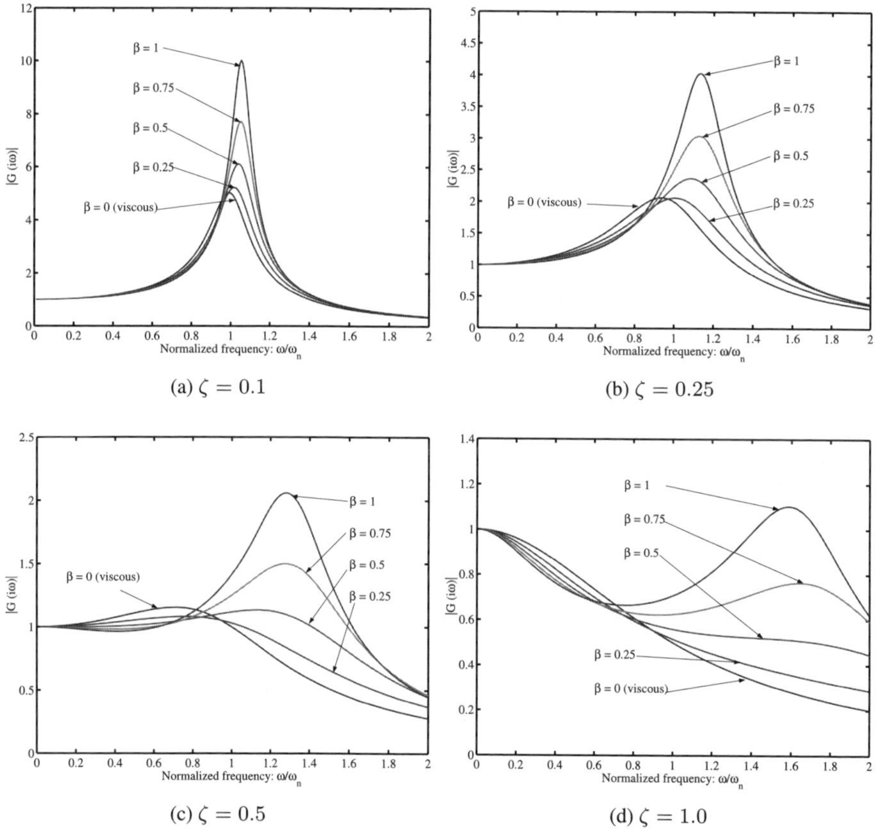

Figure 3.7. *Amplitude of the non-dimensional frequency response $|G(\mathrm{i}\omega)|$ as a function of the normalized frequency ω/ω_n for different values of ζ and β*

3.6.1. *The frequency for the maximum response amplitude*

For notational convenience, denoting

$$x = \tilde{\omega}^2 = \frac{\omega^2}{\omega_n^2} \qquad [3.94]$$

from equation [3.92], the amplitude of the dynamic response can be expressed as

$$|G|^2 = \frac{1 + \beta^2 x}{(1-x)^2 + x(2\zeta + \beta - \beta x)^2}. \qquad [3.95]$$

For the maximum value of $|G|$, we set

$$\frac{\partial |G|^2}{\partial x} = 0 \qquad [3.96]$$

or

$$\frac{2x^2\beta^2 - 2\beta^2 x - 2\beta^3\zeta x^2 - \beta^4 x^2 + \beta^4 x^3 - 1 + x + 2\zeta^2 + 2\zeta\beta - 4\zeta x\beta}{\left((1-x)^2 + x(2\zeta + \beta - \beta x)^2\right)^2} = 0. \qquad [3.97]$$

At the solution point, it is also required that

$$\frac{\partial^2 |G|^2}{\partial x^2} < 0 \qquad [3.98]$$

which in turn implies satisfying

$$\begin{aligned} &3\beta^6 x^5 + \left(9\beta^4 - 12\zeta\beta^5 - 6\beta^6\right) x^4 \\ &+ \left(9\beta^2 + 12\zeta^2\beta^4 - 18\beta^4 - 40\beta^3\zeta + 12\zeta\beta^5 + 3\beta^6\right) x^3 \\ &+ \left(60\beta^2\zeta^2 + 9\beta^4 + 3 + 60\beta^3\zeta - 24\zeta\beta - 18\beta^2\right) x^2 \\ &+ \left(9\beta^2 - 24\beta^3\zeta - 48\zeta^3\beta + 12\zeta^2 + 36\zeta\beta - 6 - 72\beta^2\zeta^2\right) x \\ &\qquad + 3 + 4\beta^3\zeta - 16\zeta^2 - 12\zeta\beta + 20\beta^2\zeta^2 + 32\zeta^3\beta + 16\zeta^4 < 0. \end{aligned} \qquad [3.99]$$

The numerator of equation [3.97] is a cubic equation in x and can be expressed as

$$x^3 + \sum_{j=0}^{2} c_j x^j = 0 \qquad [3.100]$$

where

$$c_0 = \frac{2\zeta\beta + 2\zeta^2 - 1}{\beta^4}, \quad c_1 = \frac{1 - 2\beta^2 - 4\zeta\beta}{\beta^4} \quad \text{and} \quad c_2 = \frac{2 - 2\zeta\beta - \beta^2}{\beta^2}. \qquad [3.101]$$

The three roots of equation [3.100] can either be all real or one real and one complex conjugate pair. The nature of the roots depends on the discriminant, which can be obtained from the constants

$$Q_x = \frac{3c_1 - c_2^2}{9} = -\frac{1}{9\beta^4}\left(1 + 2\zeta\beta + \beta^2\right)^2 \qquad [3.102]$$

and

$$R_x = \frac{9c_2c_1 - 27c_0 - 2c_2^3}{54} = \frac{1}{27\beta^6}\left(8\zeta^3\beta^3 + \left(12\beta^4 - 15\beta^2\right)\zeta^2 + \left(-15\beta^3 - 21\beta + 6\beta^5\right)\zeta + 3\beta^4 + 3\beta^2 + 1 + \beta^6\right) \qquad [3.103]$$

as

$$D_x = Q_x^3 + R_x^2 = -\frac{\zeta}{27\beta^{11}}\left(16\beta^4\zeta^4 + \left(13\beta^3 + 40\beta^5\right)\zeta^3 + \left(18\beta^4 + 36\beta^6 - 18\beta^2\right)\zeta^2 + \left(-13\beta - 12\beta^3 + 15\beta^5 + 14\beta^7\right)\zeta + 2 + 8\beta^2 + 12\beta^4 + 2\beta^8 + 8\beta^6\right). \qquad [3.104]$$

If $D_x > 0$, then equation [3.100] has one complex conjugate pair and only one real solution. It turns out that when $D_x > 0$, the real solution is always negative and therefore is not of interest in this study. However, when $D_x < 0$, all the roots of equation [3.100] become real. We define an angle θ as

$$\cos(\theta) = \left(R_x/\sqrt{-Q_x^3}\right) = \frac{8\zeta^3\beta^3 + \left(12\beta^4 - 15\beta^2\right)\zeta^2 + \left(6\beta^5 - 15\beta^3 - 21\beta\right)\zeta + 3\beta^4 + 3\beta^2 + 1 + \beta^6}{\left(1 + \beta^2 + 2\zeta\beta\right)^3}. \qquad [3.105]$$

Using θ, the three real solutions of equation [3.100] can be given using Dickson's formula [DIC 98] as

$$x_1 = 2\sqrt{-Q_x}\cos\left(\frac{\theta}{3}\right) - c_2/3, \qquad [3.106]$$

$$x_2 = 2\sqrt{-Q_x}\cos\left(\frac{2\pi + \theta}{3}\right) - c_2/3 \qquad [3.107]$$

$$\text{and} \quad x_3 = 2\sqrt{-Q_x}\cos\left(\frac{4\pi + \theta}{3}\right) - c_2/3. \qquad [3.108]$$

Among the above three solutions, we need to choose a positive solution that also satisfies equation [3.99]. From numerical calculations, it turns out that only x_1 in equation [3.106] satisfies the condition in equation [3.99]. Substituting Q_x from [3.102] and c_2 from [3.101] into equation [3.106], the normalized excitation frequency for which the amplitude of the frequency response function reaches its maximum value is given by

$$x_{\max} = \frac{1}{3\beta^2}\left(\left(1+2\zeta\beta+\beta^2\right)\left(2\cos\left(\theta/3\right)+1\right)-3\right). \qquad [3.109]$$

For convenience, defining the notation $\omega_{\max}$ as

$$x_{\max} = \frac{\omega_{\max}^2}{\omega_n^2} \qquad [3.110]$$

we have

$$\omega_{\max} = \frac{\omega_n}{\beta}\sqrt{\left(1+2\zeta\beta+\beta^2\right)\left(2\cos\left(\theta/3\right)+1\right)/3-1}. \qquad [3.111]$$

This is the extension of the well-known result for viscously damped systems for which $\omega_{\max} = \omega_n\sqrt{1-2\zeta^2}$.

Figure 3.8 shows the contours of $\omega_{\max}/\omega_n$ obtained from equation [3.111] as a function of ζ and β. The value of $\omega_{\max}$ is the frequency where the amplitude of the frequency response function reaches its maximum value.

For a better understanding, Figure 3.8 is divided into three regions. In region A, where $\zeta \lesssim 0.5$ and β is small, $\omega_{\max}/\omega_n < 1$. This implies that in this parameter region, the frequency at which the amplitude of the frequency response function reaches its maximum appears *below* the system natural frequency. Contour line 0 separates region B from A and C. In region B, $\beta \lesssim 0.5$ and $\zeta \gtrsim 1/\sqrt{2}$ and the amplitude of the frequency response function does not have any maximum value. This implies that within this parameter region, it is not possible to find a positive real solution of the cubic equation [3.100] and the system response decays gradually as in Figure 3.7(d) for $\beta = 0$ and $\beta = 0.25$. The shaded portion inside region B (shown in Figures 3.2 and 3.3) corresponds to the parameter region where the system cannot have any oscillatory motions. Clearly, within this overdamped region, it is not possible for the dynamic response amplitude to reach a maximum value. In region C, where $\zeta \gtrsim 1$ and $\beta \gtrsim 0.5$, observe that $\omega_{\max}/\omega_n > 1$. The contour plots in Figure 3.8 also show a general trend that $\omega_{\max}/\omega_n$ increases for increasing values of ζ and β.

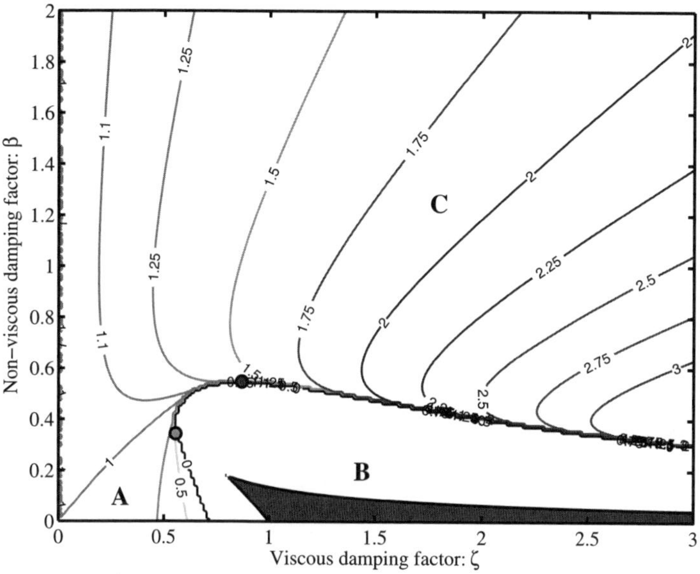

Figure 3.8. *Contours of the normalized excitation frequency corresponding to the maximum value of the amplitude of the frequency response function, $\omega_{\max}/\omega_n$, as a function of ζ and β*

An interesting contour line in Figure 3.8 is the line 1. For these parameter combinations of ζ and β, the frequency at which the amplitude of the frequency response function reaches the maximum value coincides exactly with the undamped natural frequency. This surprising observation implies that the system may be heavily damped ($\zeta > 0.5$) but still can have a peak at ω_n for some appropriate values of β. Another interesting fact observed from Figure 3.8 is that there exists a critical value of ζ, say ζ_{mL}, *below* which the amplitude of the frequency response will always have a maximum value for any values of β. Similarly, there is also a critical value of β, say β_{mU}, *above* which the amplitude of the frequency response will always have a maximum value for any values of ζ. The explanation of these observations including the derivation of the exact values of ζ_{mL} and β_{mU} is considered in the next sections.

3.6.1.1. *Critical parameter values for the maximum response amplitude*

Suppose a general complex solution of equation [3.100] is expressed as

$$x = \sigma + \mathrm{i}\psi \qquad [3.112]$$

for arbitrary $\sigma, \psi \in \mathbb{R}$. Substituting x from the above equation in [3.100] and separating the real and imaginary parts, we have

$$-2\beta^4\sigma^3 + \left(4\beta^3\zeta + 2\beta^4 - 4\beta^2\right)\sigma^2 + \left(6\beta^4\psi^2 - 2 + 4\beta^2 + 8\zeta\beta\right)\sigma \\ - 4\zeta\beta - \left(4\beta^3\zeta + 2\beta^4 - 4\beta^2\right)\psi^2 - 4\zeta^2 + 2 = 0 \quad [3.113]$$

and

$$-6\beta^4\sigma^2\psi + \left(-8\psi\beta^2 + 8\psi\beta^3\zeta + 4\psi\beta^4\right)\sigma + 2\beta^4\psi^3 + 4\psi\beta^2 + 8\psi\zeta\beta - 2\psi = 0. \quad [3.114]$$

Eliminating σ from equations [3.113] and [3.114] and substituting $\psi = 0$ (because we are interested only in the real solution) in the resulting equation, after some calculation, we have

$$\mathcal{M}(\zeta, \beta) = 0 \quad [3.115]$$

where

$$\mathcal{M}(\zeta, \beta) = 16\beta^3\zeta^3 + \left(24\beta^4 - 3\beta^2\right)\zeta^2 + \left(12\beta^5 - 15\beta - 3\beta^3\right)\zeta \\ + 2 + 6\beta^2 + 6\beta^4 + 2\beta^6. \quad [3.116]$$

The parameters ζ and β must satisfy equation [3.115] in order to have a real solution. Therefore, in view of Figure 3.8, the values of β_{mU} and ζ_{mL} can be obtained from the following optimization problems, respectively:

$$\beta_{mU}: \quad \max \beta \text{ subject to } \mathcal{M}(\zeta, \beta) = 0 \quad [3.117]$$

$$\text{and} \quad \zeta_{mL}: \quad \min \zeta \text{ subject to } \mathcal{M}(\zeta, \beta) = 0. \quad [3.118]$$

First, consider the constrained optimization problem in equation [3.117]. Using the Lagrange multiplier γ_1, we construct the Lagrangian

$$\mathcal{L}_1(\zeta, \beta) = \beta + \gamma_1 \mathcal{M}(\zeta, \beta). \quad [3.119]$$

The optimization problem shown in equation [3.117] can be solved by setting

$$\frac{\partial \mathcal{L}_1}{\partial \zeta} = 0 \quad [3.120a]$$

$$\text{and} \quad \frac{\partial \mathcal{L}_1}{\partial \beta} = 0. \quad [3.120b]$$

Differentiating the Lagrangian in equation [3.119], the above two conditions result

$$\gamma_1 \left(48\beta^3\zeta^2 + \left(-6\beta^2 + 48\beta^4\right)\zeta - 15\beta + 12\beta^5 - 3\beta^3\right) = 0 \qquad [3.121]$$

and

$$1 + \gamma_1 \left(48\beta^2\zeta^3 + \left(-6\beta + 96\beta^3\right)\zeta^2 + \left(-15 + 60\beta^4 - 9\beta^2\right)\zeta + 12\beta + 24\beta^3 + 12\beta^5\right) = 0. \qquad [3.122]$$

Because the Lagrange multiplier γ_1 cannot be zero, solving equation [3.121], we have

$$\zeta = -\frac{1+\beta^2}{2\beta} \qquad [3.123a]$$

$$\text{or} \quad \zeta = \frac{5-4\beta^2}{8\beta}. \qquad [3.123b]$$

Ignoring the first solution, which is always negative, and substituting $\zeta = \left(5-4\beta^2\right)/8\beta$ in the constraint equation [3.115] and simplifying, we have

$$\beta^4 + 2\beta^2 - 11/16 = 0. \qquad [3.124]$$

There is only one feasible solution to the above equation, which can be obtained as

$$\beta_{mU} = \frac{1}{2}\sqrt{3\sqrt{3}-4} = 0.5468. \qquad [3.125]$$

For this value of β, the value of ζ can be obtained from equation [3.123b] as

$$\zeta_{mU} = \frac{3}{4}\sqrt{(12\sqrt{3}-6)/11} = 0.8695. \qquad [3.126]$$

The point (ζ_{mU}, β_{mU}) is shown by a dot in Figure 3.8. From this plot, it can be observed that if $\beta > \beta_{mU}$, then there always exists a driving frequency for which the amplitude of the frequency response function will reach a maximum value.

The value of ζ_{mL} can be obtained from the optimization problem [3.118] by constructing the Lagrangian

$$\mathcal{L}_2(\zeta,\beta) = \zeta + \gamma_2\,\mathcal{M}\left(\zeta,\beta\right) \qquad [3.127]$$

where γ_2 is the Lagrange multiplier. Following a similar procedure, it can be shown that the optimal value of ζ is given by

$$\zeta_{mL} = \frac{1}{2}\sqrt{\sqrt{5} - 1} = 0.5559. \qquad [3.128]$$

For this value of ζ, the value of β can be obtained as

$$\beta_{mL} = \frac{1}{2}\sqrt{2\sqrt{5} - 4} = 0.3436. \qquad [3.129]$$

The point (ζ_{mL}, β_{mL}) is shown by a dot in Figure 3.8. From this plot, it can be observed that if $\zeta < \zeta_{mL}$, then there always exists a driving frequency for which the amplitude of the frequency response function will reach a maximum value. From the previous discussions, we have the following fundamental results:

THEOREM 3.12.– The amplitude of the frequency response function of a non-viscously damped system can reach a maximum value if $\zeta < \frac{1}{2}\sqrt{\sqrt{5} - 1}$ or $\beta > \frac{1}{2}\sqrt{3\sqrt{3} - 4}$.

THEOREM 3.13.– If $\zeta < \frac{1}{2}\sqrt{\sqrt{5} - 1}$ or $\beta > \frac{1}{2}\sqrt{3\sqrt{3} - 4}$, then the amplitude of the frequency response function of a non-viscously damped system reaches a maximum value when the driving frequency $\omega = \omega_n(\sqrt{(1 + 2\zeta\beta + \beta^2)(2\cos(\theta/3) + 1)/3 - 1})/\beta$.

3.6.1.2. *Parameter relationships for* $\omega_{\max} = \omega_n$

The contour line $\omega_{\max}/\omega_n = 1$ in Figure 3.8 is of special interest. For these particular parameter combinations, the maximum amplitude of the frequency response function of the damped system occurs *exactly* at the undamped natural frequency. This surprising fact occurs only in non-viscously damped systems and it is not possible for viscously damped systems. For a more detailed analysis, Figure 3.9 shows the contours of $\omega_{\max}/\omega_n$ when $\zeta \leq 1$ and $\beta \leq 1$. In Figure 3.9, when $\beta = 0$, then $\omega_{\max}/\omega_n$ can be equal to 1 if and only if $\zeta = 0$ (that is when the system is undamped). The conditions for $\omega_{\max}/\omega_n = 1$ can be obtained by enforcing $x_{\max} = 1$. Thus, substituting $x = 1$ in equation [3.100] and considering that $\zeta \neq 0$, we have

$$\beta + \beta^3 - \zeta = 0. \qquad [3.130]$$

Solving this, the required condition can be given by

$$\zeta = \beta\left(1 + \beta^2\right) \qquad [3.131]$$

when β is known, or

$$\beta = (\xi^2 - 12)/6\xi, \quad \text{where} \quad \xi = \sqrt[3]{108\zeta + 12\sqrt{12 + 81\zeta^2}} \qquad [3.132]$$

when ζ is known. Equation [3.131] is plotted in Figure 3.9. The same curve can also be obtained by plotting equation [3.132]. One interesting fact that can be observed from Figure 3.9 is that beyond certain values of ζ and β, the maximum dynamic response amplitude cannot occur at $\omega_{\max}/\omega_n = 1$. To obtain these limiting values, we substitute ζ from equation [3.131] into the condition of real solution given in equation [3.115]. After some calculation, the resulting equation becomes

$$16\beta^{12} + 72\beta^{10} + 105\beta^8 + 45\beta^6 - 15\beta^4 - 9\beta^2 + 2 = 0. \qquad [3.133]$$

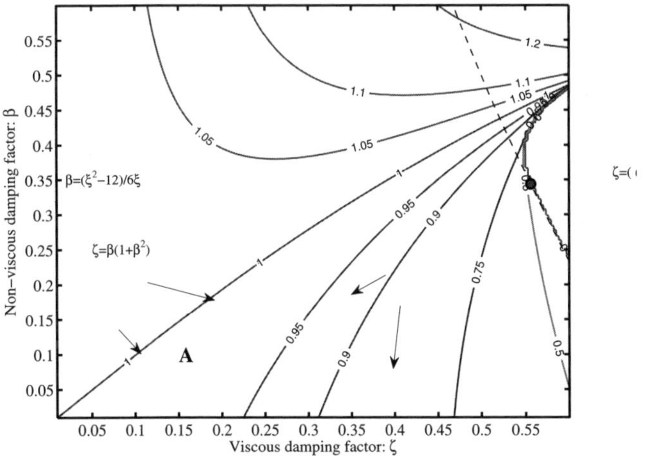

Figure 3.9. *Contours of the normalized excitation frequency corresponding to the maximum value of the amplitude of the frequency response function, $\omega_{\max}/\omega_n$, as a function of ζ and β. Equations corresponding to $\omega_{\max}/\omega_n = 0$ (dashed line) and $\omega_{\max}/\omega_n = 1$ (dotted line) are shown in the figure. The function ξ is defined in equation [3.132]*

The only positive real solution of the above equation is

$$\beta = 1/2. \qquad [3.134]$$

Substituting this value β in equation [3.131], we obtain

$$\zeta = 5/8. \qquad [3.135]$$

The point $(5/8, 1/2)$ is shown in Figure 3.9 by a dot. From this diagram, it is clear that $\omega_{\max}/\omega_n$ can be equal to 1, if and only if $\zeta < 5/8$ and $\beta < 1/2$. When $x_{\max} = 1$, the maximum value of the amplitude of the frequency response function can be obtained from equation [3.95] as

$$|G|_{x_{\max}=1} = \frac{\sqrt{1+\beta^2}}{2\zeta}. \qquad [3.136]$$

From this discussion, we have the following useful results:

THEOREM 3.14.– If it exists, the maximum amplitude of the frequency response function of a non-viscously damped system will occur below the undamped natural frequency if and only if $\zeta < 5/8$ and $\beta < 1/2$.

THEOREM 3.15.– If it exists, the maximum amplitude of the frequency response function of a non-viscously damped system will occur above the undamped natural frequency if $\zeta > 5/8$ or $\zeta < \beta\left(1+\beta^2\right)$ and $\beta > 1/2$ or $\beta > (\xi^2 - 12)/6\xi$.

Another interesting feature of Figure 3.9 is the flatness of $\omega_{\max}/\omega_n$ around the contour line 1. This implies that for a wide range of parameter combinations, it is possible to observe a damped resonance very close to the undamped natural frequency. For a viscously damped system, this can happen only if the damping is very small ($\zeta \lesssim 0.05$). But for a non-viscously damped system, this can happen even when ζ is as large as 0.6.

It was shown that the amplitude of the frequency response function cannot reach a maximum value for some combinations of ζ and β (the parameter region B in Figure 3.8 and Figure 3.9). Considering small values of ζ and β so that $\zeta \le \zeta_{mL}$ and $\beta \le \beta_{mL}$, we aim to derive a simple analytical expression for the existence of $|G|_{\max}$. Because $x = \tilde{\omega}^2$, the condition for existence of the maximum amplitude of the frequency response function can be expressed as

$$x_{\max} \ge 0. \qquad [3.137]$$

Therefore, the critical condition can be obtained by substituting $x = 0$ in equation [3.100] as

$$1 - 2\zeta\beta - 2\zeta^2 = 0. \qquad [3.138]$$

Solving this equation for ζ, the condition for existence of $|G|_{\max}$ can be expressed by

$$\zeta < \frac{1}{2}\left(\sqrt{2+\beta^2} - \beta\right) \qquad [3.139]$$

when

$$\beta \leq \frac{1}{2}\sqrt{2\sqrt{5}-4}. \tag{3.140}$$

For the special case when only viscous damping is present, substituting $\beta = 0$ in equation [3.139], we obtain the required condition as $\zeta \leq 1/\sqrt{2}$, which is well known for viscously damped systems. This condition can alternatively be expressed in terms of ζ by solving equation [3.138] for β as

$$\beta < \frac{1-2\zeta^2}{2\zeta} \tag{3.141}$$

when

$$\zeta \leq \frac{1}{2}\sqrt{\sqrt{5}-1}. \tag{3.142}$$

The validity of equations [3.139] and [3.141] can be verified from Figure 3.9. When $\zeta \leq \zeta_{mL}$ and $\beta \leq \beta_{mL}$, equations [3.139] and [3.141] match perfectly with the zero-line obtained from the expression of $x_{\max}$ in equation [3.109]. Observe that these equations become invalid when $\zeta \geq \zeta_{mL}$ and $\beta \geq \beta_{mL}$. From this discussion, we have the following result:

THEOREM 3.16.– If $\zeta \leq \frac{1}{2}\sqrt{\sqrt{5}-1}$ and $\beta \leq \frac{1}{2}\sqrt{2\sqrt{5}-4}$, the amplitude of the frequency response function of a non-viscously damped system can reach a maximum value if and only if $\zeta < (\sqrt{2+\beta^2}-\beta)/2$ or $\beta < \left(1-2\zeta^2\right)/2\zeta$.

3.6.2. *The amplitude of the maximum dynamic response*

The maximum value of the amplitude of the frequency response function is a useful quantity because it can be related to the structural failure and design. Figure 3.10 shows the contours of the maximum amplitude of the normalized frequency response function $|G|_{\max}$ as a function of ζ and β. The values of $|G|_{\max}$ are calculated from equation [3.95] by substituting $x_{\max}$ from equation [3.109] in place of x. This figure is divided into four regions for discussions. In region A, where $\zeta < \zeta_{mL}$, the amplitude of the frequency response function of the system will always have a maximum value. The values of $|G|_{\max}$ are higher for smaller values of ζ as expected. A useful fact to be noted is that for a fixed value of ζ, the value of $|G|_{\max}$ is higher for higher values of β. This can also be verified from Figure 3.7. This fact may have undesirable consequences especially if β is large. In region B, the amplitude of the frequency response function does not have a maximum value. The shaded portion inside region B (shown before in Figures 3.2 and 3.3) corresponds to the parameter region where the system cannot have any oscillatory motions. Clearly, within this overdamped region, it is not possible for the dynamic response amplitude to reach a maximum value. In region C, where $\beta > \beta_{mU}$, the amplitude of the frequency response function of the system will always have a maximum value, but the value of the maximum response is less than 1. In region D, observe that $\beta > \beta_{mU}$, but unlike region C, the value of the maximum response is more than 1. In general, for a fixed value of ζ, the values of $|G|_{\max}$ increase with the increasing values of β. The numerical values of $|G|_{\max}$ in regions C and D are, however, smaller compared to those in region A. From this discussion, we have the following general result:

THEOREM 3.17.– For a given value of ζ, if it exists, the maximum amplitude of the frequency response function of a non-viscously damped system increases with increasing values of β.

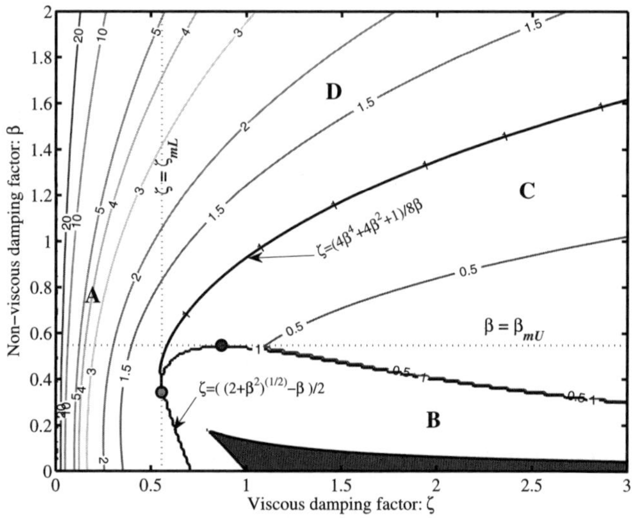

Figure 3.10. *Contours of the maximum amplitude of the normalized frequency response function $|G|_{\max}$ as a function of ζ and β*

The contour line 1 in Figure 3.10 is of special interest because $|G|_{\max} > 1$ implies that the maximum dynamic response amplitude is more than the static response. For the parameter combinations on the left side of the contour line 1, the amplitude of the maximum dynamic response is always greater than 1. The region on the right side of this line may not be of much engineering interest because the amplitude of the maximum dynamic response is less than the static response amplitude of the system. Next, we derive the exact parameter combinations for which $|G|_{\max}$ is more than 1.

Substituting $x_{\max}$ from equation [3.109] in the expression of $|G|^2$ in equation [3.95], we can obtain the expression of $|G|^2_{\max}$. Equating the resulting expression to 1 and simplifying, we have

$$\begin{aligned}&\left(8\zeta^3\beta^3 + \left(12\beta^2 + 12\beta^4\right)\zeta^2 + \left(12\beta^3 + 6\beta + 6\beta^5\right)\zeta\right.\\&\left.+3\beta^2 + 1 + \beta^6 + 3\beta^4\right)\left(8\cos^3\left(\theta/3\right) - 12\cos^2\left(\theta/3\right)\right)\\&+18\left(8\zeta^2\beta^2 + \left(-2\beta^5 + 4\beta + 4\beta^3\right)\zeta - \beta^4 - \beta^6\right)\cos\left(\theta/3\right)\\&+32\zeta^3\beta^3 + \left(48\beta^4 + 12\beta^2\right)\zeta^2 + \left(-48\beta + 6\beta^5 - 24\beta^3\right)\zeta\\&\qquad\qquad +4 + 3\beta^4 - 5\beta^6 + 12\beta^2 = 0.\end{aligned} \qquad [3.143]$$

This is a cubic equation in $\cos\left(\theta/3\right)$ and it can be solved exactly to obtain

$$\cos\left(\theta/3\right) = \frac{1 - \zeta\beta - \beta^2/2}{1 + \beta^2 + 2\zeta\beta} \qquad [3.144]$$

$$\text{or} \quad \cos\left(\theta/3\right) = \frac{\beta^2 + 2\zeta\beta + (1 \pm 3\kappa)/4}{1 + \beta^2 + 2\zeta\beta} \qquad [3.145]$$

$$\text{where } \kappa = \sqrt{4\beta^4 + 4\beta^2 - 8\zeta\beta + 1}. \qquad [3.146]$$

Among the above three solutions, any one of the two solutions given in equation [3.145] turns out be more useful. In order to obtain the relationship between ζ and β so that $|G|_{\max} = 1$, it is required to relate the expression of $\cos\left(\theta/3\right)$ in equation [3.145] to the expression of $\cos(\theta)$ in equation [3.105]. Using the identity

$$\cos(\theta) = 4\cos^3\left(\theta/3\right) - 3\cos\left(\theta/3\right) \qquad [3.147]$$

and substituting the expression of $\cos(\theta)$ from equation [3.105] and $\cos\left(\theta/3\right)$ from equation [3.145], we have

$$4\left(4\beta - \kappa\beta\right)\zeta^2 + \left(2\kappa - 4\beta^2\kappa - 8\beta^4 - 2\right)\zeta - 4\beta^3 - \beta - 2\beta^3\kappa - 4\beta^5 - \kappa\beta = 0 \qquad [3.148]$$

$$\text{or} \quad \kappa = -\frac{\left(8\beta^4 + 2\right)\zeta + 4\beta^5 + 4\beta^3 + \beta - 16\zeta^2\beta}{4\zeta^2\beta + \left(4\beta^2 - 2\right)\zeta + 2\beta^3 + \beta}. \qquad [3.149]$$

Equating the right-hand sides of equations [3.146] and [3.149] and simplifying, we have

$$\begin{aligned}&16\zeta^4\beta^2 + \left(14\beta - 8\beta^5 + 24\beta^3\right)\zeta^3 + \left(-4\beta^2 - 8\beta^4 - 2 - 16\beta^6\right)\zeta^2\\&\quad - \left(9\beta + 14\beta^3 + 8\beta^7 + 12\beta^5\right)\zeta + 1 + 5\beta^2 + 8\beta^4 + 4\beta^6 = 0\end{aligned} \qquad [3.150]$$

The two real and positive solutions of ζ of the previous equation are given by

$$\zeta = \left(\sqrt{2+\beta^2} - \beta\right)/2 \qquad [3.151]$$

$$\text{or} \quad \zeta = \left(4\beta^4 + 4\beta^2 + 1\right)/8\beta. \qquad [3.152]$$

If the expression of $\cos(\theta)$ in equation [3.144] was used in place of that in equation [3.145], then we would obtain only the condition in equation [3.151]. The expression of $\cos(\theta)$ in equation [3.145] was selected because it produces more general results. Interestingly, the condition given in equation [3.151] was also identified as the condition for the existence of the maximum value of the frequency response function in equation [3.139]. The values of ζ given in equations [3.151] and [3.152] are shown in Figure 3.10. Equation [3.151] is valid when $\beta \leq \frac{1}{2}\sqrt{2\sqrt{5}-4}$ and equation [3.152] is valid when $\beta > \frac{1}{2}\sqrt{2\sqrt{5}-4}$. From this analysis, we have the following fundamental result:

THEOREM 3.18.– The maximum amplitude of the normalized frequency response function of a non-viscously damped system will be more than 1 if and only if $\zeta < (\sqrt{2+\beta^2} - \beta)/2$ when $\beta \leq \frac{1}{2}\sqrt{2\sqrt{5}-4}$ and $\zeta < \left(4\beta^4 + 4\beta^2 + 1\right)/8\beta$ when $\beta \geq \frac{1}{2}\sqrt{2\sqrt{5}-4}$.

From this result, one practical question that naturally arises is: what is the critical value of ζ below which the maximum amplitude of the normalized frequency response function will always be more than 1? To answer this question, we look for the minimum value of ζ given by equation [3.152]. Differentiating equation [3.152] with respect to β, the optimal value can be obtained from

$$\frac{4\beta^2 + 12\beta^4 - 1}{8\beta^2} = 0. \qquad [3.153]$$

The only real and positive solution of this equation is

$$\beta = \frac{1}{\sqrt{6}}. \qquad [3.154]$$

Substituting this value of β in equation [3.152], the optimal value of ζ can be obtained as

$$\zeta = 2\sqrt{6}/9. \qquad [3.155]$$

From this discussion, we have the following theorem:

THEOREM 3.19.– The maximum amplitude of the normalized frequency response function of a non-viscously damped system will be more than 1 if $\zeta < 2\sqrt{6}/9$.

The converse statement of theorem 3.19 is, however, not always true. The value of $|G|_{\max}$ can be more than 1 even if $\zeta > 2\sqrt{6}/9$ as can be seen in region C in Figure 3.10.

3.7. Simplified analysis of the frequency response function

Dynamic characteristics of the frequency response function of a non-viscously damped SDOF system have been elucidated in the previous section. The frequency at which the amplitude of the frequency response function reaches its maximum value can be obtained from equation [3.109]. Although this is an exact expression, it is difficult to gain much physical insight due to its complexity. Here, we derive some simple expressions considering that ζ and β are small.

In Figure 3.9, it was noted that for a wide range of values of ζ and β, the amplitude of the frequency response function reaches its maximum value when the normalized excitation frequency is close to 1. For this reason, we assume that

$$x_{\max} = 1 - \delta. \qquad [3.156]$$

Substituting this in place of x in equation [3.100] and simplifying, we obtain

$$\beta^4\delta^3 + \left(-2\beta^4 + 2\beta^3\zeta - 2\beta^2\right)\delta^2 + \left(\beta^4 + 2\beta^2 - 4\beta^3\zeta - 4\zeta\beta + 1\right)\delta + 2\zeta\beta + 2\beta^3\zeta - 2\zeta^2 = 0. \qquad [3.157]$$

This is a cubic equation in δ, which needs to be solved to obtain the frequency where $|G|^2$ reaches its maximum value. Since δ is expected to be small for small values of ζ and β, neglecting the coefficients associated with δ^2 and δ^3 in equation [3.157] and solving the resulting linear equation, we obtain

$$\delta \approx \frac{2\zeta^2 - 2\zeta\beta(1+\beta^2)}{(1+\beta^2)(1+\beta^2-4\zeta\beta)}. \qquad [3.158]$$

Substituting δ in equation [3.156], the frequency corresponding to the maximum value of the amplitude of the frequency response function can be approximately obtained as

$$\tilde{\omega}_{\max} = \sqrt{x_{\max}} = \frac{\omega_{\max}}{\omega_n} \approx \sqrt{1 - \frac{2\zeta^2 - 2\zeta\beta\left(1+\beta^2\right)}{\left(1+\beta^2\right)\left(1+\beta^2-4\zeta\beta\right)}}. \qquad [3.159]$$

For the special case when only viscous damping is present, substituting $\beta = 0$ in equation [3.159], we obtain $\tilde{\omega}_{\max} = \sqrt{1 - 2\zeta^2}$, which is well known for viscously damped systems.

Substituting $x = x_{\max}$ from [3.156] into the expression of $|G|^2$ in equation [3.95] and retaining only up to quadratic terms in δ, we have

$$|G|^2_{\max} \approx \frac{1 + \beta^2 - \beta^2 \delta}{4\zeta^2 + (4\zeta\beta - 4\zeta^2)\,\delta + (\beta^2 + 1 - 4\zeta\beta)\,\delta^2}. \qquad [3.160]$$

Substituting δ from [3.158] into the previous equation and retaining only up to cubic terms in β, we have

$$|G|_{\max} \approx \frac{1}{2\zeta}\sqrt{\frac{(1 + \beta^2)}{(1 + 2\beta^2)} \frac{(1 - 4\zeta\beta + (3 - 2\zeta^2)\,\beta^2 - 6\beta^3\zeta)}{(1 - 2\zeta\beta - \zeta^2)}}. \qquad [3.161]$$

For the special case when only viscous damping is present, substituting $\beta = 0$ in equation [3.161] results in the exact corresponding expression $|G|_{\max} = 1/(2\zeta\sqrt{1 - \zeta^2})$ as given in equation [3.93]. To verify the accuracy of the approximate formulas [3.159] and [3.161], we calculate the percentage error with respect to the exact solutions obtained in the previous section. The percentage error is calculated, for example, as

$$100 \times \frac{(\tilde{\omega}_{\max})_{\text{exact}} - (\tilde{\omega}_{\max})_{\text{approx}}}{(\tilde{\omega}_{\max})_{\text{exact}}}. \qquad [3.162]$$

Figures 3.11 and 3.12, respectively, show the contours of percentage errors arising due to the use of approximate equations [3.159] and [3.161]. For $\tilde{\omega}_{\max}$ calculated from equation [3.159], the error is less than 2% when $\zeta, \beta \leq 0.5$. Error in the calculation of $|G|_{\max}$ from equation [3.161] is somewhat more. When $\zeta, \beta \approx 0.5$, the error is close to 10%. From Figure 3.12, it can be observed that the error in the calculation of $|G|_{\max}$ increases with the increasing values of β but it is relatively insensitive with respect to ζ.

The approximate expressions [3.159] and [3.161] will break down if $\zeta > \zeta_{mL}$ and $\beta > \beta_{mU}$. For such parameter values, it is not possible to extend a perturbation-type method based on viscous damped system as proposed here. Nevertheless, if a system is moderately non-viscously damped (say $\beta < 0.5$), the dynamics can be explained using the proposed approximations [3.159] and [3.161].

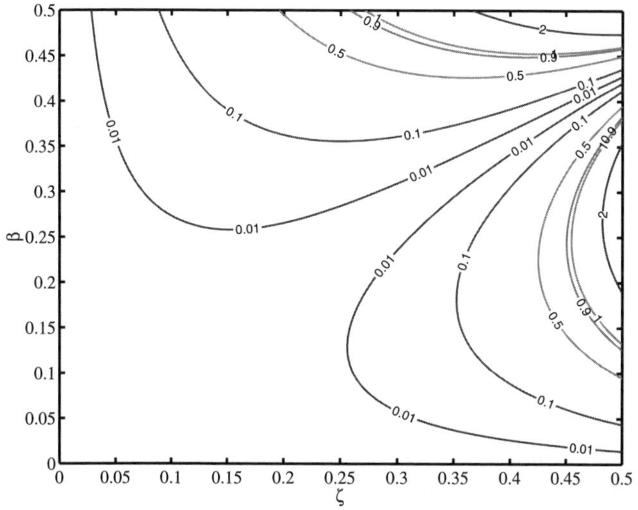

Figure 3.11. *Contours of percentage error in the approximate calculation of $\omega_{\max}/\omega_n$ from equation [3.159] as a function of ζ and β*

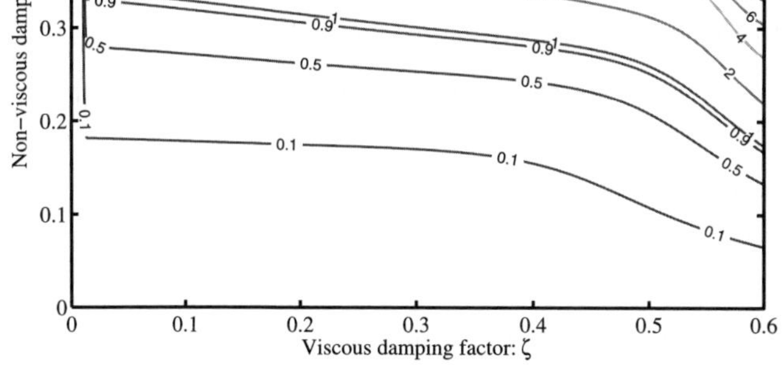

Figure 3.12. *Contours of percentage error in the approximate calculation of $|G|_{\max}$ from equation [3.161] as a function of ζ and β*

3.8. Summary

Dynamic characteristics of a non-viscously damped linear SDOF system have been discussed in this chapter. The non-viscous damping force was expressed by an exponentially fading memory kernel. It was shown that the dynamic properties of the dynamic system are governed by two non-dimensional factors, namely the viscous damping factor ζ, and the non-viscous damping factor β. The system considered reduces to the classical viscously damped system when the non-viscous damping factor is zero. Several new properties that characterize the dynamics of a non-viscously damped system have been discovered. A non-viscously damped system has three eigenvalues, one of which is always non-oscillating in nature. The conditions for occurrence of a pair of oscillating eigenvalues were derived. Characteristics of the natural frequency and decay rates were discussed in details. The conditions for the occurrence of the maximum value of the amplitude of the dynamic response were reviewed. The characteristics of the driving frequency corresponding to the maximum amplitude of the frequency response function and the value of the maximum response amplitude were discussed in detail. The main findings of the chapter are as follows:

1) A non-viscously damped system will have oscillatory motions if $\zeta < 4/(3\sqrt{3})$ or $\beta > 1/(3\sqrt{3})$.

2) If $\beta < 1/(3\sqrt{3})$, the system will have oscillatory motions if and only if $\zeta \notin [\zeta_L, \zeta_U]$. ζ_L and ζ_U given in equations [3.45] and [3.46] are the lower and upper critical damping factors, respectively.

3) If $\beta > 1/4$, the natural frequency of an exponential non-viscously damped system will be more than that of an equivalent undamped system.

4) If $\beta > 1/2$, the decay rate corresponding to the oscillating mode will be less than that of an equivalent viscously damped system.

5) If $\zeta + \beta > 1$, the normalized decay rate corresponding to the non-oscillating mode will be less than 1.

6) The amplitude of the frequency response function of a non-viscously damped system can reach a maximum value if $\zeta < \frac{1}{2}\sqrt{\sqrt{5}-1}$ or $\beta > \frac{1}{2}\sqrt{3\sqrt{3}-4}$.

7) If $\zeta < \frac{1}{2}\sqrt{\sqrt{5}-1}$ or $\beta > \frac{1}{2}\sqrt{3\sqrt{3}-4}$, then the amplitude of the frequency response function of a non-viscously damped system reaches a maximum value when the driving frequency $\omega = \omega_n(\sqrt{(1+2\zeta\beta+\beta^2)\left(2\cos\left(\theta/3\right)+1\right)/3-1})/\beta$.

8) If it exists, the maximum amplitude of the frequency response function of a non-viscously damped system will occur below the undamped natural frequency if and only if $\zeta < 5/8$ and $\beta < 1/2$.

9) If it exists, the maximum amplitude of the frequency response function of a non-viscously damped system will occur above the undamped natural frequency if $\zeta > 5/8$ or $\zeta < \beta\left(1 + \beta^2\right)$ and $\beta > 1/2$ or $\beta > (\xi^2 - 12)/6\xi$.

10) If $\zeta \leq \frac{1}{2}\sqrt{\sqrt{5} - 1}$ and $\beta \leq \frac{1}{2}\sqrt{2\sqrt{5} - 4}$, the amplitude of the frequency response function of a non-viscously damped system can reach a maximum value if and only if $\zeta < (\sqrt{2 + \beta^2} - \beta)/2$ or $\beta < \left(1 - 2\zeta^2\right)/2\zeta$.

11) For a given value of ζ, if it exists, the maximum amplitude of the frequency response function of a non-viscously damped system increases with increasing values of β.

12) The maximum amplitude of the normalized frequency response function of a non-viscously damped system will be more than 1 if and only if $\zeta < (\sqrt{2 + \beta^2} - \beta)/2$ when $\beta \leq \frac{1}{2}\sqrt{2\sqrt{5} - 4}$ and $\zeta < \left(4\beta^4 + 4\beta^2 + 1\right)/8\beta$ when $\beta \geq \frac{1}{2}\sqrt{2\sqrt{5} - 4}$.

13) The maximum amplitude of the normalized frequency response function of a non-viscously damped system will be more than 1 if $\zeta < 2\sqrt{6}/9$.

The influences of the viscous damping factor and the non-viscous damping factor on the dynamics of non-viscously damped SDOF systems are summarized in Table 3.1. The equivalent results for the case of viscous damping, that is when $\beta = 0$ are given in Table 3.1. It is clear that the addition of the non-viscous damping factor bring many additional features to the dynamics of the system. There are some results in this chapter (theorems 3.5, 3.6, 3.7, 3.11 and corollary 3.3.1) that discuss the conditions of occurrence (or non-occurrence) of particular values of natural frequency and decay rates. Using these, we can understand the nature of the eigenvalues without actually solving the eigenvalue problem. These concepts will be particularly useful in dealing with multiple-degree-of-freedom systems that are discussed in the next chapter.

Parameter values	Dynamic characteristics
$\zeta < 4/(3\sqrt{3})$ or $\beta > 1/(3\sqrt{3})$	The motion is oscillatory
$\beta < 1/(3\sqrt{3})$	Oscillatory motions if and only if $\zeta \notin [\zeta_L, \zeta_U]$. ζ_L and ζ_U given in equations [3.45] and [3.46] are the lower and upper critical damping factors, respectively
$\beta > 1/4$	The natural frequency will be more than that of an equivalent undamped system
$\beta > 1/2$	The decay rate for the oscillating mode will be less than that of an equivalent viscously damped system
$\zeta + \beta > 1$	The normalized decay rate corresponding to the non-oscillating mode will be less than 1
$\zeta < \frac{1}{2}\sqrt{\sqrt{5}-1}$ or $\beta > \frac{1}{2}\sqrt{3\sqrt{3}-4}$	The amplitude of the frequency response function can reach a maximum value
$\zeta < \frac{1}{2}\sqrt{\sqrt{5}-1}$ or $\beta > \frac{1}{2}\sqrt{3\sqrt{3}-4}$	The amplitude of the frequency response function reaches a maximum value when $\omega = \omega_n(\sqrt{(1+2\zeta\beta+\beta^2)}\,(2\cos(\theta/3)+1)/3-1)/\beta$
$\zeta < 5/8$ and $\beta < 1/2$	If it exists, the maximum amplitude of the frequency response function will occur below the undamped natural frequency
$\zeta > 5/8$ or $\zeta < \beta\left(1+\beta^2\right)$	If it exists, the maximum amplitude of the frequency response function will occur above the undamped natural frequency
$\beta > 1/2$ or $\beta > (\xi^2 - 12)/6\xi$	If it exists, the maximum amplitude of the frequency response function will occur above the undamped natural frequency
$\zeta \leq \frac{1}{2}\sqrt{\sqrt{5}-1}$ and $\beta \leq \frac{1}{2}\sqrt{2\sqrt{5}-4}$	The amplitude of the frequency response function can reach a maximum value if and only if $\zeta < (\sqrt{2+\beta^2}-\beta)/2$ or $\beta < \left(1-2\zeta^2\right)/2\zeta$
$\zeta < (\sqrt{2+\beta^2}-\beta)/2$ and $\beta \leq \frac{1}{2}\sqrt{2\sqrt{5}-4}$	The maximum amplitude of the normalized frequency response function will be more than 1
$\zeta < \left(4\beta^4+4\beta^2+1\right)/8\beta$ and $\beta \geq \frac{1}{2}\sqrt{2\sqrt{5}-4}$	The maximum amplitude of the normalized frequency response function will be more than 1
$\zeta < 2\sqrt{6}/9$, any β	The maximum amplitude of the normalized frequency response function will be more than 1

Table 3.1. *The influence of the viscous damping factor ζ and the non-viscous damping factor β on the dynamics of a non-viscously damped single-degree-of-freedom system*

Chapter 4

Non-viscously Damped Multiple-Degree-of-Freedom Systems

The previous chapter gave qualitative ideas on the dynamics of non-viscously damped single-degree-of-freedom systems. Several properties of the eigensolutions and dynamic response function have been explained. In this chapter, we devote our attention to lumped parameter linear systems with N degrees-of-freedom and non-viscous damping properties. We mainly focus on the quantitative aspects here. The system we will study can be described by an equation of motion in the form

$$\mathbf{M}\ddot{\mathbf{q}}(t) + \mathbf{C}_0\dot{\mathbf{q}}(t) + \int_0^t \boldsymbol{\mathcal{G}}(t-\tau)\,\dot{\mathbf{q}}(\tau)\,\mathrm{d}\tau + \mathbf{K}\mathbf{q}(t) = \mathbf{f}(t) \qquad [4.1]$$

together with the initial conditions

$$\mathbf{q}(t=0) = \mathbf{q}_0 \in \mathbb{R}^N \quad \text{and} \quad \dot{\mathbf{q}}(t=0) = \dot{\mathbf{q}}_0 \in \mathbb{R}^N. \qquad [4.2]$$

In this formulation, $\mathbf{q}(t) \in \mathbb{R}^N$ is the vector of displacements for the N degrees-of-freedom, $\mathbf{f}(t) \in \mathbb{R}^N$ is the vector of applied forcing, $\mathbf{M} \in \mathbb{R}^{N\times N}$ is the mass matrix, $\mathbf{K} \in \mathbb{R}^{N\times N}$ is the stiffness matrix, $\mathbf{C}_0 \in \mathbb{R}^{N\times N}$ is the viscous damping matrix and $\boldsymbol{\mathcal{G}}(t-\tau)$ is the matrix of damping kernel functions. The mass, stiffness and the viscous damping matrices are symmetric non-negative definite matrices. The mass and the stiffness matrices can be obtained using the finite element method [ZIE 91]. In general, it is difficult to obtain the viscous damping matrix $\mathbf{C}_0$ and the matrix of damping kernel functions analytically. The identification of damping matrices will be discussed in [ADH 14b], Chapters 2 and 3. The main purpose of this chapter is to introduce analytical methods for the solution of the coupled integro-differential equations [4.1].

Using this formulation, we will show how generic non-viscous damping models can be obtained for linear systems. If the systems were undamped or proportionally damped, then classical modal analysis could be used to solve the equation of motion. In this chapter, we extend the classical modal analysis to non-viscously damped systems.

For further analytical studies, it is often useful to transform the equation of motion into the Laplace domain. Taking the Laplace transform of equation [4.1] together with the initial conditions in [4.2], we obtain

$$s^2\mathbf{M}\bar{\mathbf{q}}(s) - s\mathbf{M}\mathbf{q}_0 - \mathbf{M}\dot{\mathbf{q}}_0 + s\left[\mathbf{C}_0 + \mathbf{G}(s)\right]\bar{\mathbf{q}}(s) - \left[\mathbf{C}_0 + \mathbf{G}(s)\right]\mathbf{q}_0 + \mathbf{K}\bar{\mathbf{q}}(s) = \bar{\mathbf{f}}(s) \quad [4.3]$$

or $\left[s^2\mathbf{M} + s[\mathbf{C}_0 + \mathbf{G}(s)] + \mathbf{K}\right]\bar{\mathbf{q}}(s) = \bar{\mathbf{p}}(s)$ [4.4]

where

$$\bar{\mathbf{p}}(s) = \bar{\mathbf{f}}(s) + \mathbf{M}\dot{\mathbf{q}}_0 + \left[s\mathbf{M} + \mathbf{C}_0 + \mathbf{G}(s)\right]\mathbf{q}_0 \quad [4.5]$$

is the equivalent forcing function in the Laplace domain. In the above equations, $\bar{\mathbf{q}}(s) = \mathcal{L}\left[\mathbf{q}(t)\right] \in \mathbb{C}^N$, $\bar{\mathbf{f}}(s) = \mathcal{L}\left[\mathbf{f}(t)\right] \in \mathbb{C}^N$ and $\mathcal{L}\left[\bullet\right]$ denotes the Laplace transform. The matrix $\mathbf{G}(s) \in \mathbb{C}^{N\times N}$ is the Laplace transform of $\boldsymbol{\mathcal{G}}(t)$. In the analysis of structural dynamics, the Laplace parameter is usually replaced by $s = \mathrm{i}\omega$, where $\mathrm{i} = \sqrt{-1}$ and $\omega \in \mathbb{R}^+$ denotes the frequency.

In this chapter, we develop a state-space approach from which we can study both the free vibration problem and the forced vibration response of the system. The eigenvalue analysis leads to two sets of vibration modes for the system. First, N pairs of complex conjugate *elastic* modes exist, which can be related to the normal structural vibration modes which occur in the viscous damping problem. In addition, a number of purely dissipative *non-viscous* modes exist, which together with the elastic modes form a basis for the dynamic response of the system. After an initial discussion on the choice of kernel functions, in the rest of the chapter we will use the so-called *exponential damping model*, as is relevant to the widest range of engineering applications. Different possible choices of damping kernel functions have been discussed in section 4.1. Based on this discussion, the exponential damping model is selected in section 4.2. An extended state-space formulation for exponentially damped linear system is discussed in section 4.3. The eigenvalue problem associated with the extended state-space model is discussed in section 4.4. The problem of forced vibration using the eigensolutions is discussed in section 4.5. In section 4.6, the proposed theory is illustrated using some numerical examples. Direct time-domain analysis of linear systems with exponentially decaying damping memory kernels is considered in section 4.7. The method is based on an extended state-space representation of the equations of motion. Finally, a summary of the studies taken in this chapter is given in section 4.8.

4.1. Choice of the kernel function

Equation [4.1] represents a generalization of the non-viscous damping model presented in Chapter 3 extended to multi-degree-of-freedom (MDOF) systems. The kernel functions in the elements of the matrix $\mathcal{G}(t)$ are also known as retardation functions, heredity functions, after-effect functions or relaxation functions for different applications. They can be chosen to model a range of different non-viscous damping effects. To illustrate this in a simple manner, suppose that $\mathcal{G}(t-\tau) = \bar{\mathbf{C}} g_{(j)}(t-\tau)$. Here, the scalar functions $g_{(j)}(\bullet)$ are the damping kernel functions. Let us consider the following four types of kernel functions to illustrate the nature of non-viscous damping effects:

Model 1:

$$g_{(1)}(\hat{t}) = \mu_1 \exp[-\mu_1 \hat{t}] \qquad [4.6]$$

Model 2:

$$g_{(1)}(\hat{t}) = 2\sqrt{\frac{\mu_2}{\pi}} \exp[-\mu_2 \hat{t}^2] \qquad [4.7]$$

Model 3:

$$g_{(2)}(\hat{t}) = \begin{cases} 1/\mu_3 & (0 < \hat{t} < \mu_3) \\ 0 & (\hat{t} > \mu_3) \end{cases} \qquad [4.8]$$

Model 4:

$$g_{(4)}(\hat{t}) = \begin{cases} \frac{1}{\mu_4}\left[1 + \cos\left(\frac{\pi \hat{t}}{\mu_4}\right)\right] & (0 < \hat{t} < \mu_4) \\ 0 & (\hat{t} > \mu_4) \end{cases} \qquad [4.9]$$

Here, $\hat{t}$ is an arbitrary independent time variable and $\mu_i, i = 1, \cdots, 4$ are real and positive constants. Any physically realistic damping model must satisfy the condition of positive energy dissipation at all frequencies.

It is convenient to normalize the damping functions to make comparisons between models meaningful. All the functions have already been scaled so as to have unit area when integrated to infinity. This makes them directly comparable with the viscous model in which the corresponding damping function would be a unit delta function, $g_{(0)}(\hat{t}) = \delta(t)$, and the coefficient matrix $\bar{\mathbf{C}}$ would be the usual damping matrix. It is also convenient to define a characteristic time constant θ_j for each damping function, via the first moment of $g_{(j)}(\hat{t})$

$$\theta_j = \int_0^\infty \hat{t}\, g_{(j)}(\hat{t})\, \mathrm{d}\hat{t}. \qquad [4.10]$$

For the four damping models considered here, evaluating the above integral gives

$$\theta_1 = \frac{1}{\mu_1} \qquad [4.11]$$

$$\theta_2 = \frac{1}{\sqrt{\pi \mu_2}} \qquad [4.12]$$

$$\theta_3 = \frac{\mu_3}{2} \qquad [4.13]$$

$$\text{and} \quad \theta_2 = \frac{\left(\pi^2 - 4\right)\mu_4}{2\pi^2} \qquad [4.14]$$

For viscous damping, $\theta_0 = 0$. The characteristic time constant of a damping function gives a convenient measure of "width": if it is close to zero, the damping behavior will be near-viscous, and vice versa. To establish an equivalence between the four damping models, we can choose that they have the same time constant so that $\frac{1}{\mu_1} = \frac{1}{\sqrt{\pi\mu_2}} = \frac{\mu_3}{2} = \frac{(\pi^2-4)\mu_4}{2\pi^2}$. These four models are shown in Figure 4.1 for two values of θ_j. The damping kernel functions in Figure 4.1(a) are closer to the viscous damping than those in Figure 4.1(b). As seen from these graphs, the convolution integral together with the kernel function represents the non-viscous contribution to the dynamics. This type of convolution integral approach for modeling material properties can be traced back to the early works on viscoelastic materials (see, for example, [BIO 58]). The main difference between equation [4.1] and corresponding equations for viscoelastic structures is that the kernel function is associated with damping rather than stiffness. In the limit when $\mathcal{G}(t-\tau) = \mathbf{C}\,\delta(t-\tau)$, where $\delta(t)$ is the Dirac-delta function, equation [4.1] reduces to the case of viscous damping, whereas for the viscoelastic case the equivalent would be the usual elastic constant (stiffness matrix).

Clearly, the specific form of $\mathcal{G}(\bullet)$ is application dependent and a wide variety of mathematical expressions, in addition to the four functions described here, could be used. For the kernel functions to represent damping effects, the main restriction is that the rate of energy dissipation should be non-negative. A summary of the main functional forms of $\mathcal{G}(\cdot)$, which have been used to model damping, is shown in Table 1.1. More detailed discussions regarding the individual models are given in Chapter 1.

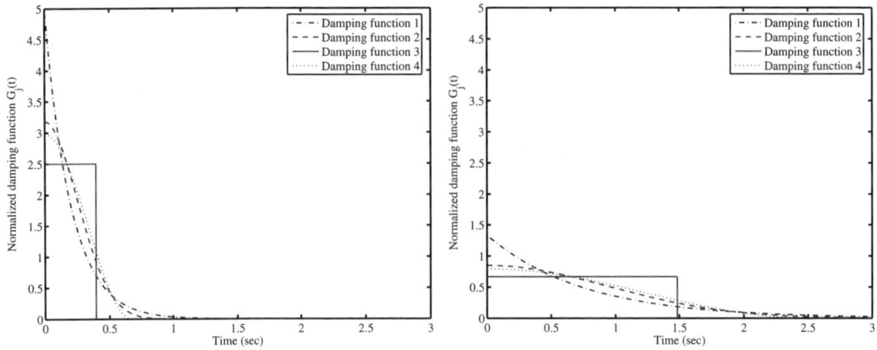

(a) $\theta_j = 0.20$ (b) $\theta_j = 0.75$

Figure 4.1. *Four models for the damping kernel functions*

4.2. The exponential model for MDOF non-viscously damped systems

In this chapter, we will focus on studying the so-called exponential damping model. We assume that the kernel function matrix has the special form

$$\mathcal{G}(\hat{t}) = \sum_{k=1}^{n} \mu_k e^{-\mu_k \hat{t}} \mathbf{C}_k \tag{4.15}$$

or in the Laplace domain

$$\mathbf{G}(s) = \sum_{k=1}^{n} \frac{\mu_k}{s + \mu_k} \mathbf{C}_k. \tag{4.16}$$

The constants $\mu_k \in \mathbb{R}^+$ are known as the relaxation parameters and n denotes the number of relaxation parameters used to describe the damping behavior. This type of model is representative of models 1, 4 and 5 in Table 1.1 and will serve to illustrate the key concepts.

A typical scenario where the exponential damping model could be used is a large engineering structure, where the damping behavior varies across different parts of the structure. For example, various members of a space-frame (see Figure 4.2) have different damping properties due to differing geometric and/or material properties. To apply an exponential damping model in the context of finite element analysis (FEA), the space-frame would first be divided into appropriate elements, each of which could be characterized by a relaxation parameter μ_k. Then the associated coefficient matrix

$\mathbf{C}_k$ would have non-zero blocks, corresponding to the relevant elements for that μ_k only. The kth member of the space-frame shown in Figure 4.2 is further divided into number of finite elements. Therefore, only the elements corresponding to the degrees-of-freedom of the nodes used in the discretization will have non-zero elements in the global $\mathbf{C}_k$ matrix. Assembling all the element matrices associated with relaxation parameter μ_k, we can obtain the global damping matrix $\mathbf{C}_k$ as

$$\mathbf{C}_k = \sum_{e=1}^{n_k} \mathbf{C}_k^{(e)}. \qquad [4.17]$$

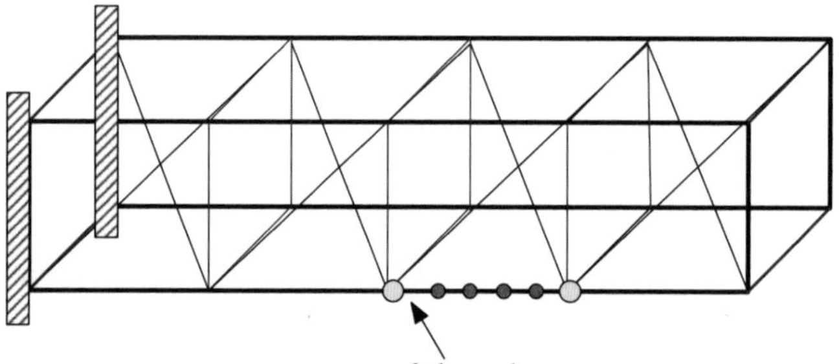

Figure 4.2. *A three-dimensional frame structure with different non-viscous damping properties in different elements*

Here, n_k is the number of finite elements sharing the same relaxation parameter μ_k and the summation is over the degrees-of-freedom corresponding these elements. To obtain the elements of the $\mathbf{C}_k^{(e)}$ matrix, an energy-based finite element technique (similar to that used for the mass and stiffness elements) would be required– see [ADH 14b], Chapter 2 for further details of these techniques. The relaxation parameters μ_k should be identified from material testing on vibration experiments [ADH 01d]. This procedure may be repeated for all damping types present in the structure to obtain μ_k and $\mathbf{C}_k$ for all k. In essence, the exponential damping model, via equation [4.15], gives significant additional flexibility to fit measured data obtained from modal testing over a standard viscous damping model. In fact, in certain cases of non-viscous damping, it has been noted that trying to fit a viscous damping model can lead to a physically unrealizable model, for example, a non-symmetric coefficient matrix [ADH 01c].

A mathematical rationalization of this model can be given in terms of the Laplace transform of $\mathcal{G}(\hat{t})$. The matrix $\mathbf{G}(s)$ in equation [4.16] is, in general, a matrix of complex functions. From the theory of complex variables, it is well known that a wide range of complex functions can be represented in the "pole-residue" form (see [KRE 06], Part D). From equation [4.16], it can be deduced that μ_k and $\mathbf{C}_k$ are directly related to the poles and residues of $\mathbf{G}(s)$. Therefore, many damping models (except the fractional derivative model – model number 3 in Table 1.1 – which has branch-points due to the fractional powers) can essentially be represented in the form of [4.15].

As most vibration analysis textbooks, finite element packages and modal analysis software only allow viscous damping, it is useful to relate the exponential damping model with the viscous damping model. From equation [4.15], let us observe that in the limit when $\mu_k \to \infty, \forall\, k$, the exponential model reduces to a viscous damping model with an equivalent viscous damping matrix

$$\mathbf{C} = \mathbf{C}_0 + \sum_{k=1}^{n} \mathbf{C}_k. \qquad [4.18]$$

In the remainder of this chapter, we will develop a state-space-based approach analogous to viscously damped systems discussed in Chapter 3, with a view to treating the exponentially damped system as a simple extension of the familiar viscously damped system.

4.3. The state-space formulation

State-space methods are commonly used in a wide range of analytical techniques for dynamics and control applications. The idea is to represent the second-order governing equation of motion – in this case equation [4.1] – as a first-order system with an extended state vector. State-space methods have been used extensively in the literature for viscously damped systems with non-proportional damping (see [NEW 87, NEW 89], for example). Here, we extend this approach to the more general non-viscously damped systems. To achieve this for second-order integro-differential equations, such as equation [4.1], we use the concept of *internal variables*. This type of approach has been well developed in the study of viscoelastic structures (see [GOL 85, MCT 93, MUR 97b, MUR 98a] for a detailed discussion on this). In this section, the state-space formulation is presented for two cases: (1) when all $\mathbf{C}_k$ matrices are of full rank and (2) when $\mathbf{C}_k$ matrices have rank deficiency.

4.3.1. *Case A: all coefficient matrices are of full rank*

This case arises when all elements of the structure are associated with one relaxation parameter μ_k. We consider the case when

$$\text{rank}\,(\mathbf{C}_k) = N, \quad \forall\, k = 1, \ldots, n. \qquad [4.19]$$

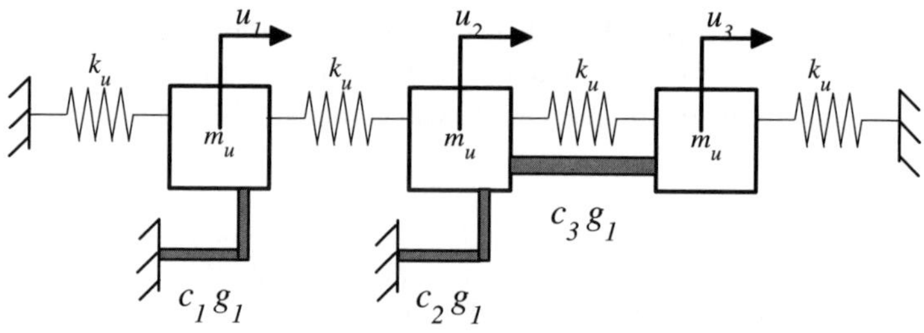

Figure 4.3. *A three DOF model system with non-viscous damping; the shaded bars represent the non-viscous damping elements.* $c_i, i = 1, 2, 3$ *are the damping constants and* $g_1(t-\tau) = \mu_1 e^{-\mu_1(t-\tau)}$ *is the non-viscous damping function*

Consider Figure 4.3 as an example, which shows a three DOF system with non-viscous damping. Three masses, each of mass m_u, are connected by springs of stiffness k_u. The non-viscous damping elements of the system are shown in the figure. The equation of motion of this model system can be represented by equation [4.1]. The mass and the stiffness matrices of the system are given by

$$\mathbf{M} = \begin{bmatrix} m_u & 0 & 0 \\ 0 & m_u & 0 \\ 0 & 0 & m_u \end{bmatrix} \qquad [4.20]$$

and $$\mathbf{K} = \begin{bmatrix} 2k_u & -k_u & 0 \\ -k_u & 2k_u & -k_u \\ 0 & -k_u & 2k_u \end{bmatrix}. \qquad [4.21]$$

The matrix of the damping function can be expressed by equation [4.15] with $n = 1$, and the coefficient matrix is given by

$$\mathbf{C}_1 = \begin{bmatrix} c_1 & 0 & 0 \\ 0 & c_2 + c_3 & -c_3 \\ 0 & -c_3 & c_3 \end{bmatrix}. \qquad [4.22]$$

From equation [4.22], we observe that $\text{rank}\,(\mathbf{C}_1) = 3$.

The exponential function is an eigenfunction in the sense that the application of the operator $d/d\hat{t}$ to the exponential term $e^{\mu\hat{t}}$ produces $\mu e^{\mu\hat{t}}$. Therefore, we introduce the internal variables $\mathbf{y}_k(t) \in \mathbb{R}^N, \forall\, k = 1, \ldots, n$ (dropping the hat notation) through the following relationship:

$$\mathbf{y}_k(t) = \int_0^t \mu_k e^{-\mu_k(t-\tau)} \dot{\mathbf{q}}(\tau) d\tau. \qquad [4.23]$$

The vectors $\mathbf{y}_k(t)$ are also known as the dissipation coordinates. Applying Leibniz's rule for differentiation of an integral [KRE 06, Part D] to equation [4.23], we obtain

$$\dot{\mathbf{y}}_k(t) = \int_0^t -\mu_k^2 e^{-\mu_k(t-\tau)} \dot{\mathbf{q}}(\tau) d\tau + \mu_k \dot{\mathbf{q}}(t). \tag{4.24}$$

Multiplying [4.23] by the relaxation parameter μ_k and then adding it to [4.24] yields the so-called evolution equation (see, for example, [CAP 01])

$$\dot{\mathbf{y}}_k(t) + \mu_k \mathbf{y}_k(t) = \mu_k \dot{\mathbf{q}}(t). \tag{4.25}$$

Substituting the kernel function matrix [4.15] into equation [4.1] gives a governing equation of motion of the form

$$\mathbf{M}\ddot{\mathbf{q}}(t) + \mathbf{C}_0\dot{\mathbf{q}}(t) + \sum_{k=1}^{n} \mathbf{C}_k \left\{ \int_0^t \mu_k e^{-\mu_k(t-\tau)} \dot{\mathbf{q}}(\tau)\, d\tau \right\} + \mathbf{K}\mathbf{q}(t) = \mathbf{f}(t). \tag{4.26}$$

Substituting the internal variables defined in equation [4.23], the equation of motion can be expressed as

$$\mathbf{M}\ddot{\mathbf{q}}(t) + \mathbf{C}_0\dot{\mathbf{q}}(t) + \sum_{k=1}^{n} \mathbf{C}_k \mathbf{y}_k(t) + \mathbf{K}\mathbf{q}(t) = \mathbf{f}(t). \tag{4.27}$$

This is the general second-order form of the equation of motion with internal variables $\mathbf{y}_k$, $k = 1, 2, 3...n$. Now, by rearranging [4.25] and then substituting $\mathbf{y}_k$ from [4.25] into equation [4.27], we obtain

$$\mathbf{M}\ddot{\mathbf{q}}(t) + \mathbf{C}_0\dot{\mathbf{q}}(t) + \sum_{k=1}^{n} \left[\mathbf{C}_k \dot{\mathbf{q}}(t) - \frac{1}{\mu_k} \mathbf{C}_k \dot{\mathbf{y}}_k(t) \right] + \mathbf{K}\mathbf{q}(t) = \mathbf{f}(t) \tag{4.28}$$

$$\text{or} \quad \mathbf{M}\ddot{\mathbf{q}}(t) + \left(\mathbf{C}_0 + \sum_{k=1}^{n} \mathbf{C}_k \right) \dot{\mathbf{q}}(t) - \sum_{k=1}^{n} \frac{1}{\mu_k} \mathbf{C}_k \dot{\mathbf{y}}_k(t) = -\mathbf{K}\mathbf{q}(t) + \mathbf{f}(t). \tag{4.29}$$

Premultiplying equation [4.25] by $\mathbf{C}_k$, dividing by μ_k^2 and rearranging, we get

$$-\frac{1}{\mu_k} \mathbf{C}_k \dot{\mathbf{q}}(t) + \frac{1}{\mu_k^2} \mathbf{C}_k \dot{\mathbf{y}}_k(t) = -\frac{1}{\mu_k} \mathbf{C}_k \mathbf{y}_k(t), \quad \forall\, k. \tag{4.30}$$

Now, using additional state-variables

$$\mathbf{v}(t) = \dot{\mathbf{q}}(t) \qquad [4.31]$$

equation [4.29] and equation [4.30] can be represented in the first-order form as

$$\mathbf{B}\,\dot{\mathbf{u}}(t) = \mathbf{A}\,\mathbf{u}(t) + \mathbf{r}(t) \qquad [4.32]$$

where

$$\mathbf{B} = \begin{bmatrix} \mathbf{C}_0 + \sum_{k=1}^{n} \mathbf{C}_k & \mathbf{M} & -\mathbf{C}_1/\mu_1 & \cdots & -\mathbf{C}_n/\mu_n \\ \mathbf{M} & \mathbf{O} & \mathbf{O} & \mathbf{O} & \mathbf{O} \\ -\mathbf{C}_1/\mu_1 & \mathbf{O} & \mathbf{C}_1/\mu_1^2 & \mathbf{O} & \mathbf{O} \\ \vdots & \mathbf{O} & \mathbf{O} & \ddots & \mathbf{O} \\ -\mathbf{C}_n/\mu_n & \mathbf{O} & \mathbf{O} & \mathbf{O} & \mathbf{C}_n/\mu_n^2 \end{bmatrix} \in \mathbb{R}^{m \times m} \qquad [4.33]$$

$$\mathbf{A} = \begin{bmatrix} -\mathbf{K} & \mathbf{O} & \mathbf{O} & \mathbf{O} & \mathbf{O} \\ \mathbf{O} & \mathbf{M} & \mathbf{O} & \mathbf{O} & \mathbf{O} \\ \mathbf{O} & \mathbf{O} & -\mathbf{C}_1/\mu_1 & \mathbf{O} & \mathbf{O} \\ \mathbf{O} & \mathbf{O} & \mathbf{O} & \ddots & \mathbf{O} \\ \mathbf{O} & \mathbf{O} & \mathbf{O} & \mathbf{O} & -\mathbf{C}_n/\mu_n \end{bmatrix} \in \mathbb{R}^{m \times m} \qquad [4.34]$$

$$\mathbf{r}(t) = \begin{Bmatrix} \mathbf{f}(t) \\ \mathbf{0} \\ \mathbf{0} \\ \vdots \\ \mathbf{0} \end{Bmatrix} \in \mathbb{R}^{m} \qquad [4.35]$$

$$\text{and} \quad \mathbf{u}(t) = \begin{Bmatrix} \mathbf{q}(t) \\ \mathbf{v}(t) \\ \mathbf{y}_1(t) \\ \vdots \\ \mathbf{y}_n(t) \end{Bmatrix} \in \mathbb{R}^{m}. \qquad [4.36]$$

In the above set of equations [4.32], $\mathbf{u}(t)$ is the extended state-vector, $\mathbf{A}$ and $\mathbf{B}$ are the system matrices in the extended state-space, $\mathbf{r}(t)$ is the force vector in the extended state-space and $\mathbf{O}$ is an $N \times N$ null matrix. Clearly, the order of the system, m, is given by

$$m = 2N + nN. \qquad [4.37]$$

Because it is assumed that $\mathbf{M}, \mathbf{K}$ and $\mathbf{C}_k, \forall\ k$ are symmetric matrices, $\mathbf{B}$ is a symmetric matrix and $\mathbf{A}$ is a block-diagonal and therefore, also a symmetric matrix. Note that matrix $\mathbf{B}$ is not positive definite and its rank is less than m. As a consequence, the generalized eigenvalue problem involving $\mathbf{A}$ and $\mathbf{B}$ in general has complex eigensolutions. Since $\mathbf{A}$ and $\mathbf{B}$ are real matrices, the complex eigensolutions always appear in complex pairs. The eigenvalue problem will be discussed in detail in section 4.4.

The equation of motion in the state-space can also be written in the form of a single-system matrix as

$$\dot{\mathbf{u}}(t) = \widehat{\mathbf{A}}\, \mathbf{u}(t) + \widehat{\mathbf{r}}(t). \qquad [4.38]$$

Here, $m \times m$ is the real non-symmetric matrix

$$\widehat{\mathbf{A}} = \begin{bmatrix} \mathbf{O} & \mathbf{I} & \mathbf{O} & \mathbf{O} & \cdots & \mathbf{O} \\ -\mathbf{M}^{-1}\mathbf{K} & -\mathbf{M}^{-1}\mathbf{C}_0 & -\mathbf{M}^{-1}\mathbf{C}_1 & -\mathbf{M}^{-1}\mathbf{C}_2 & \cdots & -\mathbf{M}^{-1}\mathbf{C}_n \\ \mathbf{O} & \mu_1\mathbf{I} & -\mu_1\mathbf{I} & \mathbf{O} & \cdots & \mathbf{O} \\ \mathbf{O} & \mu_2\mathbf{I} & \mathbf{O} & -\mu_2\mathbf{I} & \cdots & \mathbf{O} \\ \vdots & \vdots & \vdots & \vdots & \ddots & \vdots \\ \mathbf{O} & \mu_n\mathbf{I} & \mathbf{O} & \mathbf{O} & \mathbf{O} & -\mu_n\mathbf{I} \end{bmatrix} \in \mathbb{R}^{m\times m} \qquad [4.39]$$

and the m-dimensional forcing vector

$$\widehat{\mathbf{r}}(t) = \begin{Bmatrix} \mathbf{0} \\ \mathbf{M}^{-1}\mathbf{f}(t) \\ \mathbf{0} \\ \mathbf{0} \\ \vdots \\ \mathbf{0} \end{Bmatrix} \in \mathbb{R}^m. \qquad [4.40]$$

This type of representation is often used in the control literature. In this book, however, we use the symmetric representation in equation [4.32].

It is useful to consider the viscous damping limit at this stage. When $\mu_k \to \infty, \forall k$, dividing equation [4.25] by μ_k, it can be shown that

$$\mathbf{y}_k(t) = \dot{\mathbf{q}}(t), \quad \forall\, k. \tag{4.41}$$

This implies that in the viscous damping limit, *all* the internal variables reduce to the velocity vector. For this reason, the $n \times N$ equations after the first $2N$ rows in [4.32] become redundant and can be omitted from the formulation. Under these conditions, the equation of motion [4.32] reduces to the standard Duncan form [MEI 80] for viscously damped systems, as discussed in section 2.5.1.1

$$\begin{aligned} \mathbf{B} &= \begin{bmatrix} \mathbf{C} & \mathbf{M} \\ \mathbf{M} & \mathbf{O} \end{bmatrix}, \quad \mathbf{A} = \begin{bmatrix} -\mathbf{K} & \mathbf{O} \\ \mathbf{O} & \mathbf{M} \end{bmatrix} \\ \mathbf{r}(t) &= \begin{Bmatrix} \mathbf{f}(t) \\ \mathbf{0} \end{Bmatrix} \quad \text{and} \quad \mathbf{u}(t) = \begin{Bmatrix} \mathbf{q}(t) \\ \dot{\mathbf{q}}(t) \end{Bmatrix} \end{aligned} \tag{4.42}$$

where $\mathbf{C}$ is given by equation [4.18]. This shows that the representation of the equation of motion by [4.32] is a natural generalization of the standard state-space formulation for viscously damped systems.

4.3.2. *Case B: coefficient matrices are rank deficient*

In this section, we assume that, in general, the rank of $\mathbf{C}_k$ matrices is less than N. To give an example of this case, again consider a three DOF system with non-viscous damping as shown in Figure 4.4. The equation of motion of this model system can be represented by equation [4.1] with mass and stiffness matrices given by equations [4.20] and [4.21]. The matrix of the damping function can be expressed by equation [4.15] with $n = 2$. The coefficient matrices for the double-exponential model are given by

$$\mathbf{C}_1 = \begin{bmatrix} c_1 & 0 & 0 \\ 0 & c_1 & 0 \\ 0 & 0 & 0 \end{bmatrix} \quad \text{and} \quad \mathbf{C}_2 = \begin{bmatrix} 0 & 0 & 0 \\ 0 & c_2 & -c_2 \\ 0 & -c_2 & c_2 \end{bmatrix}. \tag{4.43}$$

Both the matrices have rank deficiency because we can easily verify that

$$r_1 = \text{rank}\,(\mathbf{C}_1) = 2 \le 3 \tag{4.44}$$

$$\text{and} \quad r_2 = \text{rank}\,(\mathbf{C}_2) = 1 \le 3. \tag{4.45}$$

Motivated by this type of physically realistic problems, we assume that in general

$$r_k = \text{rank}\,(\mathbf{C}_k) \le N, \quad \forall\, k = 1, \ldots, n. \tag{4.46}$$

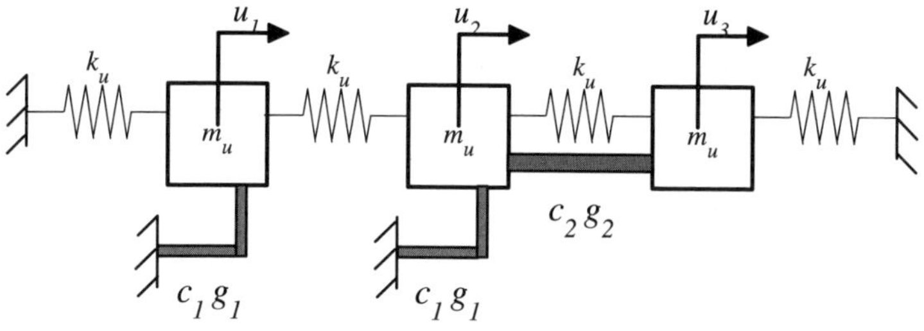

Figure 4.4. *A three DOF model system with non-viscous damping; the shaded bars represent the non-viscous damping elements.* $c_i, i = 1, 2$ *are the damping constants and* $g_i(t-\tau) = \mu_i e^{-\mu_i(t-\tau)}, i = 1, 2$ *are the non-viscous damping functions*

This implies that the number of non-zero eigenvalues of $\mathbf{C}_k$ is r_k. It is useful to decompose the $\mathbf{C}_k$ matrices into full-rank matrices of smaller dimensions. Because $\mathbf{C}_k$ is a symmetric matrix, there exists an orthogonal matrix $\mathbf{\Psi}_k \in \mathbb{R}^{N\times N}$ whose columns are the eigenvectors of $\mathbf{C}_k$ such that

$$\mathbf{\Psi}_k^T \mathbf{C}_k \mathbf{\Psi}_k = \begin{bmatrix} \mathbf{d}_k & \mathbf{O}_{1k} \\ \mathbf{O}_{1k}^T & \mathbf{O}_{2k} \end{bmatrix}. \qquad [4.47]$$

In the above equation, $\mathbf{d}_k \in \mathbb{R}^{r_k\times r_k}$ is a diagonal matrix consisting of only the non-zero eigenvalues of $\mathbf{C}_k$ and $\mathbf{O}_{1k} \in \mathbb{R}^{r_k\times(N-r_k)}$ and $\mathbf{O}_{2k} \in \mathbb{R}^{(N-r_k)\times(N-r_k)}$ are the null matrices. For convenience, partition $\mathbf{\Psi}_k$ as

$$\mathbf{\Psi}_k = [\mathbf{\Psi}_{1k} | \mathbf{\Psi}_{2k}] \qquad [4.48]$$

where the columns of $\mathbf{\Psi}_{1k} \in \mathbb{R}^{N\times r_k}$ are the eigenvectors corresponding to the non-zero block $\mathbf{d}_k$ and the columns of $\mathbf{\Psi}_{2k} \in \mathbb{R}^{N\times(N-r_k)}$ are the eigenvectors corresponding to the rest $(N - r_k)$ number of zero eigenvalues. Defining a rectangular transformation matrix

$$\mathbf{R}_k = \mathbf{\Psi}_{1k} \in \mathbb{R}^{N\times r_k} \qquad [4.49]$$

it is easy to show that

$$\mathbf{R}_k^T \mathbf{C}_k \mathbf{R}_k = \mathbf{d}_k. \qquad [4.50]$$

Therefore, the matrix $\mathbf{R}_k$ in equation [4.49] transforms the originally rank-deficient matrix $\mathbf{C}_k$ into a full-rank (diagonal) matrix of rank r_k. It may be noted that the choice of $\mathbf{R}_k$ can be arbitrary and, in general, $\mathbf{R}_k$ may be expressed as

$$\mathbf{R}_k = \mathbf{\Psi}_k \times \begin{bmatrix} \mathbf{Q}_{r_k} \\ \mathbf{O}_{3k} \end{bmatrix} \tag{4.51}$$

where $\mathbf{Q}_{r_k} \in \mathbb{R}^{r_k \times r_k}$ is *any* orthogonal matrix and $\mathbf{O}_{3k} \in \mathbb{R}^{(N-r_k)\times r_k}$ is a null matrix. The matrix $\mathbf{R}_k$ appearing in equation [4.49] can be obtained as a special case when $\mathbf{Q}_{r_k}$ is an identity (therefore orthogonal) matrix.

Now, define a set of internal variables of reduced dimension $\widetilde{\mathbf{y}}_k(t) \in \mathbb{R}^{r_k}$ using the rectangular transformation matrix $\mathbf{R}_k$ as

$$\mathbf{y}_k(t) = \mathbf{R}_k, \widetilde{\mathbf{y}}_k(t). \tag{4.52}$$

From this equation, it immediately follows that

$$\dot{\mathbf{y}}_k(t) = \mathbf{R}_k \dot{\widetilde{\mathbf{y}}}_k(t) \tag{4.53}$$

where $\mathbf{y}_k(t)$ is defined in equation [4.23]. Using these relationships, equations [4.29] and [4.30] can be expressed as

$$\mathbf{M}\ddot{\mathbf{q}}(t) + \left(\mathbf{C}_0 + \sum_{k=1}^{n} \mathbf{C}_k\right)\dot{\mathbf{q}}(t) - \sum_{k=1}^{n} \frac{1}{\mu_k}\mathbf{C}_k\mathbf{R}_k\dot{\widetilde{\mathbf{y}}}_k(t) = -\mathbf{K}\mathbf{q}(t) + \mathbf{f}(t) \tag{4.54}$$

and $$-\frac{1}{\mu_k}\mathbf{C}_k\dot{\mathbf{q}}(t) + \frac{1}{\mu_k^2}\mathbf{C}_k\mathbf{R}_k\dot{\widetilde{\mathbf{y}}}_k(t) = -\frac{1}{\mu_k}\mathbf{C}_k\mathbf{R}_k\widetilde{\mathbf{y}}_k(t). \tag{4.55}$$

Because equation [4.55] still represents a set of N equations, we premultiply this by $\mathbf{R}_k^T$ to obtain a reduced set of r_k equations

$$-\frac{1}{\mu_k}\mathbf{R}_k^T\mathbf{C}_k\dot{\mathbf{q}}(t) + \frac{1}{\mu_k^2}\mathbf{R}_k^T\mathbf{C}_k\mathbf{R}_k\dot{\widetilde{\mathbf{y}}}_k(t) = -\frac{1}{\mu_k}\mathbf{R}_k^T\mathbf{C}_k\mathbf{R}_k\widetilde{\mathbf{y}}_k(t), \quad \forall k. \tag{4.56}$$

Now equations [4.54] and [4.56] can be combined into a first-order form as

$$\widetilde{\mathbf{B}}\,\dot{\widetilde{\mathbf{u}}}(t) = \widetilde{\mathbf{A}}\,\widetilde{\mathbf{u}}(t) + \widetilde{\mathbf{r}}(t) \tag{4.57}$$

where

$$\widetilde{\mathbf{B}} = \begin{bmatrix} \mathbf{C}_0 + \sum\limits_{k=1}^{n} \mathbf{C}_k & \mathbf{M} & -\mathbf{C}_1\mathbf{R}_1/\mu_1 & \cdots & -\mathbf{C}_n\mathbf{R}_n/\mu_n \\ \mathbf{M} & \mathbf{O}_{N,N} & \mathbf{O}_{N,r_1} & \cdots & \mathbf{O}_{N,r_n} \\ -\mathbf{R}_1^T\mathbf{C}_1/\mu_1 & \mathbf{O}_{N,r_1}^T & \mathbf{R}_1^T\mathbf{C}_1\mathbf{R}_1/\mu_1^2 & \cdots & \mathbf{O}_{r_1,r_n} \\ \vdots & \vdots & \vdots & \ddots & \vdots \\ -\mathbf{R}_1^T\mathbf{C}_n/\mu_n & \mathbf{O}_{N,r_n}^T & \mathbf{O}_{r_1,r_n}^T & \cdots & \mathbf{R}_n^T\mathbf{C}_n\mathbf{R}_n/\mu_n^2 \end{bmatrix} \in \mathbb{R}^{m\times m} \qquad [4.58]$$

$$\widetilde{\mathbf{A}} = \begin{bmatrix} -\mathbf{K} & \mathbf{O}_{N,N} & \mathbf{O}_{N,r_1} & \cdots & \mathbf{O}_{N,r_n} \\ \mathbf{O}_{N,N} & \mathbf{M} & \mathbf{O}_{N,r_1} & \cdots & \mathbf{O}_{N,r_n} \\ \mathbf{O}_{N,r_1}^T & \mathbf{O}_{N,r_1}^T & -\mathbf{R}_1^T\mathbf{C}_1\mathbf{R}_1/\mu_1 & \cdots & \mathbf{O}_{r_1,r_n} \\ \vdots & \vdots & \vdots & \ddots & \vdots \\ \mathbf{O}_{N,r_n}^T & \mathbf{O}_{N,r_n}^T & \mathbf{O}_{r_1,r_n}^T & \cdots & -\mathbf{R}_n^T\mathbf{C}_n\mathbf{R}_n/\mu_n \end{bmatrix} \in \mathbb{R}^{m\times m} \qquad [4.59]$$

$$\widetilde{\mathbf{r}}(t) = \begin{Bmatrix} \mathbf{f}(t) \\ \mathbf{0}_N \\ \mathbf{0}_{r_1} \\ \vdots \\ \mathbf{0}_{r_n} \end{Bmatrix} \in \mathbb{R}^{m} \qquad [4.60]$$

$$\text{and} \quad \widetilde{\mathbf{u}}(t) = \begin{Bmatrix} \mathbf{q}(t) \\ \mathbf{v}(t) \\ \widetilde{\mathbf{y}}_1(t) \\ \vdots \\ \widetilde{\mathbf{y}}_n(t) \end{Bmatrix} \in \mathbb{R}^{m}. \qquad [4.61]$$

In the above equations

$$m = 2N + \sum_{k=1}^{n} r_k \qquad [4.62]$$

is the order of the system, $\mathbf{O}_{ij}$ are $i \times j$ null matrices and $\mathbf{0}_j$ are vectors of j zeros. The term $\widetilde{(\bullet)}$ corresponds to the term $(\bullet)$ defined in equation [4.32] for the full rank case. Again note that the system matrices $\widetilde{\mathbf{A}}$ and $\widetilde{\mathbf{B}}$ are symmetric. When all $\mathbf{C}_k$ matrices are of full rank, that is, when $r_k = N, \forall\ k$, then one can choose each $\mathbf{R}_k$ matrix as the identity matrix and equation [4.57] reduces to equation [4.32]. Using this symmetric state-space representation, the system response can be obtained easily by the mode superposition method in a similar way to that normally used for undamped or viscously damped systems.

4.4. The eigenvalue problem

The free vibration characteristics of the system are governed by the eigenvalue problem associated with the equation of motion [4.1]. Substituting the expression of $\mathbf{G}(s)$ from equation [4.16] into equation [4.3], we have

$$\left[s^2\mathbf{M} + s\left\{\mathbf{C}_0 + \sum_{k=1}^{n} \frac{\mu_k}{s+\mu_k}\mathbf{C}_k\right\} + \mathbf{K}\right]\bar{\mathbf{q}}(s) = \bar{\mathbf{p}}(s). \qquad [4.63]$$

Considering the free vibration from equation [4.63], the eigenvalue problem can be represented by

$$\left[s_j^2\mathbf{M} + s_j\mathbf{C}_0 + s_j\sum_{k=1}^{n} \frac{\mu_k}{s_j+\mu_k}\mathbf{C}_k + \mathbf{K}\right]\mathbf{z}_j = \mathbf{0}, \quad j = 1, 2, \cdots, m. \qquad [4.64]$$

Here, $s_j \in \mathbb{C}$ is the jth eigenvalue and $\mathbf{z}_j \in \mathbb{C}^N$ is the jth eigenvector. The number of eigenvalues, m, depends on the rank of the $\mathbf{C}_k$ matrices and has been derived in the previous section.

For lightly damped systems, among the m eigenvalues, $2N$ appears in complex conjugate pairs and the rest are purely real and negative. The N pairs of complex conjugate eigenvalues can be related to the N (complex) modes of structural vibration. These modes are therefore called *elastic modes*. The rest $(m - 2N)$ purely dissipative modes appear due to non-viscous nature of the damping model and are therefore called *non-viscous modes*. Non-viscous modes, or those similar to these, are known by different names in the literature of different subjects, for example, "wet modes" in the context of ship dynamics [BIS 79] and "damping modes" in the context of viscoelastic structures [MCT 93]. For convenience, we construct and partition the following matrices:

$$\mathbf{S} = [s_1, s_2, \ldots, s_m] \in \mathbb{C}^{m\times m} = \mathrm{diag}\left[\mathbf{S}_e, \mathbf{S}_e^*, \mathbf{S}_{nv}\right] \qquad [4.65]$$

$$\text{and} \quad \mathbf{Z} = [\mathbf{z}_1, \mathbf{z}_2, \ldots, \mathbf{z}_m] \in \mathbb{C}^{n\times m} = \left[\mathbf{Z}_e, \mathbf{Z}_e^*, \mathbf{Z}_{nv}\right]. \qquad [4.66]$$

Here, $(\bullet)^*$ denotes complex conjugation, the subscript $(\bullet)_e$ corresponds to elastic modes and the subscript $(\bullet)_{nv}$ corresponds to non-viscous modes. Next, we analyze the structure of the modal matrices arising from the state-space approach and relate them to the eigenvector of the original system.

4.4.1. *Case A: all coefficient matrices are of full rank*

The eigenvalue problem associated with equation [4.32] can be expressed as

$$\mathbf{A}\boldsymbol{\phi}_j = s_j\mathbf{B}\boldsymbol{\phi}_j \quad j = 1, 2, \ldots, (2+n)N \qquad [4.67]$$

where $\boldsymbol{\phi}_j \in \mathbb{C}^{(2+n)N}$ is the jth eigenvector in the extended state-space. Because **A** and **B** are real matrices, the eigenvalues can only be real or, if complex, they must appear in complex conjugate pairs. We construct the "modal matrix"

$$\boldsymbol{\Phi} = \left[\boldsymbol{\phi}_1, \boldsymbol{\phi}_2, \ldots, \boldsymbol{\phi}_{(2+n)N}\right] \in \mathbb{C}^{(2+n)N \times (2+n)N}. \qquad [4.68]$$

Due to the symmetry of **A** and **B**, it can be shown that the eigenvectors satisfy an orthogonality relationship, that is

$$\boldsymbol{\phi}_j^T \mathbf{B} \boldsymbol{\phi}_k = 0 \quad \text{and} \quad \boldsymbol{\phi}_j^T \mathbf{A} \boldsymbol{\phi}_k = 0, \quad \forall\, j \neq k. \qquad [4.69]$$

We also normalize the eigenvectors such that

$$\boldsymbol{\Phi}^T \mathbf{B} \boldsymbol{\Phi} = \mathbf{I}_{(2+n)N} \qquad [4.70]$$

$$\text{and} \quad \boldsymbol{\Phi}^T \mathbf{A} \boldsymbol{\Phi} = \mathbf{S}. \qquad [4.71]$$

This normalization is equivalent to the mass normalization of the classical modes.

4.4.1.1. *The structure of the modal matrix*

From the definition of $\mathbf{u}(t)$ in equation [4.36], the right-eigenvectors in the extended state-space can be related to the right-eigenvectors in the original space by

$$\boldsymbol{\phi}_j = \begin{Bmatrix} \mathbf{z}_j \\ s_j \mathbf{z}_j \\ \boldsymbol{y}_{1j} \\ \boldsymbol{y}_{2j} \\ \vdots \\ \boldsymbol{y}_{nj} \end{Bmatrix} \qquad [4.72]$$

where $\boldsymbol{y}_{1j}, \boldsymbol{y}_{2j}, \ldots, \boldsymbol{y}_{nj}$ are components of the jth eigenvector corresponding to the internal variables $\mathbf{y}_1(t), \mathbf{y}_2(t), \ldots, \mathbf{y}_n(t)$. The vectors $\boldsymbol{y}_{kj} \in \mathbb{C}^N, \forall\, k = 1, 2, \ldots, n$ can be related to $\mathbf{z}_j$ using equation [4.25]. Taking the Laplace transform of equation [4.25] results

$$s\, \bar{\mathbf{y}}_k + \mu_k\, \bar{\mathbf{y}}_k = s \mu_k \bar{\mathbf{q}} \qquad [4.73]$$

where $\bar{\mathbf{y}}_k$ is the Laplace transform of $\mathbf{y}_k(t)$. For the jth eigenvalue, we obtain

$$s_j\, \boldsymbol{y}_{kj} + \mu_k\, \boldsymbol{y}_{kj} = s_j\, \mu_k\, \mathbf{z}_j \quad \text{or} \quad (s_j + \mu_k)\, \boldsymbol{y}_{kj} = s_j\, \mu_k\, \mathbf{z}_j. \qquad [4.74]$$

Provided $s_j \neq -\mu_k$, from the preceding equation

$$\boldsymbol{y}_{kj} = \frac{s_j\, \mu_k}{s_j + \mu_k}\mathbf{z}_j, \quad \forall\, k = 1, 2, \ldots, n;\ \forall\, j = 1, 2, \ldots, (2+n)N. \qquad [4.75]$$

Using equations [4.72] and [4.75], the eigenvectors in the state-space $\boldsymbol{\phi}_j$ can be related to the eigenvectors in the original space $\mathbf{z}_j$. It is useful to represent this relationship in the matrix form. Define the matrix

$$\boldsymbol{Y}_k = \left[\boldsymbol{y}_{k1}, \boldsymbol{y}_{k2}, \ldots, \boldsymbol{y}_{k(2+n)N}\right] \in \mathbb{C}^{N\times(2+n)N}. \qquad [4.76]$$

For $j = 1, 2, \ldots, (2+n)N$, equation [4.74] can be written in the matrix form as

$$\boldsymbol{Y}_k\mathbf{S} + \mu_k\boldsymbol{Y}_k = \mu_k\mathbf{Z}\mathbf{S}. \qquad [4.77]$$

Dividing this equation by μ_k, we obtain

$$\boldsymbol{Y}_k = \mathbf{Z}\mathbf{S}\left[\mathbf{S}/\mu_k + \mathbf{I}_{(2+n)N}\right]^{-1}. \qquad [4.78]$$

Using this expression, the matrix of eigenvectors in the state-space, given by equation [4.68], can be expressed as

$$\boldsymbol{\Phi} = \begin{bmatrix} \mathbf{Z} \\ \mathbf{Z}\mathbf{S} \\ \mathbf{Z}\mathbf{S}\left[\mathbf{S}/\mu_1 + \mathbf{I}_{(2+n)N}\right]^{-1} \\ \mathbf{Z}\mathbf{S}\left[\mathbf{S}/\mu_2 + \mathbf{I}_{(2+n)N}\right]^{-1} \\ \vdots \\ \mathbf{Z}\mathbf{S}\left[\mathbf{S}/\mu_n + \mathbf{I}_{(2+n)N}\right]^{-1} \end{bmatrix}. \qquad [4.79]$$

In view of the partitions shown in equations [4.65] and [4.66], the preceding equation can be conveniently partitioned as

$$\boldsymbol{\Phi} = \begin{bmatrix} \mathbf{Z}_e & \mathbf{Z}_e^* & \mathbf{Z}_{nv} \\ \mathbf{Z}_e\mathbf{S}_e & \mathbf{Z}_e^*\mathbf{S}_e^* & \mathbf{Z}_{nv}\mathbf{S}_{nv} \\ \mathbf{Z}_e\mathbf{S}_e\left[\mathbf{S}_e/\mu_1 + \mathbf{I}_N\right]^{-1} & \mathbf{Z}_e^*\mathbf{S}_e^*\left[\mathbf{S}_e^*/\mu_1 + \mathbf{I}_N\right]^{-1} & \mathbf{Z}_{nv}\mathbf{S}_{nv}\left[\mathbf{S}_{nv}/\mu_1 + \mathbf{I}_{nN}\right]^{-1} \\ \mathbf{Z}_e\mathbf{S}_e\left[\mathbf{S}_e/\mu_2 + \mathbf{I}_N\right]^{-1} & \mathbf{Z}_e^*\mathbf{S}_e^*\left[\mathbf{S}_e^*/\mu_2 + \mathbf{I}_N\right]^{-1} & \mathbf{Z}_{nv}\mathbf{S}_{nv}\left[\mathbf{S}_{nv}/\mu_2 + \mathbf{I}_{nN}\right]^{-1} \\ \vdots & \vdots & \vdots \\ \mathbf{Z}_e\mathbf{S}_e\left[\mathbf{S}_e/\mu_n + \mathbf{I}_N\right]^{-1} & \mathbf{Z}_e^*\mathbf{S}_e^*\left[\mathbf{S}_e^*/\mu_n + \mathbf{I}_N\right]^{-1} & \mathbf{Z}_{nv}\mathbf{S}_{nv}\left[\mathbf{S}_{nv}/\mu_n + \mathbf{I}_{nN}\right]^{-1} \end{bmatrix}. \qquad [4.80]$$

This equation completely defines the structure of the modal matrix in the state-space. The modal matrix of a viscously damped system consists of only $2N \times 2N$ block in the top left corner of this expression.

4.4.2. *Case B: coefficient matrices are rank deficient*

The eigenvalue problem associated with equation [4.57] can be expressed as

$$\widetilde{\mathbf{A}}\widetilde{\boldsymbol{\phi}}_j = s_j\widetilde{\mathbf{B}}\widetilde{\boldsymbol{\phi}}_j \quad j = 1, 2, \ldots, m \tag{4.81}$$

where $\widetilde{\boldsymbol{\phi}}_j \in \mathbb{C}^m$ is the jth eigenvector and the order of the system m is defined in equation [4.62]. Again we construct the modal matrix

$$\widetilde{\boldsymbol{\Phi}} = \left[\widetilde{\boldsymbol{\phi}}_1, \widetilde{\boldsymbol{\phi}}_2, \ldots, \widetilde{\boldsymbol{\phi}}_m\right] \in \mathbb{C}^{m\times m}. \tag{4.82}$$

This modal matrix also satisfies the orthogonality property defined in equations [4.70] and [4.71].

4.4.2.1. *The structure of the modal matrix*

From the definition of $\widetilde{\mathbf{u}}(t)$ in equation [4.61], the eigenvectors in the extended state-space can be related to the eigenvectors in the original space by

$$\widetilde{\boldsymbol{\phi}}_j = \begin{Bmatrix} \mathbf{z}_j \\ s_j\mathbf{z}_j \\ \widetilde{\boldsymbol{y}}_{1_j} \\ \widetilde{\boldsymbol{y}}_{2_j} \\ \vdots \\ \widetilde{\boldsymbol{y}}_{n_j} \end{Bmatrix} \tag{4.83}$$

where $\widetilde{\boldsymbol{y}}_{1_j} \in \mathbb{C}^{r_1}, \widetilde{\boldsymbol{y}}_{2_j} \in \mathbb{C}^{r_2}, \ldots, \widetilde{\boldsymbol{y}}_{n_j} \in \mathbb{C}^{r_n}$ are components of the jth eigenvector corresponding to the internal variables $\widetilde{\mathbf{y}}_1(t), \widetilde{\mathbf{y}}_2(t), \ldots, \widetilde{\mathbf{y}}_n(t)$. From equation [4.52], we may obtain

$$\boldsymbol{y}_{kj} = \mathbf{R}_k\,\widetilde{\boldsymbol{y}}_{k_j}. \tag{4.84}$$

where $\boldsymbol{y}_{kj}$ is defined in equation [4.75]. To obtain $\widetilde{\boldsymbol{y}}_{k_j}$ from the preceding equation, we need the Moore–Penrose generalized inverse [RAO 71] of the rectangular matrix $\mathbf{R}_k$. Therefore

$$\widetilde{\boldsymbol{y}}_{k_j} = \mathbf{T}_k\,\boldsymbol{y}_{kj} \tag{4.85}$$

$$\mathbf{T}_k = \mathbf{R}_k^\dagger = \left[\mathbf{R}_k^T \mathbf{R}_k\right]^{-1} \mathbf{R}_k^T \in \mathbb{R}^{r_k \times N}, \quad \forall\, k = 1, \cdots, n. \qquad [4.86]$$

Substituting $\widetilde{\boldsymbol{y}}_{k_j}$ in equation [4.83] for $j = 1, 2, \ldots, m$, the matrix of eigenvectors in the extended state-space can be obtained as

$$\widetilde{\mathbf{\Phi}} = \begin{bmatrix} \mathbf{Z}_e & \mathbf{Z}_e^* & \mathbf{Z}_{nv} \\ \mathbf{Z}_e\mathbf{S}_e & \mathbf{Z}_e^*\mathbf{S}_e^* & \mathbf{Z}_{nv}\mathbf{S}_{nv} \\ \mathbf{T}_1\mathbf{Z}_e\mathbf{S}_e\left[\mathbf{S}_e/\mu_1 + \mathbf{I}_N\right]^{-1} & \mathbf{T}_1\mathbf{Z}_e^*\mathbf{S}_e^*\left[\mathbf{S}_e^*/\mu_1 + \mathbf{I}_N\right]^{-1} & \mathbf{T}_1\mathbf{Z}_{nv}\mathbf{S}_{nv}\left[\mathbf{S}_{nv}/\mu_1 + \mathbf{I}_{nN}\right]^{-1} \\ \mathbf{T}_2\mathbf{Z}_e\mathbf{S}_e\left[\mathbf{S}_e/\mu_2 + \mathbf{I}_N\right]^{-1} & \mathbf{T}_2\mathbf{Z}_e^*\mathbf{S}_e^*\left[\mathbf{S}_e^*/\mu_2 + \mathbf{I}_N\right]^{-1} & \mathbf{T}_2\mathbf{Z}_{nv}\mathbf{S}_{nv}\left[\mathbf{S}_{nv}/\mu_2 + \mathbf{I}_{nN}\right]^{-1} \\ \vdots & \vdots & \vdots \\ \mathbf{T}_n\mathbf{Z}_e\mathbf{S}_e\left[\mathbf{S}_e/\mu_n + \mathbf{I}_N\right]^{-1} & \mathbf{T}_n\mathbf{Z}_e^*\mathbf{S}_e^*\left[\mathbf{S}_e^*/\mu_n + \mathbf{I}_N\right]^{-1} & \mathbf{T}_n\mathbf{Z}_{nv}\mathbf{S}_{nv}\left[\mathbf{S}_{nv}/\mu_n + \mathbf{I}_{nN}\right]^{-1} \end{bmatrix}. \qquad [4.87]$$

Note that only the first N rows of this matrix correspond to the eigenvectors of the original system.

4.5. Forced vibration response

The purpose of this section is to obtain the system response using the mode superposition method, commonly used for undamped or proportionally damped systems. One obvious way to obtain the response $\bar{\mathbf{q}}(s)$ is to invert the coefficient matrix (also known as the dynamic stiffness matrix) associated with $\bar{\mathbf{q}}(s)$ in equation [4.4], that is,

$$\bar{\mathbf{q}}(s) = \left[s^2\mathbf{M} + s\left[\mathbf{C}_0 + \mathbf{G}(s)\right] + \mathbf{K}\right]^{-1} \bar{\mathbf{f}}(s). \qquad [4.88]$$

The inverse of the dynamic stiffness matrix is known as the transfer function matrix. The direct inversion of the dynamic stiffness matrix for each frequency point is numerically very expensive for large systems and is normally avoided in engineering practice. Therefore, the focus is on performing the inversion after diagonalizing the system using the orthogonality property of the modal matrix in the extended state-space. It must be noted that although the eigenvectors in the extended state-space ϕ_j satisfy an orthogonality relationship, the eigenvectors of the original system $\mathbf{z}_j$ do not enjoy such simple orthogonality relationship. Further discussions on the general properties of $\mathbf{z}_j$ are taken up in section 5.6. Transfer function matrices of general non-viscously damped systems can be obtained using the mode superposition method. We will discuss this in the next chapter. In this chapter, state-space eigensolutions are used to obtain the system response.

4.5.1. *Frequency domain analysis*

We begin by assuming that all the initial conditions are zero. Taking the Laplace transform of equation [4.32] or [4.57], we obtain

$$s\mathbf{B}\,\bar{\mathbf{u}}(s) = \mathbf{A}\,\bar{\mathbf{u}}(s) + \bar{\mathbf{r}}(s). \qquad [4.89]$$

where $\bar{\mathbf{u}}(s)$ and $\bar{\mathbf{r}}(s)$ are the Laplace transforms of $\mathbf{u}(t)$ and $\mathbf{r}(t)$ or $\widetilde{\mathbf{u}}(t)$ and $\widetilde{\mathbf{r}}(t)$, respectively. We consider that the modal matrix in equation [4.80], or equation [4.94] for the rank-deficient case, can be used to decouple the system of order m by means of the modal transformation

$$\mathbf{u}(t) = \mathbf{\Phi}\,\mathbf{q}(t). \qquad [4.90]$$

Taking the Laplace transform of [4.90], we obtain

$$\bar{\mathbf{u}}(s) = \mathbf{\Phi}\,\bar{\mathbf{q}}(s). \qquad [4.91]$$

Substituting $\bar{\mathbf{u}}(s)$ into equation [4.89], premultiplying by $\mathbf{\Phi}^T$ and using the orthogonality relationship [4.69] yields

$$(s\mathbf{I} - \mathbf{S})\,\bar{\mathbf{q}}(s) = \mathbf{\Phi}^T\bar{\mathbf{r}}(s) \quad \text{or} \quad \bar{\mathbf{q}}(s) = (s\mathbf{I} - \mathbf{S})^{-1}\mathbf{\Phi}^T\bar{\mathbf{r}}(s). \qquad [4.92]$$

From [4.91] and [4.92], we have

$$\bar{\mathbf{u}}(s) = \mathbf{\Phi}\,(s\mathbf{I} - \mathbf{S})^{-1}\mathbf{\Phi}^T\bar{\mathbf{r}}(s) = \sum_{j=1}^{m} \frac{\boldsymbol{\phi}_j\boldsymbol{\phi}_j^T}{s - s_j}\bar{\mathbf{r}}(s). \qquad [4.93]$$

Using the expression of $\boldsymbol{\phi}_j$ in [4.72] and recalling from equation [4.35] that only the first N rows of $\bar{\mathbf{r}}(s)$ are non-zero, we have $\mathbf{\Phi}^T\bar{\mathbf{r}}(s) = \mathbf{z}_j^T\bar{\mathbf{f}}(s)$. Using this, equation [4.93] can be rewritten as

$$\bar{\mathbf{u}}(s) = \sum_{j=1}^{m} \frac{\mathbf{z}_j^T\bar{\mathbf{f}}(s)}{s - s_j}\boldsymbol{\phi}_j. \qquad [4.94]$$

From [4.36], it may be noted that we are interested only in the first N rows of the preceding equation as the aim is to determine $\bar{\mathbf{q}}(s)$. Therefore, taking only the first N rows of equation [4.94] and using [4.72], we obtain

$$\bar{\mathbf{q}}(s) = \sum_{j=1}^{m} \frac{\mathbf{z}_j^T\bar{\mathbf{f}}(s)}{s - s_j}\mathbf{z}_j. \qquad [4.95]$$

In view of the partitions in equations [4.65] and [4.66], the preceding expression can be conveniently expressed as

$$\bar{\mathbf{q}}(s) = \sum_{j=1}^{N} \left[\frac{\mathbf{z}_{e_j}^T \bar{\mathbf{f}}(s)}{s - s_j} \mathbf{z}_{e_j} + \frac{\mathbf{z}_{e_j}^{*T} \bar{\mathbf{f}}(s)}{s - s_j^*} \mathbf{z}_{e_j}^* \right] + \sum_{j=2N+1}^{m} \frac{\mathbf{z}_{nv_j}^T \bar{\mathbf{f}}(s)}{s - s_{nv_j}} \mathbf{z}_{nv_j}. \qquad [4.96]$$

This expression of the dynamic response is in terms of the eigenvalues and eigenvectors of the system in the N space. The last term of equation [4.96] is the contribution of the non-viscous modes to the global dynamic response. This part is not present for viscously damped systems, and consequently the response of a viscously damped system can be obtained only form the first part of the right-hand side of equation [4.96].

In the presence of non-zero initial conditions, one simply replaces the vector $\bar{\mathbf{f}}(s)$ by $\bar{\mathbf{p}}(s)$ defined in equation [4.5]. The complete solution in the Laplace domain is given by

$$\bar{\mathbf{q}}(s) = \sum_{j=1}^{m} \frac{1}{s - s_j} \left(\mathbf{z}_j^T \bar{\mathbf{f}}(s) + \mathbf{z}_j^T \mathbf{M} \dot{\mathbf{q}}_0 + \mathbf{z}_j^T \mathbf{C}_0 \mathbf{q}_0 + s\, \mathbf{z}_j^T \mathbf{M} \mathbf{q}_0 + \sum_{k=1}^{n} \mu_k \frac{\mathbf{z}_j^T \mathbf{C}_k \mathbf{q}_0}{s + \mu_k} \right) \mathbf{z}_j. \qquad [4.97]$$

The results in the frequency domain can be obtained by substituting $s = \mathrm{i}\omega$ in the above expressions.

4.5.2. *Time-domain analysis*

In the time domain, the response can be obtained by taking the inverse Laplace transform of equation [4.97]

$$\begin{aligned}
\mathbf{q}(t) &= \mathcal{L}^{-1}[\bar{\mathbf{q}}(s)] \\
&= \mathcal{L}^{-1} \left[\sum_{j=1}^{m} \frac{1}{s-s_j} \left(\mathbf{z}_j^T \bar{\mathbf{f}}(s) + \sum_{k=1}^{n} \frac{1}{s+\mu_k} \mu_k \left(\mathbf{z}_j^T \mathbf{C}_k \mathbf{q}_0 \right) \right) \right. \\
&\qquad \left. + \frac{\mathbf{z}_j^T \mathbf{M} \dot{\mathbf{q}}_0 + \mathbf{z}_j^T \mathbf{C}_0 \mathbf{q}_0}{s-s_j} + \left(\frac{s}{s-s_j} \right) \mathbf{z}_j^T \mathbf{M} \mathbf{q}_0 \right] \mathbf{z}_j \\
&= \sum_{j=1}^{m} \left(\int_0^t e^{s_j(t-\tau)} \mathbf{z}_j^T \mathbf{f}(\tau)\, \mathrm{d}\tau + \sum_{k=1}^{n} \int_0^t \mu_k e^{s_j t - (s_j + \mu_k)\tau} \mathbf{z}_j^T \mathbf{C}_k \mathbf{q}_0\, \mathrm{d}\tau \right) \mathbf{z}_j \\
&\qquad + e^{s_j t} \left(\mathbf{z}_j^T \mathbf{M} \dot{\mathbf{q}}_0 + \mathbf{z}_j^T \mathbf{C}_0 \mathbf{q}_0 \right) \mathbf{z}_j + \left\{ s_j e^{s_j t} + \delta(t) \right\} \left(\mathbf{z}_j^T \mathbf{M} \mathbf{q}_0 \right) \mathbf{z}_j.
\end{aligned} \qquad [4.98]$$

For $t > 0$, the preceding equation may be rewritten as

$$\mathbf{q}(t) = \sum_{j=1}^{m} \left\{ \int_0^t e^{s_j(t-\tau)} \mathbf{z}_j^T \mathbf{f}(\tau)\,\mathrm{d}\tau + a_j(t) \right\} \mathbf{z}_j \qquad [4.99]$$

where

$$\begin{aligned} a_j(t) &= e^{s_j t}\left(\mathbf{z}_j^T \mathbf{M}\dot{\mathbf{q}}_0 + \mathbf{z}_j^T \mathbf{C}_0 \mathbf{q}_0\right) + s_j e^{s_j t}\left(\mathbf{z}_j^T \mathbf{M}\mathbf{q}_0\right) \\ &\quad + \sum_{k=1}^{n} \mu_k \frac{\left(e^{s_j t} - e^{-\mu_k t}\right)}{s_j + \mu_k} \mathbf{z}_j^T \mathbf{C}_k \mathbf{q}_0. \end{aligned} \qquad [4.100]$$

In the viscous damping limit, when $\mu_k \to \infty, \forall\, k, m = 2N$ and equation [4.97] or [4.98] reduces exactly to the expression of dynamic response of a viscously damped system with an equivalent viscous damping matrix given by [4.18]. It is interesting to note that the expression of $a_j(t)$ is independent of the ranks of the $\mathbf{C}_k$ matrices. The ranks of the $\mathbf{C}_k$ matrices only effect the number of terms (m) to be added in equation [4.99] to obtain $\mathbf{q}(t)$.

4.6. Numerical examples

4.6.1. *Example 1: SDOF system with non-viscous damping*

Consider a simple SDOF system with non-viscous damping. The equation of motion is given by

$$m\,\ddot{u}(t) + f_d + k\,u = f(t) \qquad [4.101]$$

where f_d, the damping force due to the non-viscous element, is assumed to be of the double-exponential form

$$f_d = \int_0^t \left\{ c_1\,\mu_1 e^{-\mu_1(t-\tau)} + c_2\,\mu_2 e^{-\mu_2(t-\tau)} \right\} \dot{u}(\tau)\,\mathrm{d}\tau. \qquad [4.102]$$

This damping model is also known as the double-exponential model. The equation of motion can be transformed into the first-order form as

$$\begin{bmatrix} c_1+c_2 & m & -c_1/\mu_1 & -c_2/\mu_2 \\ m & 0 & 0 & 0 \\ -c_1/\mu_1 & 0 & c_1/\mu_1^2 & 0 \\ -c_2/\mu_2 & 0 & 0 & c_1/\mu_1^2 \end{bmatrix} \begin{Bmatrix} \dot{u}(t) \\ \ddot{u}(t) \\ \dot{y}_1(t) \\ \dot{y}_2(t) \end{Bmatrix} = \begin{bmatrix} -k & 0 & 0 & 0 \\ 0 & m & 0 & 0 \\ 0 & 0 & -c_1/\mu_1 & 0 \\ 0 & 0 & 0 & -c_1/\mu_1 \end{bmatrix} \begin{Bmatrix} u(t) \\ \dot{u}(t) \\ y_1(t) \\ y_2(t) \end{Bmatrix} + \begin{Bmatrix} f(t) \\ 0 \\ 0 \\ 0 \end{Bmatrix}. \qquad [4.103]$$

The eigenvalue problem associated with the preceding equation can be easily solved. Numerical values of the parameters appearing in equations [4.101] and [4.102] are assumed to be $m = 3$ kg, $k = 20$ N/m, $c_1 = 2.5$ Ns/m, $c_2 = 1.5$ Ns/m, $\mu_1 = 0.2$ s^{-1} and $\mu_2 = 15.0$ s^{-1}. For these parameter values, the four eigenvalues of the SDOF system are obtained as

$$\begin{aligned} s_1 &= -0.4297 + 2.6580\text{i}, \quad s_2 = -0.4297 - 2.6580\text{i}, \\ s_3 &= -0.1950, \quad s_4 = -14.1455. \end{aligned} \qquad [4.104]$$

Note that s_1 and s_2 appear in a complex conjugate pair. The eigenvalues of the corresponding undamped system are given by $\omega_{1,2} = \pm 2.5820\text{i}$. The eigenvalues s_3 and s_4 are purely real, negative and they appear due to non-viscous damping mechanism.

4.6.2. *Example 2: a rank-deficient system*

We consider the rank-deficient system introduced in Figure 4.4 with the double-exponential. For the numerical calculation, we have assumed $m_u = 3.0$ kg, $k_u = 2.0$ N/m, $c_1 = 0.6$ Ns/m, $c_2 = 0.2$ Ns/m, $\mu_1 = 1.0$ s^{-1} and $\mu_2 = 5.0$ s^{-1}. The order of the system matrices in the state-space, m, expressed by equation [4.62], can be obtained using r_1 and r_2 given by equations [4.44] and [4.45]. From equation [4.62], it can be easily shown that $m = 2 \times 3 + (2 + 1) = 9$. The transformation matrices

$\mathbf{R}_k, k = 1, 2$ appearing in equations [4.58] and [4.59] can be obtained from equation [4.49]

$$\mathbf{R}_1 = \begin{bmatrix} 1 & 0 \\ 0 & 1 \\ 0 & 0 \end{bmatrix} \quad \text{and} \quad \mathbf{R}_2 = \begin{bmatrix} 0 \\ -1/\sqrt{2} \\ 1/\sqrt{2} \end{bmatrix}. \qquad [4.105]$$

Using these, the system matrices in the state-space, given by equations [4.58] and [4.59], are obtained as

$$\widetilde{\mathbf{B}} = \begin{bmatrix} c_1 & 0 & 0 & m_u & 0 & 0 & 0 & -c_1/\mu_1 & 0 \\ 0 & c_1+c_2 & -c2 & 0 & m_u & 0 & -c_1/\mu_1 & 0 & c_2\sqrt{2}/\mu_2 \\ 0 & -c_2 & c2 & 0 & 0 & m_u & 0 & 0 & -c_2\sqrt{2}/\mu_2 \\ m_u & 0 & 0 & 0 & 0 & 0 & 0 & 0 & 0 \\ 0 & m_u & 0 & 0 & 0 & 0 & 0 & 0 & 0 \\ 0 & 0 & m_u & 0 & 0 & 0 & 0 & 0 & 0 \\ 0 & -c_1/\mu_1 & 0 & 0 & 0 & 0 & c_1/\mu_1^2 & 0 & 0 \\ -c_1/\mu_1 & 0 & 0 & 0 & 0 & 0 & 0 & c_1/\mu_1^2 & 0 \\ 0 & c_2\sqrt{2}/\mu_2 & -c_2\sqrt{2}/\mu_2 & 0 & 0 & 0 & 0 & 0 & 2c_2/\mu_2^2 \end{bmatrix} \qquad [4.106]$$

and

$$\widetilde{\mathbf{A}} = \begin{bmatrix} -2k_u & k_u & 0 & 0 & 0 & 0 & 0 & 0 & 0 \\ k_u & -2k_u & k_u & 0 & 0 & 0 & 0 & 0 & 0 \\ 0 & k_u & -2k_u & 0 & 0 & 0 & 0 & 0 & 0 \\ 0 & 0 & 0 & m_u & 0 & 0 & 0 & 0 & 0 \\ 0 & 0 & 0 & 0 & m_u & 0 & 0 & 0 & 0 \\ 0 & 0 & 0 & 0 & 0 & m_u & 0 & 0 & 0 \\ 0 & 0 & 0 & 0 & 0 & 0 & -c_1/\mu_1 & 0 & 0 \\ 0 & 0 & 0 & 0 & 0 & 0 & 0 & -c_1/\mu_1 & 0 \\ 0 & 0 & 0 & 0 & 0 & 0 & 0 & 0 & -2c_2/\mu_2 \end{bmatrix}. \qquad [4.107]$$

The diagonal matrix containing the eigenvalues may be expressed in terms of three diagonal block matrices as

$$\mathbf{S} = \text{diag}\left[\boldsymbol{\lambda}_e, \boldsymbol{\lambda}_e^*, \boldsymbol{\lambda}_{nv}\right] \qquad [4.108]$$

where the subscript $(\bullet)_e$ corresponds to elastic modes and the subscript $(\bullet)_{nv}$ corresponds to non-viscous modes. Similarly, the modal matrix $\boldsymbol{\Phi}$ can also be expressed as

$$\boldsymbol{\Phi} = \left[\boldsymbol{\Phi}_e, \boldsymbol{\Phi}_e^*, \boldsymbol{\Phi}_{nv}\right]. \qquad [4.109]$$

For this example, there are three elastic modes appearing in complex conjugate pairs and, coincidentally, the number of non-viscous modes is also 3, two corresponding to μ_1 and one corresponding to μ_2. The eigenvalues and the eigenvectors corresponding to the elastic modes are

$$\boldsymbol{\lambda}_e = \text{diag}\left[-0.0559 + 0.6628\text{i}, -0.0402 + 1.1838\text{i}, -0.0680 + 1.5569\text{i}\right] \qquad [4.110]$$

and

$$\boldsymbol{\Phi}_e = \begin{bmatrix} -0.1659 + 0.1946\text{i} & -0.2027 + 0.1848\text{i} & -0.1050 + 0.1318\text{i} \\ -0.2451 + 0.2726\text{i} & -0.0173 + 0.0121\text{i} & 0.1716 - 0.1679\text{i} \\ -0.2018 + 0.1839\text{i} & 0.1726 - 0.2007\text{i} & -0.1160 + 0.1066\text{i} \\ -0.1197 - 0.1208\text{i} & -0.2106 - 0.2474\text{i} & -0.1981 - 0.1724\text{i} \\ -0.1669 - 0.1777\text{i} & -0.0136 - 0.0210\text{i} & 0.2498 + 0.2786\text{i} \\ -0.1106 - 0.1441\text{i} & 0.2307 + 0.2124\text{i} & -0.1581 - 0.1878\text{i} \\ -0.1451 - 0.0261\text{i} & -0.2131 + 0.0051\text{i} & -0.1376 + 0.0449\text{i} \\ -0.2070 - 0.0429\text{i} & -0.0163 - 0.0017\text{i} & 0.2024 - 0.0392\text{i} \\ 0.0427 + 0.0183\text{i} & 0.2023 + 0.1181\text{i} & -0.3618 - 0.2201\text{i} \end{bmatrix}. \qquad [4.111]$$

The eigenvalues and the eigenvectors corresponding to the non-viscous modes are

$$\boldsymbol{\lambda}_{nv} = \text{diag}\left[-0.8649, -0.9324, -4.8744\right] \qquad [4.112]$$

and

$$\boldsymbol{\Phi}_{nv} = \begin{bmatrix} -0.1220 & 0.0684 & -0.0008 \\ -0.1467 & -0.0572 & -0.0286 \\ -0.0435 & -0.0159 & 0.0289 \\ 0.1055 & -0.0638 & 0.0037 \\ 0.1268 & 0.0533 & 0.1395 \\ 0.0376 & 0.0148 & -0.1408 \\ 0.7807 & -0.9443 & -0.0009 \\ 0.9386 & 0.7894 & -0.0360 \\ -0.0763 & -0.0335 & -7.8912 \end{bmatrix}. \qquad [4.113]$$

The eigenvectors in [4.111] and [4.113] are normalized according to equation [4.70].

Now attention is given to the dynamic response of the system. It is assumed that the system is subjected to a unit initial displacement at the first DOF, that is, $\bar{\mathbf{f}}(\text{i}\omega) = \mathbf{0}$, $\mathbf{q}_0 = \{1, 0, 0\}^T$ and $\dot{\mathbf{q}}_0 = \mathbf{0}$. First, the response in the frequency domain is considered. The responses of the three masses are obtained using the mode superposition approach from equation [4.97] and shown in Figure 4.5. The peaks in this plot corresponding to the elastic modes $\boldsymbol{\lambda}_e$ are given in equation [4.110]. In the time domain, responses of the three masses are obtained using the mode superposition approach from equation [4.98] and shown in Figure 4.6.

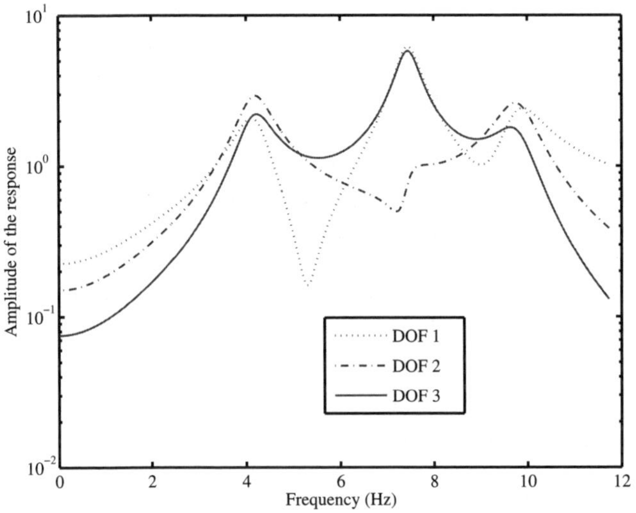

Figure 4.5. *Absolute value of the frequency domain response of the three masses due to unit initial displacement at the first DOF*

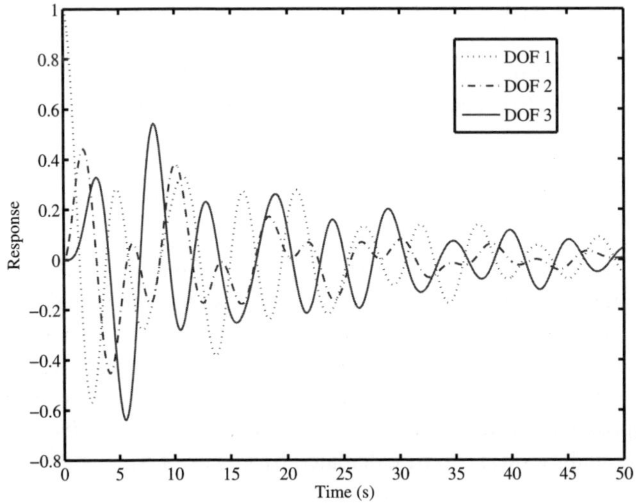

Figure 4.6. *Time-domain response of the three masses due to unit initial displacement at the first DOF*

4.7. Direct time-domain approach

Different methods for the analysis of non-viscously damped systems have been discussed in section 1.3. Earlier in this chapter, we have shown a state-space approach for the analysis of linear systems with exponential damping. This method is based on representing equation [4.1] in terms of two symmetric matrices in an augmented state-space. The dynamic response of the system was obtained by mode superposition method using the state-space eigensolutions. Although the method gives exact results, it requires significant computation because the size of the eigenvalue problem in the extended state-space is usually very large. In this section, an alternative approach based on direct integration in the time domain has been proposed. This approach uses the state-space representation proposed earlier in section 4.3. Again, the two cases, namely, (1) when all the damping coefficient matrices are of full rank and (2) when the damping coefficient matrices are rank deficient, have been presented. The proposed method is illustrated by numerical examples.

4.7.1. *Integration in the time domain*

First, consider the full-rank case. The linear approximations

$$\mathbf{q}(t) = \mathbf{q}_j \left(1 - \frac{t}{h}\right) + \mathbf{q}_{j+1}\frac{t}{h} \qquad [4.114]$$

and $$\mathbf{v}(t) = \mathbf{v}_j \left(1 - \frac{t}{h}\right) + \mathbf{v}_{j+1}\frac{t}{h} \qquad [4.115]$$

for the displacements $\mathbf{q}(t)$ and the velocities $\mathbf{v}(t)$ are used. In addition, the internal variables $\mathbf{y}_k(t)$ have also been interpolated in a linear manner

$$\mathbf{y}_k(t) = \mathbf{y}_{k_j} \left(1 - \frac{t}{h}\right) + \mathbf{y}_{k_{j+1}}\frac{t}{h}, \quad k = 1, 2, \ldots, n_d. \qquad [4.116]$$

Here, the step size $h = T/N_d$, where T is the time up to which the response calculation is required, and N_d is the number of divisions in the time axis. To simplify the formulas, only constant time-steps across the whole time axis are used. The next and the last step concerns the integration of the system [4.32] with respect to the time interval with $t/h \in [j, j+1]$. This integration, exact where it is possible and by using the linear approximation where it is necessary, leads to a totally discretized one-step formulation that allows $\mathbf{u}_{j+1}$ to be calculated by means of $\mathbf{u}_j$

$$\mathbf{B}\left(\mathbf{u}_{j+1} - \mathbf{u}_j\right) = \frac{h}{2}\mathbf{A}\left(\mathbf{u}_{j+1} + \mathbf{u}_j\right) + \mathbf{i}_r, \quad bi_r = \int_{j\,h}^{(j+1)\,h} \mathbf{r}(t)\, dt \qquad [4.117]$$

or

$$\left[\mathbf{B} - \frac{h}{2}\mathbf{A}\right]\mathbf{u}_{j+1} = \left[\mathbf{B} + \frac{h}{2}\mathbf{A}\right]\mathbf{u}_j + bi_r \qquad [4.118]$$

$$\text{where} \quad \mathbf{u}_0^T = \left[\ \mathbf{q}_0^T, \mathbf{v}_0^T,\ \mathbf{0}^T,\ \cdots,\ \mathbf{0}^T\ \right]. \qquad [4.119]$$

The initial values for the internal variables follow from their definition in equation [4.23]. Equation [4.118] can now be used to compute $\mathbf{u}(t)$ at discrete points. The result in equation [4.118] is also identical with the transition process based on a $\mathbf{P}_{11}$ Padé-expansion of an exponential representation proposed by Ruge and Wagner [RUG 00]. A result similar to equation [4.118] can also be obtained for the rank-deficient case.

4.7.2. *Numerical realization*

From equation [4.118], note that factorization (e.g. LU factorization) of the matrix

$$\mathbf{F} = \mathbf{B} - \frac{h}{2}\mathbf{A} \qquad [4.120]$$

is required to compute each $\mathbf{u}_j$. The order of the matrix $\mathbf{F}$ is often too large to do efficient numerical computations. For example, for a three DOF system with four exponential terms, factorization of a 18×18 matrix may be required at each time step. Therefore, although the numerical method outlined in the previous section is very simple, it is required to reduce the computational effort in order to put the method into practice. Following Ruge [RUG 01], the most important properties required for an effective numerical scheme are as follows:

– The coefficient matrix $\mathbf{F}$ of the time-stepping scheme $\mathbf{Fx} = \mathbf{p}$ to finally be solved should contain the original system matrices $\mathbf{M}, \mathbf{C}_k, \mathbf{K}$ in their original form without any mass scaling form $\mathbf{M}^{-1}\mathbf{K}$ or products like $\mathbf{MK}$.

– Any special properties of $\mathbf{M}, \mathbf{C}_k, \mathbf{K}$ like sparsity or symmetry should be preserved when generating $\mathbf{F}$.

– The column $\mathbf{x}$ of unknowns to finally be solved should contain nothing but the N nodal quantities $\mathbf{q}$. Thus, $\mathbf{F}$ should be of order $N \times N$.

An essential advantage of the $\mathbf{P}_{11}$-scheme used here is the possibility of satisfying these requirements. In the following section, we outline how one can achieve them. It may be noted that in some problems the use of variable time-steps may be desired. Variable time-stepping algorithms have not been considered in this section, and our discussions are confined to fixed time-stepping algorithm only.

4.7.2.1. *Case A: all $\mathbf{C}_k$ matrices are of full rank*

The second row in [4.118] is used to eliminate the velocities

$$\mathbf{v}_{j+1} = \frac{2}{h}\left[\mathbf{q}_{j+1} - \mathbf{q}_j\right] - \mathbf{v}_j. \qquad [4.121]$$

All succeeding rows in [4.118] are used to express the internal variables in terms of the displacements as

$$\frac{\mathbf{C}_k}{\mu_k}\mathbf{y}_{k,j+1} = \frac{2\mu_k}{2+\mu_k h}\left[\frac{\mathbf{C}_k}{\mu_k}\left(\mathbf{q}_{j+1} - \mathbf{q}_j\right) + \frac{\mathbf{C}_k}{2\mu_k^2}\left(2-\mu_k h\right)\mathbf{y}_{k,j}\right]. \qquad [4.122]$$

Substituting [4.121] and [4.122] into the first row of equation [4.118] leads to a totally discretized one-step formulation of order N as

$$\left[\frac{2}{h}\mathbf{M} + \sum_{k=1}^{n}\frac{2\mu_k}{2+h\mu_k}\mathbf{C}_k + \frac{h}{2}\mathbf{K}\right]\mathbf{q}_{j+1}$$

$$= \left[\frac{2}{h}\mathbf{M} + \sum_{k=1}^{n}\frac{2\mu_k}{2+h\mu_k}\mathbf{C}_k - \frac{h}{2}\mathbf{K}\right]\mathbf{q}_j + 2\mathbf{M}\mathbf{v}_j - 2h\sum_{k=1}^{n}\frac{\mathbf{C}_k}{2+\mu_k h}\mathbf{y}_{k,j} + \int_{jh}^{(j+1)h}\mathbf{f}(t)dt.$$

$$[4.123]$$

Simplifying equation [4.122], we also have

$$\frac{2+\mu_k h}{2\mu_k^2}\mathbf{C}_k\mathbf{y}_{k,j+1} = \frac{2-\mu_k h}{2\mu_k^2}\mathbf{C}_k\mathbf{y}_{k,j} + \frac{1}{\mu_k}\mathbf{C}_k\left[\mathbf{q}_{j+1} - \mathbf{q}_j\right] \qquad [4.124]$$

or

$$\mathbf{y}_{k,j+1} = \frac{2-\mu_k h}{2+\mu_k h}\mathbf{y}_{k,j} + \frac{2\mu_k}{2+\mu_k h}\left[\mathbf{q}_{j+1} - \mathbf{q}_j\right]. \qquad [4.125]$$

Equation [4.123] together with equations [4.121] and [4.125] allow, $\mathbf{q}_{j+1}$ to be calculated by means of $\mathbf{q}_j$, $\mathbf{y}_{k,j}$ $\forall k = 1, \ldots, n$. Note that this scheme meets all the requirements for an effective numerical scheme just mentioned.

4.7.2.2. *Case B:* $\mathbf{C}_k$ *matrices are rank deficient*

We rewrite equation [4.118] for the rank-deficient case as

$$\left[\widetilde{\mathbf{B}} - \frac{h}{2}\widetilde{\mathbf{A}}\right]\widetilde{\mathbf{u}}_{j+1} = \left[\widetilde{\mathbf{B}} + \frac{h}{2}\widetilde{\mathbf{A}}\right]\widetilde{\mathbf{u}}_j + \widetilde{\mathbf{i}}_r, \quad [4.126]$$

$$\text{where} \quad \widetilde{\mathbf{u}}_0^T = \left[\ \mathbf{q}_0^T, \mathbf{v}_0^T,\ \mathbf{0}_{r_1}^T,\ \cdots,\ \mathbf{0}_{r_n}^T\ \right] \quad [4.127]$$

$$\text{and} \quad \widetilde{\mathbf{i}}_r = \int_{j\,h}^{(j+1)\,h} \widetilde{\mathbf{r}}(t)\,\mathrm{d}t. \quad [4.128]$$

From the second row in [4.126], the velocities can be eliminated by using [4.121]. All succeeding rows in [4.126] are used to express the internal variables in terms of the displacements

$$\frac{\mathbf{C}_k\mathbf{R}_k}{\mu_k}\widetilde{\mathbf{y}}_{k,j+1} = \frac{2\mu_k}{2+\mu_k h}\left[\frac{\mathbf{C}_k}{\mu_k}\left(\mathbf{q}_{j+1} - \mathbf{q}_j\right) + \frac{\mathbf{C}_k\mathbf{R}_k}{2\mu_k^2}\left(2-\mu_k h\right)\widetilde{\mathbf{y}}_{k,j}\right]. \quad [4.129]$$

Substituting [4.121] and [4.129] into the first row of [4.126] again leads to a totally discretized one-step formulation of order N, which allows $\mathbf{q}_{j+1}$ to be calculated by means of $\mathbf{q}_j$, $\mathbf{y}_{k,j}$ $\forall k = 1, 2, \ldots, n_d$ as

$$\left[\frac{2}{h}\mathbf{M} + \sum_{k=1}^{n}\frac{2\mu_k}{2+h\mu_k}\mathbf{C}_k + \frac{h}{2}\mathbf{K}\right]\mathbf{q}_{j+1} = \left[\frac{2}{h}\mathbf{M} + \sum_{k=1}^{n}\frac{2\mu_k}{2+h\mu_k}\mathbf{C}_k - \frac{h}{2}\mathbf{K}\right]\mathbf{q}_j$$
$$+2\mathbf{M}\mathbf{v}_j - 2h\sum_{k=1}^{n}\frac{\mathbf{C}_k\mathbf{R}_k}{2+\mu_k h}\widetilde{\mathbf{y}}_{k,j} + \int_{jh}^{(j+1)h}\mathbf{f}(t)dt \quad [4.130]$$

with

$$\frac{2+\mu_k h}{2\mu_k^2}\mathbf{R}_k^T\mathbf{C}_k\mathbf{R}_k\,\widetilde{\mathbf{y}}_{k,j+1} = \frac{2-\mu_k h}{2\mu_k^2}\mathbf{R}_k^T\mathbf{C}_k\mathbf{R}_k\widetilde{\mathbf{y}}_{k,j} + \frac{1}{\mu_k}\mathbf{R}_k^T\mathbf{C}_k\left[\mathbf{q}_{j+1} - \mathbf{q}_j\right] \quad [4.131]$$

$$\text{or} \quad \left[\mathbf{R}_k^T\mathbf{C}_k\mathbf{R}_k\right]\widetilde{\mathbf{y}}_{k,j+1} = \frac{2-\mu_k h}{2+\mu_k h}\left[\mathbf{R}_k^T\mathbf{C}_k\mathbf{R}_k\right]\widetilde{\mathbf{y}}_{k,j} + \frac{2\mu_k}{2+\mu_k h}\mathbf{R}_k^T\mathbf{C}_k\left[\mathbf{q}_{j+1} - \mathbf{q}_j\right] \quad [4.132]$$

$$\text{and} \quad \mathbf{v}_{j+1} = \frac{2}{h}\left[\mathbf{q}_{j+1} - \mathbf{q}_j\right] - \mathbf{v}_j. \quad [4.133]$$

It should be noted that at the most n additional factorizations regarding the internal variables $\mathbf{R}_k^T\mathbf{C}_k\mathbf{R}_k$ need to be computed when $r_k = \text{rank}(\mathbf{C}_k) < N,\ \forall k = 1,\dots,n$. Nevertheless, the order of those matrices is always less than N.

Recall that the computation of each $\mathbf{R}_k$ matrix requires the solution of an eigenvalue problem involving the matrix $\mathbf{C}_k$. This is a significant computation for systems with many exponential terms. Now it will be shown that these additional decompositions can be avoided. We start with the first step $t \in [0, h]$

$$\left[\frac{2}{h}\mathbf{M} + \sum_{k=1}^{n}\frac{2\mu_k}{2+h\mu_k}\mathbf{C}_k + \frac{h}{2}\mathbf{K}\right]\mathbf{q}_1 = \left[\frac{2}{h}\mathbf{M} + \sum_{k=1}^{n}\frac{2\mu_k}{2+h\mu_k}\mathbf{C}_k - \frac{h}{2}\mathbf{K}\right]\mathbf{q}_0 + 2\mathbf{M}\mathbf{v}_0 - 2h\sum_{k=1}^{n}\frac{\mathbf{C}_k\mathbf{R}_k}{2+\mu_k h}\widetilde{\mathbf{y}}_{k,0} + \int_0^h \mathbf{f}(t)dt \qquad [4.134]$$

$$\frac{2+\mu_k h}{2\mu_k^2}\mathbf{R}_k^T\mathbf{C}_k\mathbf{R}_k\widetilde{\mathbf{y}}_{k,1} = \frac{2-\mu_k h}{2\mu_k^2}\mathbf{R}_k^T\mathbf{C}_k\mathbf{R}_k\widetilde{\mathbf{y}}_{k,0} + \frac{1}{\mu_k}\mathbf{R}_k^T\mathbf{C}_k\left[\mathbf{q}_1 - \mathbf{q}_0\right]$$

$$\text{alternatively}\quad \mathbf{C}_k\mathbf{R}_k\widetilde{\mathbf{y}}_{k,1} = \frac{2-\mu_k h}{2+\mu_k h}\mathbf{C}_k\mathbf{R}_k\widetilde{\mathbf{y}}_{k,0} + \frac{2\mu_k}{2+\mu_k h}\mathbf{C}_k\left[\mathbf{q}_1 - \mathbf{q}_0\right] \qquad [4.135]$$

$$\text{and}\quad \mathbf{v}_1 = \frac{2}{h}\left[\mathbf{q}_1 - \mathbf{q}_0\right] - \mathbf{v}_0. \qquad [4.136]$$

Since $\mathbf{y}_{k,0} = \mathbf{0}\quad \forall\, k$, equations [4.134] and [4.135] are simplified to

$$\left[\frac{2}{h}\mathbf{M} + \sum_{k=1}^{n}\frac{2\mu_k}{2+h\mu_k}\mathbf{C}_k + \frac{h}{2}\mathbf{K}\right]\mathbf{q}_1 = \left[\frac{2}{h}\mathbf{M} + \sum_{k=1}^{n}\frac{2\mu_k}{2+h\mu_k}\mathbf{C}_k - \frac{h}{2}\mathbf{K}\right]\mathbf{q}_0 + 2\mathbf{M}\mathbf{v}_0 + \int_0^h \mathbf{f}(t)dt \qquad [4.137]$$

and

$$\mathbf{C}_k\mathbf{R}_k\widetilde{\mathbf{y}}_{k,1} = \frac{2\mu_k}{2+\mu_k h}\mathbf{C}_k\left[\mathbf{q}_1 - \mathbf{q}_0\right] =: \mathbf{s}_{k,1} \qquad [4.138]$$

where auxiliary quantities $\mathbf{s}_k$ are introduced conveniently. All succeeding steps $t \in [jh, (j+1)h]$, $j \geq 1$:

$$\left[\frac{2}{h}\mathbf{M}+\sum_{k=1}^{n}\frac{2\mu_k}{2+h\mu_k}\mathbf{C}_k+\frac{h}{2}\mathbf{K}\right]\mathbf{q}_{j+1} = \left[\frac{2}{h}\mathbf{M}+\sum_{k=1}^{n}\frac{2\mu_k}{2+h\mu_k}\mathbf{C}_k-\frac{h}{2}\mathbf{K}\right]\mathbf{q}_j +2\mathbf{M}\mathbf{v}_j - 2h\sum_{k=1}^{n}\frac{\mathbf{s}_{k,j}}{2+\mu_k h}+\int_{jh}^{(j+1)h}\mathbf{f}(t)dt \quad [4.139]$$

$$\mathbf{s}_{k,j+1} = \frac{2-\mu_k h}{2+\mu_k h}\mathbf{s}_{k,j}+\frac{2\mu_k}{2+\mu_k h}\mathbf{C}_k\left[\mathbf{q}_{j+1}-\mathbf{q}_j\right] \quad [4.140]$$

and $$\mathbf{v}_{j+1} = \frac{2}{h}\left[\mathbf{q}_{j+1}-\mathbf{q}_j\right]-\mathbf{v}_j. \quad [4.141]$$

From this analysis, it can be seen that it is not necessary to compute the matrices $\mathbf{R}_k$ unless one is interested in $\widetilde{\mathbf{y}}_k$ explicitly. If, and only if, that is the case, then additional decompositions $\mathbf{R}_k^T\mathbf{C}_k\mathbf{R}_k$ must be computed and $\widetilde{\mathbf{y}}_k$ should be obtained from [4.131]. The full-rank scenario follows directly from [4.130] and [4.131] by substituting $\mathbf{R}_k = \mathbf{I}$. Finally, because the modulus of the eigenvalues of the associated amplification matrix is 1, the linear interpolation scheme proposed here is unconditionally stable.

4.7.3. *Summary of the method*

Following the procedure outlined in the previous sections, the time-domain response of an exponentially damped linear system can be obtained in an efficient manner. Here, we briefly summarize the steps to be followed:

1) Select a sufficiently small step size $h = T/N_d$ and construct the symmetric matrices

$$\mathbf{F}_1 = \left[\frac{2}{h}\mathbf{M}+\sum_{k=1}^{n}\frac{2\mu_k}{2+h\mu_k}\mathbf{C}_k+\frac{h}{2}\mathbf{K}\right] \quad [4.142]$$

and $$\mathbf{F}_2 = \left[\frac{2}{h}\mathbf{M}+\sum_{k=1}^{n}\frac{2\mu_k}{2+h\mu_k}\mathbf{C}_k-\frac{h}{2}\mathbf{K}\right] = \mathbf{F}_1 - h\mathbf{K}. \quad [4.143]$$

2) Rank deficient [yes/no]: if yes step 3, else step 6.

3) Explicit computation of $\widetilde{\mathbf{y}}_k$ [yes/no]: if yes step 4, else step 5.

4) Explicit computation of $\widetilde{\mathbf{y}}_k$:

i) Solve the eigenvalue problem $\mathbf{C}_k\mathbf{U}_k = \mathbf{U}_k \begin{bmatrix} \mathbf{d}_k & \mathbf{O}_{1k} \\ \mathbf{O}_{1k}^T & \mathbf{O}_{2k} \end{bmatrix}$ for all k. Normalize the eigenvector matrix $\mathbf{U}_k$ so that $\mathbf{U}_k^T\mathbf{U}_k = \mathbf{I}$ and partition $\mathbf{U}_k = [\mathbf{U}_{1k}|\mathbf{U}_{2k}]$. Select $\mathbf{R}_k = \mathbf{U}_{1k} \in \mathbb{R}^{N\times r_k}$.

ii) For all j, solve the displacements from

$$\mathbf{F}_1\mathbf{q}_{j+1} = \mathbf{F}_2\mathbf{q}_j + 2\mathbf{M}\mathbf{v}_j - 2h\sum_{k=1}^{n}\frac{\mathbf{C}_k\mathbf{R}_k}{2+\mu_k h}\widetilde{\mathbf{y}}_{k,j} + \int_{jh}^{(j+1)h}\mathbf{f}(t)dt. \qquad [4.144]$$

iii) Velocities from

$$\mathbf{v}_{j+1} = \frac{2}{h}\left[\mathbf{q}_{j+1} - \mathbf{q}_j\right] - \mathbf{v}_j. \qquad [4.145]$$

iv) Solve for the internal variables

$$\left[\mathbf{R}_k^T\mathbf{C}_k\mathbf{R}_k\right]\widetilde{\mathbf{y}}_{k,j+1} = \frac{2-\mu_k h}{2+\mu_k h}\left[\mathbf{R}_k^T\mathbf{C}_k\mathbf{R}_k\right]\widetilde{\mathbf{y}}_{k,j} + \frac{2\mu_k}{2+\mu_k h}\mathbf{R}_k^T\mathbf{C}_k\left[\mathbf{q}_{j+1} - \mathbf{q}_j\right]. \qquad [4.146]$$

5) Explicit computation of $\widetilde{\mathbf{y}}_k$ is omitted. For all j:

i) Solve for displacements

$$\mathbf{F}_1\mathbf{q}_{j+1} = \mathbf{F}_2\mathbf{q}_j + 2\mathbf{M}\mathbf{v}_j - 2h\sum_{k=1}^{n}\frac{\mathbf{s}_{k,j}}{2+\mu_k h} + \int_{jh}^{(j+1)h}\mathbf{f}(t)dt. \qquad [4.147]$$

ii) Velocities

$$\mathbf{v}_{j+1} = \frac{2}{h}\left[\mathbf{q}_{j+1} - \mathbf{q}_j\right] - \mathbf{v}_j. \qquad [4.148]$$

iii) Auxiliary variables

$$\mathbf{s}_{k,j+1} = \frac{2-\mu_k h}{2+\mu_k h}\mathbf{s}_{k,j} + \frac{2\mu_k}{2+\mu_k h}\mathbf{C}_k\left[\mathbf{q}_{j+1} - \mathbf{q}_j\right] \qquad [4.149]$$

where $\mathbf{s}_{k,1}$ are defined in equation [4.138].

6) Full-rank scenario: for all j

i) Solve for displacements

$$\mathbf{F}_1\mathbf{q}_{j+1} = \mathbf{F}_2\mathbf{q}_j + 2\mathbf{M}\mathbf{v}_j - 2h\sum_{k=1}^{n}\frac{\mathbf{C}_k\mathbf{y}_{k,j}}{2+\mu_k h} + \int\limits_{jh}^{(j+1)h}\mathbf{f}(t)dt. \qquad [4.150]$$

ii) Velocities

$$\mathbf{v}_{j+1} = \frac{2}{h}\left[\mathbf{q}_{j+1} - \mathbf{q}_j\right] - \mathbf{v}_j. \qquad [4.151]$$

iii) Solve for the internal variables

$$\mathbf{y}_{k,j+1} = \frac{2-\mu_k h}{2+\mu_k h}\mathbf{y}_{k,j} + \frac{2\mu_k}{2+\mu_k h}\left[\mathbf{q}_{j+1} - \mathbf{q}_j\right]. \qquad [4.152]$$

Numerical efficiency is mainly governed by the special properties of $\mathbf{F}_1 \in \mathbb{R}^{N\times N}$. If $\mathbf{F}_1$ is symmetric positive definite, a Cholesky decomposition $\mathbf{F}_1 = \mathbf{L}\mathbf{L}^T$ is used. It is a factor of 2 faster than alternative methods for solving linear equations. Once $\mathbf{F}_1$ is decomposed, the triangular factor $\mathbf{L}$ can be used to solve [4.144], [4.147] and [4.150] by backsubstitution. However, if time-step adaptation is considered, the decomposition $\mathbf{F}_1$ has to be done at each time when h is modified. Finally, recall that as the viscous damping model is a special case of the exponential damping model, the proposed algorithm is also applicable to linear systems with non-proportional viscous damping.

4.7.4. *Numerical examples*

4.7.4.1. *Example 1: a three DOF system*

A three DOF system with double-exponential model, discussed in section 4.3.2, is again considered. The equations of motion of this model system can be represented by equation [4.26] with $n = 2$ and $\mathbf{C}_0 = \mathbf{O}$. The mass and the stiffness matrices of the system are given by equation [4.20]. For numerical calculation, we have assumed $m_u = 3.0$ kg, $k_u = 2.0$ N/m, $c_1 = 0.6$ Ns/m, $c_2 = 0.2$ Ns/m, $\mu_1 = 1.0$ s^{-1} and $\mu_2 = 5.0$ s^{-1}. Both the damping coefficient matrices have rank deficiency because $r_1 = \text{rank}(\mathbf{C}_1) = 2 \leq 3$ and $r_2 = \text{rank}(\mathbf{C}_2) = 1 \leq 3$. The order of the system matrices in the state-space, expressed by equation [4.62], can be obtained as $m = 2\times 3 + (2+1) = 9$.

Attention is given to the dynamic response of the system subjected to a unit initial displacement at the first DOF, that is, $\bar{\mathbf{f}}(\mathrm{i}\omega) = \mathbf{0}$, $\mathbf{q}_0 = \{1,0,0\}^T$ and $\dot{\mathbf{q}}_0 = \mathbf{0}$. The responses of the three masses are obtained using the direct time-domain approach outlined in section 4.7.2 with time step $h = 0.02$ s. Figure 4.7 shows the response of the first DOF as a function of time. To verify the numerical accuracy of the proposed method, the same quantity is obtained using the exact state-space mode superposition method (as in section 4.5) and shown in the same figure. It may be noted that the results obtained using both the approaches match with excellent accuracy. For numerical calculation, a total of 100 steps with step size $h = 0.5$ s have been used. The time-domain method proposed here, however, is more efficient because the computation of state-space eigensolutions is avoided.

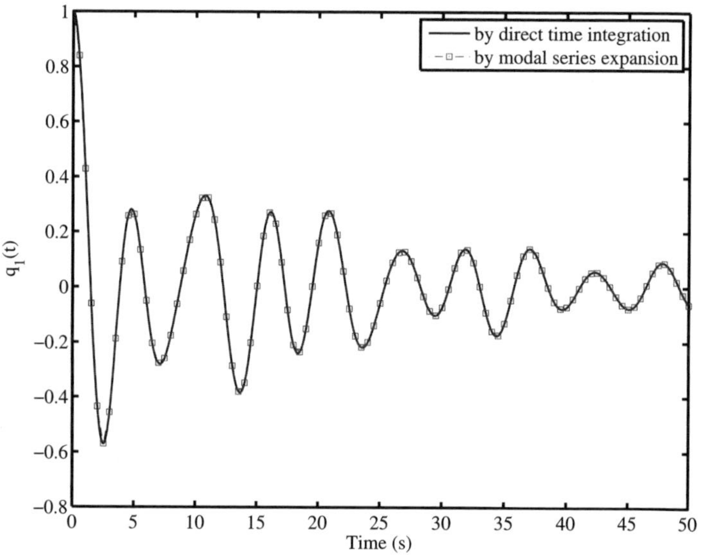

Figure 4.7. *Displacement at the first DOF*

4.7.4.2. *Example 2: axial vibration of a rod*

We consider a fixed-free axially vibrating rod with double-exponential damping. A finite element model of the rod, shown in Figure 4.8, with N linear elements is used. The element stiffness matrix and the corresponding (consistent) element mass matrix for a rod element are well known

$$\mathbf{M}_e = \frac{\rho A l_e}{6}\begin{bmatrix} 2 & 1 \\ 1 & 2 \end{bmatrix} \quad \text{and} \quad \mathbf{K}_e = \frac{EA}{l_e}\begin{bmatrix} 1 & -1 \\ -1 & 1 \end{bmatrix}. \qquad [4.153]$$

Here, ρ is the mass density, A is the area, E is the Young's modulus and $l_e = L/N$ is the length of an element. Numerical values of these quantities are assumed to be $A = 6.25 \times 10^{-4}\,\mathrm{m}^2$, $E = 2.1 \times 10^{11}\,\mathrm{N/m}^2$, $\rho = 7.8 \times 10^3\,\mathrm{kg/m}^3$ and the length of the rod $L = 4\,\mathrm{m}$. The global stiffness matrix, **K**, and the global mass matrix, **M**, can be obtained using the usual finite element assembly procedure. For numerical calculation, 80 elements have been used so that the degrees-of-freedom of the system $N = 80$. Thus, the response vector $\mathbf{q}(t) = \{u_1(t), u_2(t), \ldots, u_{80}(t)\}^T$ and the (discretized) equation of motion of this rod with double-exponential damping can be expressed as

$$\mathbf{M}\ddot{\mathbf{q}}(t) + \int_0^t \left[\mu_1 e^{-\mu_1(t-\tau)}\mathbf{C}_1 + \mu_2 e^{-\mu_2(t-\tau)}\mathbf{C}_2\right] \dot{\mathbf{q}}(\tau)\,\mathrm{d}\tau + \mathbf{K}\mathbf{q}(t) = \mathbf{f}(t). \quad [4.154]$$

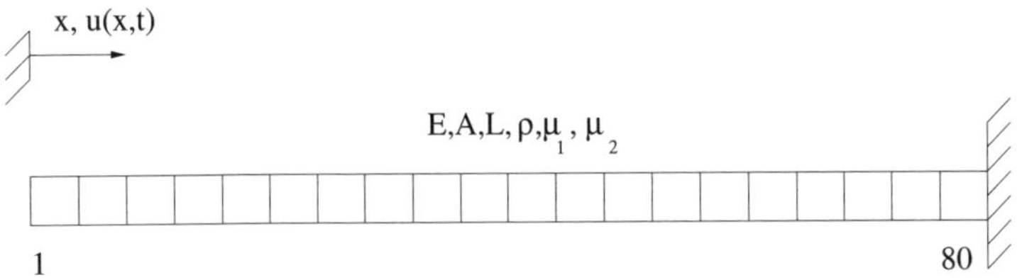

Figure 4.8. *Fixed-free axially vibrating rod with double-exponential damping*

Here, numerical values of the relaxation parameters μ_1 and μ_2 are conveniently selected as

$$\mu_1 = \frac{1}{T_{min}} \quad \text{and} \quad \mu_2 = \frac{1}{2T_{min}} \quad [4.155]$$

where T_{min} is the minimum time period given by $T_{min} = 2\pi/\omega_{max}$. For an axially vibrating rod, the highest undamped natural frequency $\omega_{max} = \sqrt{\frac{E}{\rho}}\,\frac{2\,N-1}{2\,L}\pi$. The global damping coefficient matrices $\mathbf{C}_1$ and $\mathbf{C}_2$ are assumed to be proportional to the mass and stiffness matrices, respectively

$$\mathbf{C}_1 = \alpha_1\mathbf{M} \quad \text{and} \quad \mathbf{C}_2 = \alpha_2\mathbf{K}. \quad [4.156]$$

The proportionality constants α_1 and α_2 are assumed to be

$$\alpha_1 = 2\zeta\frac{\omega_1\omega_2}{\omega_1+\omega_2} \quad \text{and} \quad \alpha_2 = 2\zeta\frac{1}{\omega_1+\omega_2} \quad [4.157]$$

where the damping factor $\zeta = 0.05$ (i.e. 5% damping) and the jth undamped natural frequency $\omega_j = \sqrt{\frac{E}{\rho}}\,\frac{2\,j-1}{2\,L}\pi$.

The objective here is to obtain the time-domain response of the free end of the rod subjected to a unit initial velocity at the same end. For this problem, the forcing vectors $\mathbf{f}(t) = \mathbf{0}$ and the initial conditions $\mathbf{q}_0 = \mathbf{0}$, $\mathbf{v}_0 = \mathbf{e}_1$ (unit vector). The problem is solved by the time-stepping procedure developed in this section. Because we have assumed proportional damping and $\mathbf{M}, \mathbf{K}$ are symmetric positive definite, the matrices $\mathbf{C}_1$ and $\mathbf{C}_2$ are of full rank. Figures 4.9 and 4.10 show the tip displacement $q(x = 0, t) = q_1(t)$ and the tip velocity $\dot{q}(x = 0, t) = v_1(t)$ as a function of time. For numerical computation, a total of $8,000$ steps with step size $h = 1.5 \times 10^{-6}$ s have been used.

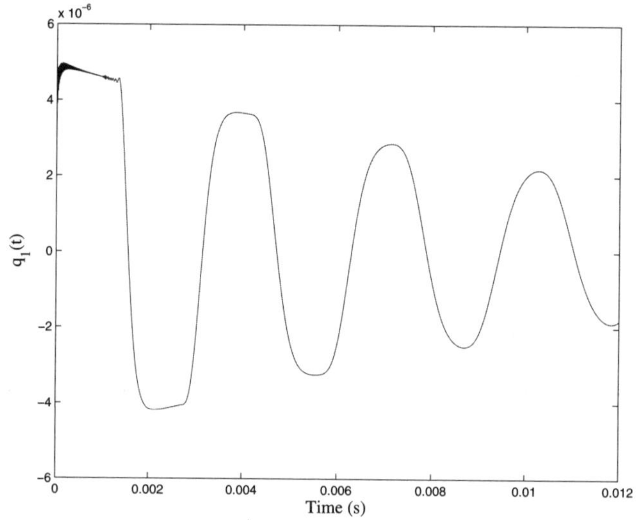

Figure 4.9. *Tip displacement of the axially vibrating rod*

4.8. Summary

In this chapter, we have considered the problem of dynamic analysis of non-viscously damped linear systems with multiple-degrees-of-freedom. The assumed non-viscous damping model is such that the damping forces depend on the past history of motion via convolution integrals over some application-dependent kernel functions. We have assumed that, in general, the mass and stiffness matrices as well as the matrix of the kernel functions cannot be simultaneously diagonalized by any linear transformation. The analysis is, however, restricted to systems with non-repetitive eigenvalues and non-singular mass matrices.

This approach, in contrast to the usual state-vector for viscously damped systems, uses a set of internal variables in addition to the displacement and the velocity as the state vector. Analysis for two physically realistic cases, (1) when all the damping coefficient matrices are of full rank and (2) when the damping coefficient matrices have rank deficiency, has been presented. It was shown that for both cases, the equation of motion can be represented in terms of two symmetric matrices. The eigenvalues and the corresponding eigenvectors of the system were obtained by solving the state-space eigenvalue problem. Unlike, the viscously damped systems, the number of eigensolutions is more than $2N$ and depends on the rank of the damping coefficient matrices. The modes were grouped into two types: (1) elastic modes and (2) non-viscous modes. In the analysis developed here, the elastic modes appear in complex conjugate pairs, that is, they are subcritically damped. The elastic modes, which consist of N eigenvectors together with their complex conjugate pairs, correspond to N modes of vibration of the structural system.

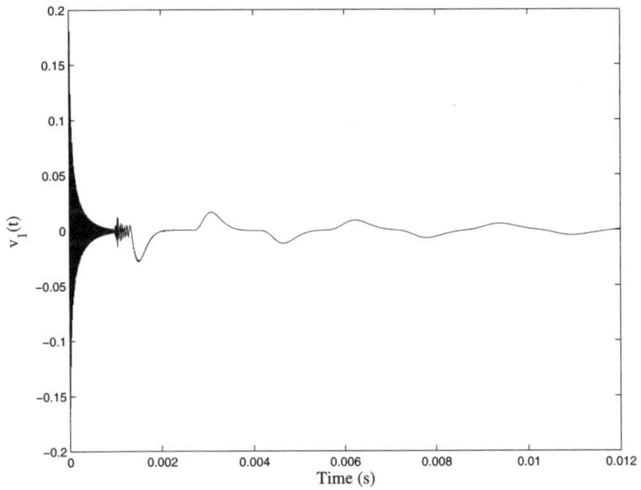

Figure 4.10. *Tip velocity of the axially vibrating rod*

The eigenanalysis allowed exact closed-form expressions of the response in the time and frequency domains due to arbitrary forcing functions and initial conditions to be obtained in terms of the state-space eigensolutions. The response of the non-viscously damped MDOF system can then be expressed as a sum of two parts, one that arises in viscously damped systems and the other that occurs due to the non-viscous nature of the damping forces.

Direct time-domain analysis of linear systems with exponentially decaying damping memory kernels has been considered. The proposed method is based on an extended state-space representation of the equations of motion. The advantage of the proposed method is that the coefficient matrix of the time-stepping scheme finally to be solved contains nothing but a linear combination of the original system matrices $\mathbf{M}, \mathbf{C}_k$ and $\mathbf{K}$. This is useful because we can benefit from the special properties of

$\mathbf{M}, \mathbf{C}_k$ and $\mathbf{K}$, like sparsity or symmetry. Thus, although the problem is originally formulated in the extended state-space, in effect the actual solution is carried out in the original reduced space. This implies that the extra effort needed to handle the non-viscous damping model like the one considered here is small compared to a viscous damping model. The proposed algorithm also extends to the physically realistic case when some or all damping coefficient matrices $\mathbf{C}_k$ are rank deficient. If the interest is limited to the evolution of displacements and velocities only (not the internal variables), the proposed algorithm allows us to bypass the expensive process of factorization of the $\mathbf{C}_k$ matrices for the rank-deficient case. The proposed algorithm allows only uniform time steps. In principle, it can be extended to variable time steps but at additional computational cost.

One possible drawback of the eigensolution approach proposed in the chapter is that the solution of the eigenvalue problem in the extended state-space is necessary. For large systems, this can be computationally expensive. This issue is addressed in the next chapter in the context of general non-viscous damping models.

Chapter 5

Linear Systems with General Non-Viscous Damping

In Chapter 4, a symmetric state-space approach for exponentially damped linear systems was proposed. It was shown that the dynamic response of the system can be obtained in a manner similar to undamped or viscously damped systems provided that the eigenvalues and eigenvectors are available. Calculation of the eigensolutions in the state-space is computationally expensive as the dimension of the system can be very large due to the additional internal variables. The purpose of this chapter is to consider methods based on the configuration space. We also consider general non-viscous damping model for analytical developments. The exponential model discussed in Chapter 4 can be considered as a special case. We express the equation of motion of forced vibration of an N degrees-of-freedom linear system with general non-viscous damping as

$$\mathbf{M}\ddot{\mathbf{q}}(t) + \int_0^t \boldsymbol{\mathcal{G}}(t-\tau)\dot{\mathbf{q}}(\tau)\mathrm{d}\tau + \mathbf{K}\mathbf{q}(t) = \mathbf{f}(t). \qquad [5.1]$$

The *initial conditions* associated with the above equation are

$$\begin{aligned} \mathbf{q}(0) &= \mathbf{q}_0 \in \mathbb{R}^N \\ \text{and} \quad \dot{\mathbf{q}}(0) &= \dot{\mathbf{q}}_0 \in \mathbb{R}^N. \end{aligned} \qquad [5.2]$$

In section 1.3, currently available methods for analyzing such systems were discussed. The majority of the available methods employ additional dissipation coordinates and then use the state-space formalism. This approach is not only computationally more involved but also physical insight is lost in such approach. For this reason, in this chapter, we develop procedures that avoid this approach and are consistent with traditional modal analysis.

First, we discuss the idea of proportional damping in the context of general non-viscously damped systems in section 5.1. In section 5.2, the nature of the eigenvalues and eigenvectors is discussed under certain simplified but physically realistic assumptions on the system matrices and damping kernel functions. A series expansion method for the determination of complex eigenvectors is proposed. The transfer function matrix of the system is derived in terms of these eigenvectors in section 5.3. Exact closed-form expressions are derived for the transient response and the response due to non-zero initial conditions in section 5.4. Applications of the proposed method and related numerical issues are explained using a non-viscously damped three degrees-of-freedom system in section 5.5. Some general mathematical relationship involving the eigensolutions of non-viscously damped systems was derived in section 5.6. Orthogonality and normalization properties of complex modes of generally damped systems are introduced in section 5.7. The concept of the Rayleigh quotient for non-viscously damped systems is discussed. Three new Rayleigh quotients are proposed and their stationary properties are investigated. A summary of the results obtained in this chapter is given in section 5.8.

5.1. Existence of classical normal modes

It is required to find out the conditions when there exists a non-zero matrix $\mathbf{X} \in \mathbb{R}^{N\times N}$ that diagonalizes equation [5.1] by a congruence transformation. Unlike the viscously damped case where all the system matrices were constant and symmetric, here the system dynamics is characterized by two constant symmetric matrices and one symmetric matrix containing real functions. The problem of simultaneous diagonalization of Hermitian matrices with functional entries through constant complex transformations has been discussed in [CHA 78]. In view of their results and our earlier result (theorem 2.1) on viscously damped systems, the conditions for existence of classical normal modes in system [5.1] can be described as follows:

THEOREM 5.1.– A non-viscously damped system [5.1] possesses classical normal modes if and only if $\mathbf{K}\mathbf{M}^{-1}\boldsymbol{\mathcal{G}}(\tau) = \boldsymbol{\mathcal{G}}(\tau)\mathbf{M}^{-1}\mathbf{K}, \forall\,\tau$.

A proof of this theorem can be constructed following the proof of theorem 2.1 together with the results established in [CHA 78]. Intuitively, if $\boldsymbol{\mathcal{G}}(\tau)$ is a sufficiently smooth matrix function, then we can obtain a sequence $\boldsymbol{\mathcal{G}}_j = \boldsymbol{\mathcal{G}}(\tau_j) \in \mathbb{R}^{N\times N}, \forall j = 1, 2, \cdots, \infty$, where all $\boldsymbol{\mathcal{G}}_j$ are symmetric. Use of theorem 2.1 for all $j = 1, 2, \cdots, \infty$ essentially leads to the result of this theorem.

EXAMPLE 5.1.– Consider a system whose equation of motion can be described by [5.1] with

$$\mathbf{M} = \begin{bmatrix} 2 & -1 \\ -1 & 3 \end{bmatrix}, \quad \mathbf{K} = \begin{bmatrix} 1 & -2 \\ -2 & 5 \end{bmatrix} \qquad [5.3]$$

and

$$\boldsymbol{\mathcal{G}}(\hat{\tau}) = \begin{bmatrix} 1.0177e^{-\mu_1\hat{\tau}} + 0.3517e^{-\mu_2\hat{\tau}^2/2} & -1.7724e^{-\mu_1\hat{\tau}} - 0.5364e^{-\mu_2\hat{\tau}^2/2} \\ -1.7724e^{-\mu_1\hat{\tau}} - 0.5364e^{-\mu_2\hat{\tau}^2/2} & 4.4749e^{-\mu_1\hat{\tau}} + 1.3688e^{-\mu_2\hat{\tau}^2/2} \end{bmatrix},$$
$$\text{where} \quad \hat{\tau} = t - \tau \qquad [5.4]$$

and μ_1, μ_2 are constants. Note that **M**, **K** and $\boldsymbol{\mathcal{G}}(\tau)$ are positive definite. It may be easily verified that the condition outlined in theorem 5.1 is satisfied. For this reason, the undamped modal matrix

$$\mathbf{X} = \begin{bmatrix} 0.7710 & -0.0743 \\ 0.3124 & 0.5499 \end{bmatrix} \qquad [5.5]$$

which diagonalizes **M** and **K** also diagonalizes $\boldsymbol{\mathcal{G}}(\tau)$ as

$$\mathbf{X}^T \boldsymbol{\mathcal{G}}(\hat{\tau})\mathbf{X} = \begin{bmatrix} 0.1879e^{-\mu_1\hat{\tau}} + 0.0843e^{-\mu_2\hat{\tau}^2/2} & 0.0 \\ 0.0 & 1.5037e^{-\mu_1\hat{\tau}} + 0.4597e^{-\mu_2\hat{\tau}^2/2} \end{bmatrix} \qquad [5.6]$$

is a diagonal matrix $\forall \tau$.

5.1.1. *Generalization of proportional damping*

We can now follow the approach similar to the viscously damped case, as in section 2.4.2, for obtaining an expression of the damping function in terms of **M** and **K** so that system [5.1] possesses classical normal modes. From theorem 5.1, the conditions for existence of classical normal modes can be expressed as

$$\mathbf{A}\boldsymbol{\mathcal{B}}(\tau) = \boldsymbol{\mathcal{B}}(\tau)\mathbf{A} \quad \forall \tau \qquad [5.7]$$

where $\mathbf{A} = \mathbf{M}^{-1}\mathbf{K}$ and $\boldsymbol{\mathcal{B}}(\tau) = \mathbf{M}^{-1}\boldsymbol{\mathcal{G}}(\tau)$. In view of this commutative condition, we say system [5.1] possesses classical normal modes if and only if

$$\boldsymbol{\mathcal{G}}(\tau) = \mathbf{M}\mathcal{F}_1\left(\mathbf{M}^{-1}\mathbf{K}, \tau\right) + \mathbf{K}\mathcal{F}_2\left(\mathbf{K}^{-1}\mathbf{M}, \tau\right) \qquad [5.8]$$

where $\mathcal{F}_i(z, \tau)$, $i = 1, 2$ are any real analytic functions such that $\left|\int_{-\infty}^{\infty} \mathcal{F}_i(z, \tau)\, \mathrm{d}\tau\right| < \infty$, $\forall z$. This representation of damping is the right-functional form. Following a similar approach to that outlined for the viscous damping case in section 2.4.2, we can also have the left-functional form as

$$\mathcal{G}(\tau) = \mathcal{F}_3\left(\mathbf{K}\mathbf{M}^{-1}, \tau\right)\mathbf{M} + \mathcal{F}_4\left(\mathbf{M}\mathbf{K}^{-1}, \tau\right)\mathbf{K}. \qquad [5.9]$$

This is possibly the most general class of damping that can be treated within the scope of classical modal analysis. The representation of $\mathbf{C}$ in equations [2.109] and [2.110] can be obtained as a special case when $\mathcal{F}_i(z, \tau) = f_i(z)\delta(\tau)$. This, in turn, relates equations [5.8] and [5.9] to the Rayleigh's representation and Caughey series.

EXAMPLE 5.2.– It will be shown that the linear dynamic system satisfying the following equation of motion of free vibration

$$\mathbf{M}\ddot{\mathbf{q}}(t) + \int_{-\infty}^{t} \left[\mathbf{M}(\mathbf{K}^{-1}\mathbf{M})^{-\pi/4} e^{-\mu(t-\tau)} + e^{-\sqrt{\mathbf{K}\mathbf{M}^{-1}}}\mathbf{K}\delta(t-\tau)\right] \dot{\mathbf{q}}(\tau)\, \mathrm{d}\tau + \mathbf{K}\mathbf{q}(t) = \mathbf{0}$$

[5.10]

possesses classical normal modes. Here, $\mathbf{M}$ and $\mathbf{K}$ are the same as in example 2.1 and μ is a constant.

Introduce the modal transformation $\mathbf{q}(t) = \mathbf{X}\mathbf{y}(t)$ where the mass normalized modal matrix $\mathbf{X}$ is given as before in equation [2.104]. From direct calculation, the equation of motion in the modal coordinates may be obtained as

$$\begin{aligned}\mathbf{I}\ddot{\mathbf{y}}(t) \; &+ \int_{-\infty}^{t} \begin{bmatrix} 0.2695 & 0.0 & 0.0 \\ 0.0 & 0.9732 & 0.0 \\ 0.0 & 0.0 & 5.2433 \end{bmatrix} e^{-\mu(t-\tau)}\dot{\mathbf{y}}(\tau)\, \mathrm{d}\tau \\ &+ \begin{bmatrix} 0.1220 & 0.0 & 0.0 \\ 0.0 & 0.3615 & 0.0 \\ 0.0 & 0.0 & 0.4668 \end{bmatrix} \dot{\mathbf{y}}(t) + \begin{bmatrix} 0.1883 & 0.0 & 0.0 \\ 0.0 & 0.9660 & 0.0 \\ 0.0 & 0.0 & 8.2457 \end{bmatrix} \mathbf{y}(t) = \mathbf{0}.\end{aligned}$$

[5.11]

The above equations are uncoupled and can be solved in the frequency domain adopting similar procedures to those used for viscously damped systems.

The concept of proportional damping was extended to non-viscously damped systems. The condition for existence of proportional damping in general non-viscously damped systems was derived. It is clear that this condition is purely mathematical in nature and there is no reason why a general linear system should obey such a condition. Thus, in general, non-viscously damped systems are non-proportionally damped. The central theme of the remainder of the chapter is to analyze general non-viscously damped multiple-degrees-of-freedom linear systems with non-proportional damping.

5.2. Eigenvalues and eigenvectors

Considering the free vibration, that is $\mathbf{f}(t) = \mathbf{q}_0 = \dot{\mathbf{q}}_0 = \mathbf{0}$, and taking the Laplace transform of the equations of motion [5.1], we have

$$s^2\mathbf{M}\bar{\mathbf{q}}(s) + s\mathbf{G}(s)\bar{\mathbf{q}}(s) + \mathbf{K}\bar{\mathbf{q}}(s) = \mathbf{0}. \qquad [5.12]$$

Here, $\bar{\mathbf{q}}(s) = \mathcal{L}\left[\mathbf{q}(t)\right] \in \mathbb{C}^N$, $\mathbf{G}(s) = \mathcal{L}\left[\mathcal{G}(t)\right] \in \mathbb{C}^{N\times N}$ and $\mathcal{L}\left[\bullet\right]$ denotes the Laplace transform. In the context of structural dynamics, $s = \mathrm{i}\omega$, where $\mathrm{i} = \sqrt{-1}$ and $\omega \in \mathbb{R}^+$ denotes the frequency. It is assumed that (1) $\mathbf{M}^{-1}$ exists and (2) *all* the eigenvalues of $\mathbf{M}^{-1}\mathbf{K}$ are distinct and positive. Because $\mathcal{G}(t)$ is a real function, $\mathbf{G}(s)$ is also a real function of the parameter s. We assume that $\mathbf{G}(s)$ is such that the motion is dissipative. Conditions that $\mathbf{G}(s)$ must satisfy in order to produce dissipative motion were given in [GOL 85]. Several physically realistic mathematical forms of the elements of $\mathbf{G}(s)$ were given in Table 1.1. For the linear viscoelastic case, it can be shown that (see [BLA 60, MUR 97a]), in general, the elements of $\mathbf{G}(s)$ can be represented as

$$G_{jk}(s) = \frac{p_{jk}(s)}{q_{jk}(s)} \qquad [5.13]$$

where $p_{jk}(s)$ and $q_{jk}(s)$ are finite-order polynomials in s. Here, we do not assume any specific functional form of $G_{jk}(s)$ but assume that $|G_{jk}(s)| < \infty$ when $s \to \infty$. This, in turn, implies that the elements of $\mathbf{G}(s)$ are at the most of order $1/s$ in s or constant, as in the case of viscous damping. The eigenvalues, s_j, associated with equation [5.12] are roots of the characteristic equation

$$\det\left[s^2\mathbf{M} + s\mathbf{G}(s) + \mathbf{K}\right] = 0. \qquad [5.14]$$

If the elements of $\mathbf{G}(s)$ have simple forms, for example, as in equation [5.13], then the characteristic equation becomes a polynomial equation of finite order. In other cases, the characteristic equation can be expressed as a polynomial equation by expanding $\mathbf{G}(s)$ in a Taylor series. However, the order of the equation will be infinite in those cases. For practical purposes, the Taylor expansion of $\mathbf{G}(s)$ can be truncated to a finite series to make the characteristic equation as a polynomial equation of finite order. Such equations can be solved using standard numerical methods, see [PRE 92] ([ADH 14b], Chapter 3). Advanced computational methods for solving nonlinear eigenvalue equation [5.14] are discussed in Chapter 6. Suppose the order of the characteristic polynomial is m. In general, m is more than $2N$, that is $m = 2N + p; p \geq 0$. Thus, although the system has N degrees-of-freedom, the number of eigenvalues is more than $2N$. This is a major difference between the non-viscously damped systems and the viscously damped systems where the number of eigenvalues is exactly $2N$, including any multiplicities.

A general analysis on the nature of the eigenvalues of non-viscously damped systems is beyond the scope of this chapter. It is assumed that *all* m eigenvalues are distinct. We further restrict our attention to a special case when, among the m eigenvalues, $2N$ appears in complex conjugate pairs. Because $\mathcal{G}(t)$ is real, $\mathbf{G}^*(s) = \mathbf{G}(s^*)$. Using this and taking complex conjugation of [5.12], it is clear that ${s^*}^2\mathbf{M}\bar{\mathbf{q}} + s^*\mathbf{G}(s^*)\bar{\mathbf{q}} + \mathbf{K}\bar{\mathbf{q}} = \mathbf{0}$, i.e. if s satisfies equation [5.12] then so does s^*. Under such assumptions, it is easy to show that the remaining p eigenvalues become purely real. The mathematical conditions that **M**, **K** and $\mathbf{G}(s)$ must satisfy in order to produce such eigenvalues will not be obtained but a physical justification will follow shortly. For convenience, the eigenvalues are arranged as

$$s_1, s_2, \cdots, s_N, s_1^*, s_2^*, \cdots, s_N^*, s_{2N+1}, \cdots, s_m \qquad [5.15]$$

where $(\bullet)^*$ denotes complex conjugation.

The eigenvalue problem associated with equation [5.1] can be defined from [5.12] as

$$\mathbf{D}(s_j)\mathbf{z}_j = \mathbf{0}, \quad \text{for } j = 1, \cdots, m \qquad [5.16]$$

where

$$\mathbf{D}(s_j) = s_j^2\mathbf{M} + s_j\,\mathbf{G}(s_j) + \mathbf{K} \qquad [5.17]$$

is the *dynamic stiffness matrix* corresponding to the jth eigenvalue and $\mathbf{z}_j$ is the jth eigenvector. Here, $(\bullet)^T$ denotes the matrix transpose. From equation [5.16], it is clear that, when s_j appears in complex conjugate pairs, $\mathbf{z}_j$ also appears in complex conjugate pairs, and when s_j is real $\mathbf{z}_j$ is also real. Corresponding to the $2N$ complex conjugate pairs of eigenvalues, the N eigenvectors together with their complex conjugates will be called *elastic modes*. These modes are related to the N modes of vibration of the structural system. Physically, the assumption of "$2N$ complex conjugate pairs of eigenvalues" implies that all the elastic modes are oscillatory in nature, that is, they are subcritically damped. The modes corresponding to the "additional" p eigenvalues will be called *non-viscous modes*. These modes are induced by the non-viscous effect of the damping mechanism. For stable passive systems the non-viscous modes are overcritically damped (i.e. negative real eigenvalues) and not oscillatory in nature. Non-viscous modes, or those similar to these, are known by different names in the literature of different subjects, for example "wet modes" in the context of ship dynamics [BIS 79] and "damping modes" in the context of viscoelastic structures [MCT 93]. Determination of the eigenvectors is considered next.

5.2.1. *Elastic modes*

Once the eigenvalues are known, $\mathbf{z}_j, \forall j = 1, \cdots, 2N$ can be obtained from equation [5.16] by fixing any one element and inverting the matrix $\mathbf{D}(s_j) \in \mathbb{C}^{N \times N}$. Note that inversion of an $(N-1) \times (N-1)$ complex matrix is required for the calculation of every $\mathbf{z}_j$. Although the method is exact, it is computationally expensive and does not offer much physical insight. Here, we propose an alternative method based on Neumann expansion approach that utilizes the familiar undamped eigenvectors discussed in section 2.3.1.

5.2.1.1. *Neumann expansion method*

For distinct undamped eigenvalues (ω_l^2), $\mathbf{x}_l$, $\forall\, l = 1, \cdots, N$, form a *complete* set of vectors. For this reason, $\mathbf{z}_j$ can be expanded as a complex linear combination of $\mathbf{x}_l$. Thus, an expansion of the form

$$\mathbf{z}_j = \sum_{l=1}^{N} \alpha_l^{(j)} \mathbf{x}_l \qquad [5.18]$$

may be considered. Now, without any loss of generality, we can assume that $\alpha_j^{(j)} = 1$ (normalization), which leaves us to determine $\alpha_l^{(j)}, \forall l \neq j$. Substituting the expansion of $\mathbf{z}_j$, from equation [5.16], we obtain

$$\sum_{l=1}^{N} s_j^2 \alpha_l^{(j)} \mathbf{M}\mathbf{x}_l + s_j \alpha_l^{(j)} \mathbf{G}(s_j)\mathbf{x}_l + \alpha_l^{(j)} \mathbf{K}\mathbf{x}_l = \mathbf{0}. \qquad [5.19]$$

Premultiplying the above equation by $\mathbf{x}_k^T$ and using the orthogonality property of the undamped eigenvectors described by [2.63] and [2.64], we obtain

$$s_j^2 \alpha_k^{(j)} + s_j \sum_{l=1}^{N} \alpha_l^{(j)} G'_{kl}(s_j) + \omega_k^2 \alpha_k^{(j)} = 0, \quad \forall k = 1, \cdots, N \qquad [5.20]$$

where $G'_{kl}(s_j) = \mathbf{x}_k^T \mathbf{G}(s_j)\mathbf{x}_l$. This is mathematically similar to performing a Galerkin error minimization. The jth equation of this set obtained by setting $k = j$ is a trivial case because $\alpha_j^{(j)} = 1$ has already been assumed. From the above set of equations, excluding this trivial case, we have

$$s_j^2 \alpha_k^{(j)} + s_j \left(G'_{kj}(s_j) + \alpha_k^{(j)} G'_{kk}(s_j) + \sum_{l \neq k \neq j}^{N} \alpha_l^{(j)} G'_{kl}(s_j) \right) + \omega_k^2 \alpha_k^{(j)} = 0, \quad \forall k = 1, \cdots, N; \neq j. \qquad [5.21]$$

These equations can be combined into a matrix form as

$$\left[\mathbf{P}_j - \mathbf{Q}_j\right] \mathbf{a}_j = \mathbf{b}_j. \qquad [5.22]$$

In the above equation

$$\mathbf{P}_j = \text{diag}\left[\frac{s_j^2 + s_j G'_{11}(s_j) + \omega_1^2}{-s_j}, \cdots, \{j\text{th term deleted}\}, \cdots, \frac{s_j^2 + s_j G'_{NN}(s_j) + \omega_N^2}{-s_j} \right] \in \mathbb{C}^{(N-1)\times(N-1)}, \qquad [5.23]$$

the traceless matrix

$$\mathbf{Q}_j = \begin{bmatrix} 0 & G'_{12}(s_j) & \cdots & \{j\text{th term deleted}\} & \cdots & G'_{1N}(s_j) \\ G'_{21}(s_j) & 0 & \vdots & \vdots & \vdots & G'_{2N}(s_j) \\ \vdots & \vdots & \vdots & \{j\text{th term deleted}\} & \vdots & \vdots \\ \vdots & \vdots & \vdots & \vdots & \vdots & \vdots \\ G'_{N1}(s_j) & G'_{N2}(s_j) & \cdots & \{j\text{th term deleted}\} & \cdots & 0 \end{bmatrix} \in \mathbb{C}^{(N-1)\times(N-1)}, \qquad [5.24]$$

$$\mathbf{b}_j = \left\{ G'_{1j}(s_j), G'_{2j}(s_j), \cdots, \{j\text{th term deleted}\}, \cdots, G'_{Nj}(s_j) \right\}^T \in \mathbb{C}^{(N-1)} \qquad [5.25]$$

and

$$\mathbf{a}_j = \left\{ \alpha_1^{(j)}, \alpha_2^{(j)}, \cdots, \{j\text{th term deleted}\}, \cdots, \alpha_N^{(j)} \right\}^T \in \mathbb{C}^{(N-1)} \qquad [5.26]$$

is the vector of unknown $\alpha_k^{(j)}, \forall k \neq j$. From equation [5.22], $\hat{\mathbf{a}}^{(j)}$ has to be determined by performing the associated matrix inversion and is achieved by using the Neumann expansion method. Now, using the Neumann expansion, we have

$$\begin{aligned} \mathbf{a}_j &= \left[\mathbf{I}_{N-1} - \mathbf{P}_j^{-1}\mathbf{Q}_j\right]^{-1} \{\mathbf{P}_j^{-1}\mathbf{b}_j\} \\ &= \left[\mathbf{I}_{N-1} + \mathbf{R}_j + \mathbf{R}_j^2 + \mathbf{R}_j^3 + \cdots\right] \mathbf{a}_{j_0} \end{aligned} \qquad [5.27]$$

where $\mathbf{I}_{N-1}$ is an, $(N-1) \times (N-1)$ identity matrix,

$$\mathbf{R}_j = \mathbf{P}_j^{-1}\mathbf{Q}^{(j)} \in \mathbb{C}^{(N-1)\times(N-1)} \qquad [5.28]$$

and

$$\mathbf{a}_{j_0} = \mathbf{P}_j^{-1}\mathbf{b}_j \in \mathbb{C}^{(N-1)}. \qquad [5.29]$$

Because $\mathbf{P}^{(j)}$ is a diagonal matrix, its inversion can be carried out analytically and subsequently the closed-form expressions of ${\mathbf{R}_u}^{(j)}$ and $\mathbf{a}_0^{(j)}$ can be obtained as

$$R_{j_{kl}} = \frac{-s_j G'_{kl}(s_j)\,(1-\delta_{kl})}{\omega_k^2 + s_j^2 + s_j G'_{kk}(s_j)}, \quad \forall k, l \neq j \qquad [5.30]$$

$$a_{j_{0_l}} = \frac{-s_j G'_{lj}(s_j)}{\omega_l^2 + s_j^2 + s_j G'_{ll}(s_j)}, \quad \forall l \neq j. \qquad [5.31]$$

This makes further calculations involving these quantities simpler. From equation [5.27], $\mathbf{a}^{(j)}$ can be calculated in an efficient way as we can write

$$\mathbf{a}_j = \mathbf{a}_{j_0} + \mathbf{a}_{j_1} + \mathbf{a}_{j_2} + \cdots + \mathbf{a}_{j_k} + \cdots \qquad [5.32]$$

where

$$\mathbf{a}_{j_1} = \mathbf{R}_j\mathbf{a}_{j_0},\ \mathbf{a}_{j_2} = \mathbf{R}_j\mathbf{a}_{j_1},\ \cdots\ \mathbf{a}_{j_k} = \mathbf{R}_j\mathbf{a}_{j_{k-1}}. \qquad [5.33]$$

This implies that all the $\mathbf{a}_{j_k}$ can be obtained using successive matrix-vector multiplications only. Noting that $\mathbf{a}_j$ is the vector of $\alpha_k^{(j)}, \forall k \neq j$, its substitution in equation [5.18] will give the elastic modes. It is easy to see that by taking more terms in the series represented by [5.32], we can obtain the elastic modes to any desired accuracy provided that the complex matrix power series $\mathbf{I}_{N-1} + \mathbf{R}_j + \mathbf{R}_j^2 + \mathbf{R}_j^3 + \cdots$ is convergent. Convergence of this series will be addressed later.

From the previous formulation we may verify that, corresponding to the complex conjugate pairs of the eigenvalues, the eigenvectors also appear in complex conjugate pairs. For many engineering problems, it is often observed that the damping forces are not very "big" and that by retaining only a few terms in the series expression [5.32] will result in an acceptable accuracy. Closed-form approximate expressions for the elastic modes obtained by retaining one and two terms of these series are given in section 5.2.3. These expressions might be useful whenever we find that the entries of the damping kernel-functions are small compared to those of **M** and **K**.

5.2.1.2. *Convergence of the Neumann series*

For the validity of the series expressions for $\mathbf{a}_j$ in [5.32], it is required that the series

$$\mathbf{S}_{\mathbf{u}} = \mathbf{I}_{N-1} + \mathbf{R}_j + \mathbf{R}_j^2 + \mathbf{R}_j^3 + \cdots \qquad [5.34]$$

should be convergent. We develop following conditions for convergence.

CONDITION 5.1.– The complex matrix power series $\mathbf{S}_{\mathbf{u}}$ converges if, and only if, for all the eigenvalues $\sigma_l^{(j)}$ of the matrix $\mathbf{R}_j$, the inequality $|\sigma_l^{(j)}| < 1$ holds.

Although this condition is both necessary and sufficient, checking convergence for all $j = 1, \cdots, N$ is often not feasible. So we look for a sufficient condition that is relatively easy to check and that ensures convergence for all $j = 1, \cdots, N$.

CONDITION 5.2.– The complex matrix power series $\mathbf{S}_{\mathbf{u}}$ converges for any s_j if $\mathbf{G}'(s_j)$ is a diagonally dominant matrix.

PROOF.– Since a matrix norm is always greater than or equal to its maximum eigenvalue, it follows from condition 5.1 that convergence of the series is guaranteed if $\| \mathbf{R}_j \| < 1$. Writing the sum of absolute values of entries of $\mathbf{R}_j$ results in the following inequality as the required sufficient condition for convergence

$$\sum_{\substack{k=1\\k\neq j}}^{N}\sum_{\substack{l=1\\l\neq j}}^{N}\left|\frac{s_j G'_{kl}(s_j)}{\omega_k^2 + s_j^2 + s_j G'_{kk}(s_j)}\right|(1-\delta_{lk}) < 1. \qquad [5.35]$$

Dividing both numerator and denominator by s_j, the above inequality can be written as

$$\sum_{\substack{k=1\\k\neq j}}^{N}\sum_{\substack{l=1\\l\neq i\neq k}}^{N}\frac{|G'_{kl}(s_j)|}{|1/s_j\left(\omega_k^2 + s_j^2\right) + G'_{kk}(s_j)|} < 1. \qquad [5.36]$$

Taking the maximum for all $k \neq j$, this condition can further be represented as

$$\max_{k \neq j} \frac{\sum_{\substack{l=1 \\ l \neq j,k}}^{N} |G'_{kl}(s_j)|}{|1/s_j\left(\omega_k^2 + s_j^2\right) + G'_{kk}(s_j)|} < 1. \quad [5.37]$$

It is clear that [5.37] always holds if

$$\sum_{\substack{l=1 \\ l \neq j \neq k}}^{N} |G'_{kl}(s_j)| < |G'_{kk}(s_j)|, \quad \forall k \neq j \quad [5.38]$$

which, in turn, implies that, for all $j = 1 \cdots N$, the inequality $\| \mathbf{R}_j \| < 1$ holds if $\mathbf{G}'(s_j)$ is a diagonally dominant matrix. It is important to note that the diagonal dominance of $\mathbf{G}'(s_j)$ is only a sufficient condition and the lack of it does not necessarily prevent convergence of $\mathbf{S_u}$.

5.2.2. *Non-viscous modes*

When $2N < j \leq m$, the eigenvalues become real and consequently, from equation [5.17], we observe that $\mathbf{D}(s_j) \in \mathbb{R}^{N \times N}$. The non-viscous modes can be obtained from equation [5.16] by fixing any one element of the eigenvectors. Since $\mathbf{D}(s_j) \in \mathbb{R}^{N \times N}$, from equation [5.16], it is easy to see that $\mathbf{z}_j \in \mathbb{R}^N$. We partition the vector $\mathbf{z}_j$ as

$$\mathbf{z}_j = \begin{Bmatrix} \mathbf{z_1}_j \\ \mathbf{z_2}_j \end{Bmatrix}. \quad [5.39]$$

We select $\mathbf{z_1}_j = 1$ so that $\mathbf{z_2}_j \in \mathbb{R}^{(N-1)}$ has to be determined from equation [5.16]. Further, partition the matrix $\mathbf{D}(s_j)$ as

$$\mathbf{D}(s_j) = \begin{bmatrix} \mathbf{D_{11}}(s_j) & \mathbf{D_{12}}(s_j) \\ \mathbf{D_{21}}(s_j) & \mathbf{D_{22}}(s_j) \end{bmatrix} \quad [5.40]$$

where $\mathbf{D_{11}}(s_j) \in \mathbb{R}$, $\mathbf{D_{12}}(s_j) \in \mathbb{R}^{1 \times (N-1)}$, $\mathbf{D_{21}}(s_j) \in \mathbb{R}^{(N-1) \times 1}$ and $\mathbf{D_{22}}^{(j)} \in \mathbb{R}^{(N-1) \times (N-1)}$. In view of [5.40] and recalling that $\mathbf{z_1}_j = 1$, from equation [5.16], we can have

$$\begin{aligned} \mathbf{D_{22}}(s_j)\mathbf{z_2}_j &= -\mathbf{D_{21}}(s_j) \\ \text{or} \quad \mathbf{z_2}_j &= -\left[\mathbf{D_{22}}(s_j)\right]^{-1} \mathbf{D_{21}}(s_j). \end{aligned} \quad [5.41]$$

It may be noted that determination of the non-viscous modes is computationally more demanding than the elastic modes because inversion of an $(N-1)\times(N-1)$ real matrix is associated with each eigenvector. However, for most physically realistic non-viscous damping models, it appears that the number of non-viscous modes is not very high and also their contribution to the global dynamic response is not very significant (see the example problem in section 5.2.2). For this reason, calculation of the first few non-viscous modes may be sufficient from a practical point of view.

5.2.3. *Approximations for lightly damped systems*

The expression of the elastic modes derived in section 5.2.1 is quite general. Here, we consider some special cases and approximation of this general expression that are of practical interest. For light non-proportionally damped systems, the off-diagonal entries of the modal damping function matrix are small compared to the diagonal entries, that is $G'_{kl}(s_j) \leq G'_{kk}(s_j), \forall k \neq l, s_j$. For this reason, higher order terms in the series expression [5.32] can be neglected.

The expression for the elastic modes obtained by taking one term in the series [5.32] is close to the expression obtained from the first-order perturbation analysis. Considering the jth set of equation [5.20] and neglecting the second-order terms involving $\alpha_k^{(j)}$ and $G'_{kl}(s_j), \forall k \neq l$, and also noting that $\alpha_j^{(j)} = 1$, we obtain

$$\begin{aligned} & s_j^2 + s_j G'_{jj}(s_j) + \omega_j^2 \approx 0 \\ \text{or} \quad s_j & \approx \pm \mathrm{i}\omega_j - G'_{jj}(\pm \mathrm{i}\omega_j)/2 \\ & = \mathrm{i}\omega_j - G'_{jj}(\mathrm{i}\omega_j)/2, \quad -\mathrm{i}\omega_j - G'_{jj}(-\mathrm{i}\omega_j)/2. \end{aligned} \qquad [5.42]$$

This is the first-order approximate expression for the complex eigenvalues of system [5.1] corresponding to the elastic modes. A similar result was also obtained by Woodhouse [WOO 98]. In deriving this expression, the assumption has been made that $\mathbf{G}(s_j) \approx \mathbf{G}(\mathrm{i}\omega_j)$. Since the off-diagonal elements of $\mathbf{G}(s_j)$ are assumed small, it is expected that this approximation will not result in significant errors. Note that, as $\mathcal{G}(t)$ is a real function, $G'_{jj}(\bullet)$ satisfies the property

$$G'_{jj}(-\mathrm{i}\omega_j) = G'^{*}_{jj}(\mathrm{i}\omega_j). \qquad [5.43]$$

Using this relationship, it may be confirmed that the eigenvalues corresponding to the elastic modes, approximately given by equation [5.42], appear in complex conjugate pairs.

To obtain approximate expression for the eigenvectors, we simply consider only the first term of the series [5.32] and substitute $\mathbf{a}_j$ into equation [5.18] to obtain

$$\mathbf{z}_j \approx \mathbf{x}_j - \sum_{\substack{k=1 \\ k \neq j}}^{N} \frac{s_j G'_{kj}(s_j)\mathbf{x}_k}{\omega_k^2 + s_j^2 + s_j G'_{kk}(s_j)}. \qquad [5.44]$$

This is the first-order approximation of the complex eigenvectors. Now, retaining the first two terms of the series expression [5.32] and substituting $\hat{\mathbf{a}}^{(j)}$ into equation [5.18], we obtain

$$\begin{aligned} \mathbf{z}_j \approx\ & \mathbf{x}_j - \sum_{\substack{k=1 \\ k \neq j}}^{N} \frac{s_j G'_{kj}(s_j)\mathbf{x}_k}{\omega_k^2 + s_j^2 + s_j G'_k(s_j)} \\ & + \sum_{\substack{k=1 \\ k \neq j}}^{N} \sum_{\substack{l=1 \\ l \neq j \neq k}}^{N} \frac{s_j^2 G'_{kl}(s_j) G'_{lj}(s_j)\mathbf{x}_k}{\left(\omega_k^2 + s_j^2 + s_j G'_{kk}(s_j)\right)\left(\omega_l^2 + s_j^2 + s_j G'_{ll}(s_j)\right)}. \end{aligned} \qquad [5.45]$$

The above equation is second-order approximate expression for the eigenvectors corresponding to the elastic modes of system [5.1]. Finally, note that no such simple expressions can be obtained for the non-viscous modes unless some specific forms for the kernel function are assumed.

5.3. Transfer function

The transfer function (matrix) of a system completely defines its input–output relationship in *steady state*. It is well known that for any linear system, if the forcing function is harmonic, that is $\mathbf{f}(t) = \bar{\mathbf{f}} \exp[st]$ with $s = \mathrm{i}\omega$ and amplitude vector $\bar{\mathbf{f}} \in \mathbb{R}^N$, the steady-state response will also be harmonic at frequency $\omega \in \mathbb{R}^+$. So, we seek a solution of the form $\mathbf{q}(t) = \bar{\mathbf{q}} \exp[st]$, where $\bar{\mathbf{q}} \in \mathbb{C}^N$ is the response vector in the frequency domain. Substitution of $\mathbf{q}(t)$ and $\mathbf{f}(t)$ into equation [5.1] gives

$$s^2\mathbf{M}\bar{\mathbf{q}}(s) + s\,\mathbf{G}(s)\bar{\mathbf{q}}(s) + \mathbf{K}\bar{\mathbf{q}}(s) = \bar{\mathbf{f}}(s) \quad \text{or} \quad \mathbf{D}(s)\bar{\mathbf{q}} = \bar{\mathbf{f}}(s). \qquad [5.46]$$

Here, the *dynamic stiffness matrix* is

$$\mathbf{D}(s) = s^2\mathbf{M} + s\,\mathbf{G}(s) + \mathbf{K} \ \in \mathbb{C}^{N \times N}. \qquad [5.47]$$

From equation [5.46], the response vector $\bar{\mathbf{q}}(s)$ can be obtained as

$$\bar{\mathbf{q}}(s) = \mathbf{D}^{-1}(s)\bar{\mathbf{f}}(s) = \mathbf{H}(s)\bar{\mathbf{f}}(s) \qquad [5.48]$$

where

$$\mathbf{H}(s) = \mathbf{D}^{-1}(s) \in \mathbb{C}^{N\times N} \qquad [5.49]$$

is the transfer function matrix. From this equation, we further have

$$\mathbf{H}(s) = \frac{\text{adj}\,[\mathbf{D}(s)]}{\det\,[\mathbf{D}(s)]}. \qquad [5.50]$$

The *poles* of $\mathbf{H}(s)$, denoted by s_j, are the eigenvalues of the system. Because it is assumed that *all* the m eigenvalues are distinct, each pole is a *simple pole*. The matrix inversion in [5.48] is difficult to carry out in practice because of the singularities associated with the poles. Moreover, such an approach would be an expensive numerical exercise and may not offer much physical insight. For these reasons, we seek a solution analogous to the classical modal series solution of the undamped or proportionally damped systems.

From the residue theorem, it is known that any complex function can be expressed in terms of the poles and *residues*, that is the transfer function has the form

$$\mathbf{H}(s) = \sum_{j=1}^{m} \frac{\mathbf{R}_j}{s - s_j}. \qquad [5.51]$$

Here

$$\mathbf{R}_j = \operatorname*{res}_{s=s_j}\,[\mathbf{H}(s)] \stackrel{\text{def}}{=} \lim_{s\to s_j}(s - s_j)\,[\mathbf{H}(s)] \qquad [5.52]$$

is the residue of the transfer function matrix at the pole s_j. It may be noted that equation [5.51] is equivalent to expressing the right-hand side of equation [5.50] in the partial fraction form. Here, we try to obtain the residues, that is the coefficients in the partial fraction form, in terms of the system eigenvectors.

5.3.1. *Eigenvectors of the dynamic stiffness matrix*

It turns out that the eigenvectors of the dynamic stiffness matrix play an important role in determining the residues of the transfer function matrix. For any given $s \in \mathbb{C}$, the eigenvalue problem associated with the dynamic stiffness matrix can be expressed by

$$\mathbf{D}(s)\boldsymbol{\varphi}_k(s) = \nu_k(s)\boldsymbol{\varphi}_k(s), \quad \forall\, k = 1, \cdots, N. \qquad [5.53]$$

In the previous equation, the eigenvalues $\nu_k(s) \in \mathbb{C}$ are the roots of the characteristic equation

$$\det\left[\mathbf{D}(s) - \nu(s)\mathbf{I}_N\right] = 0 \qquad [5.54]$$

and $\boldsymbol{\varphi}_k(s) \in \mathbb{C}^N$ is the kth eigenvector of $\mathbf{D}(s)$. The symbols $\nu_k(s)$ and $\boldsymbol{\varphi}_k(s)$ indicate functional dependence of these quantities on the complex parameter s. Such a continuous dependence is expected whenever $\mathbf{D}(s)$ is a sufficiently smooth matrix function of s. It should be noted that because $\mathbf{D}(s)$ is an $N \times N$ complex matrix for a fixed s, the number of eigenvalues (and consequently the eigenvectors) must be N. Further, it can be shown that, for distinct eigenvalues, $\boldsymbol{\varphi}_k(s)$ also satisfies an orthogonality relationship although $\mathbf{z}_k$ does not enjoy any such simple relationship. We normalize $\boldsymbol{\varphi}_k(s)$ such that

$$\boldsymbol{\varphi}_j^T(s)\boldsymbol{\varphi}_k(s) = \delta_{kj}, \quad \forall\, k, j = 1, \cdots, N. \qquad [5.55]$$

In view of the above relationship, from equation [5.53], we have

$$\boldsymbol{\varphi}_j^T(s)\mathbf{D}(s)\boldsymbol{\varphi}_k(s) = \nu_k(s)\delta_{kj}, \quad \forall\, k, j = 1, \cdots, N \qquad [5.56]$$

or in the matrix form

$$\boldsymbol{\Phi}^T(s)\mathbf{D}(s)\boldsymbol{\Phi}(s) = \boldsymbol{\nu}(s). \qquad [5.57]$$

Here

$$\boldsymbol{\Phi}(s) = \left[\boldsymbol{\varphi}_1(s), \boldsymbol{\varphi}_2(s), \cdots, \boldsymbol{\varphi}_N(s)\right] \in \mathbb{C}^{N \times N}, \qquad [5.58]$$

$$\text{and} \quad \boldsymbol{\nu}(s) = \text{diag}\left[\nu_1(s), \nu_2(s), \cdots, \nu_N(s)\right] \in \mathbb{C}^{N \times N}. \qquad [5.59]$$

It is possible to establish the relationships between the original eigenvalue problem of the system defined by equation [5.16] and that by equation [5.53]. Consider the case when the parameter s approaches any one of the system eigenvalues, say s_j. Since *all* the $\nu_k(s)$ are assumed to be distinct, for non-trivial eigenvectors, comparing equations [5.16] and [5.53] we can conclude that one and only one of the $\nu_k(s)$ must be zero when $s \to s_j$, see [YAN 98]. Suppose that the rth eigenvalue of the eigenvalue problem [5.53] is zero when $s \to s_j$. It is also clear that the eigenvector in [5.53] corresponding to the rth eigenvalue also approaches the eigenvector in [5.16] as $s \to s_j$. Thus, when $s = s_j$, we have

$$\nu_r(s_j) = 0 \quad \text{and} \quad \nu_k(s_j) \neq 0, \forall k = 1, \cdots, N; \neq r \qquad [5.60]$$

and also

$$\boldsymbol{\varphi}_r(s_j) = \mathbf{z}_j. \qquad [5.61]$$

These equations completely relate the eigensolutions of [5.16] with [5.53]. Now, these relationships will be utilized to obtain the residues of the transfer function matrix.

5.3.2. *Calculation of the residues*

From equation [5.57], we have

$$\mathbf{D}^{-1}(s) = \boldsymbol{\Phi}(s)\boldsymbol{\nu}^{-1}(s)\boldsymbol{\Phi}^T(s). \qquad [5.62]$$

Using the expression of the transfer function in equation [5.49] and noting that $\boldsymbol{\nu}(s)$ is a diagonal matrix, we may expand the right-hand side of the above equation to obtain

$$\mathbf{H}(s) = \mathbf{D}^{-1}(s) = \sum_{k=1}^{N} \frac{\boldsymbol{\varphi}_k(s)\boldsymbol{\varphi}_k^T(s)}{\nu_k(s)}. \qquad [5.63]$$

Separation of the rth term in the above sum yields

$$\mathbf{H}(s) = \frac{\boldsymbol{\varphi}_r(s)\boldsymbol{\varphi}_r^T(s)}{\nu_r(s)} + \left[\sum_{\substack{k=1\\k\neq r}}^{N} \frac{\boldsymbol{\varphi}_k(s)\boldsymbol{\varphi}_k^T(s)}{\nu_k(s)}\right]. \qquad [5.64]$$

Clearly, when $s \to s_j$, the second term of the right-hand side of equation [5.64] is analytic because according to equation [5.60] $\nu_k(s_j) \neq 0, \forall k = 1, \cdots, N; \neq r$. Now, from equation [5.52], the residue at $s = s_j$ may be obtained as

$$\begin{aligned}
\mathbf{R}_j &\stackrel{\text{def}}{=} \lim_{s \to s_j} (s - s_j) \left\{ \frac{\boldsymbol{\varphi}_r(s)\boldsymbol{\varphi}_r^T(s)}{\nu_r(s)} \left[\sum_{\substack{k=1 \\ k \neq r}}^{N} \frac{\boldsymbol{\varphi}_k(s)\boldsymbol{\varphi}_k^T(s)}{\nu_k(s)} \right] \right\} \\
&= \lim_{s \to s_j} (s - s_j) \frac{\boldsymbol{\varphi}_r(s)\boldsymbol{\varphi}_r^T(s)}{\nu_r(s)} \\
&= \frac{\boldsymbol{\varphi}_r(s)\boldsymbol{\varphi}_r^T(s)|_{s=s_j}}{\frac{\partial \nu_r(s)}{\partial s}|_{s=s_j}} + \lim_{s \to s_j} \frac{(s - s_j)\frac{\partial}{\partial s}\left[\boldsymbol{\varphi}_k(s)\boldsymbol{\varphi}_k^T(s)\right]}{\frac{\partial \nu_r(s)}{\partial s}} \\
&\qquad \text{(using l'Hôspital's rule)} \\
&= \frac{\mathbf{z}_j \mathbf{z}_j^T}{\frac{\partial \nu_r(s)}{\partial s}|_{s=s_j}} \qquad \text{(by equation [5.61]).}
\end{aligned} \qquad [5.65]$$

The denominator in the above expression for the residues, $\frac{\partial \nu_r(s)}{\partial s}|_{s=s_j}$, is still unknown. Now, consider the rth eigenvalue problem associated with the dynamic stiffness matrix. Differentiation of equation [5.53] for $k = r$ with respect to s yields

$$\frac{\partial \mathbf{D}(s)}{\partial s}\boldsymbol{\varphi}_r(s) + \mathbf{D}(s)\frac{\partial \boldsymbol{\varphi}_r(s)}{\partial s} = \frac{\partial \nu_r(s)}{\partial s}\boldsymbol{\varphi}_r(s) + \nu_r(s)\frac{\partial \boldsymbol{\varphi}_r(s)}{\partial s}. \qquad [5.66]$$

Premultiplying the above equation by $\boldsymbol{\varphi}_r^T(s)$ and rearranging, we obtain

$$\boldsymbol{\varphi}_r^T(s)\frac{\partial \mathbf{D}(s)}{\partial s}\boldsymbol{\varphi}_r(s) + \left[\boldsymbol{\varphi}_r^T(s)\mathbf{D}(s) - \boldsymbol{\varphi}_r^T(s)\nu_r(s)\right]\frac{\partial \boldsymbol{\varphi}_r(s)}{\partial s} = \boldsymbol{\varphi}_r^T(s)\frac{\partial \nu_r(s)}{\partial s}\boldsymbol{\varphi}_r(s). \qquad [5.67]$$

Taking transpose of equation [5.53], it follows that the second term of the left-hand side of the above equation is zero. Using the normalizing condition in [5.55] and letting $s \to s_j$, from equation [5.67], we have

$$\frac{\partial \nu_r(s)}{\partial s}|_{s=s_j} = \mathbf{z}_j^T \frac{\partial \mathbf{D}(s)}{\partial s}|_{s=s_j}\mathbf{z}_j = \mathbf{z}_j^T \frac{\partial \mathbf{D}(s_j)}{\partial s_j}\mathbf{z}_j. \qquad [5.68]$$

The term $\frac{\partial \mathbf{D}(s_j)}{\partial s_j}$ can be obtained by differentiating equation [5.47] as

$$\frac{\partial \mathbf{D}(s_j)}{\partial s_j} = 2s_j\mathbf{M} + \mathbf{G}(s_j) + s_j\frac{\partial \mathbf{G}(s_j)}{\partial s_j}. \qquad [5.69]$$

Using [5.65] and [5.68], we finally obtain the residue as

$$\mathbf{R}_j = \frac{\mathbf{z}_j \mathbf{z}_j^T}{\mathbf{z}_j^T \frac{\partial \mathbf{D}(s_j)}{\partial s_j} \mathbf{z}_j}. \qquad [5.70]$$

The above equation completely relates the transfer function residues to the eigenvalues and eigenvectors of the system. Recalling that, among the m eigenvalues $2N$ appear in complex conjugate pairs, from equation [5.51], the transfer function matrix may be obtained as

$$\mathbf{H}(\mathrm{i}\omega) = \sum_{j=1}^{N} \left[\frac{\gamma_j \mathbf{z}_j \mathbf{z}_j^T}{\mathrm{i}\omega - s_j} + \frac{\gamma_j^* \mathbf{z}_j^* \mathbf{z}_j^{*T}}{\mathrm{i}\omega - s_j^*} \right] + \sum_{j=2N+1}^{m} \frac{\gamma_j \mathbf{z}_j \mathbf{z}_j^T}{\mathrm{i}\omega - s_j} \qquad [5.71]$$

where

$$\gamma_j = \frac{1}{\mathbf{z}_j^T \frac{\partial \mathbf{D}(s_j)}{\partial s_j} \mathbf{z}_j}. \qquad [5.72]$$

The transfer function matrix has two parts: the first part is due to the elastic modes and the second part is due to the non-viscous modes. Using a first-order perturbation method, Woodhouse [WOO 98] has obtained an expression of the transfer functions similar to equation [5.71]. However, the non-viscous part of the transfer functions was not included in [WOO 98].

5.3.3. *Special cases*

The expression for the transfer function matrix in equation [5.71] is a natural generalization for the familiar expressions for the transfer function matrix of undamped or viscously damped systems. Transfer functions for several useful special cases may be obtained from [5.71] as follows:

– *Undamped systems*: in this case, $\mathbf{G}(s) = 0$ results the order of the characteristic polynomial $m = 2N$; s_j is purely imaginary so that $s_j = \mathrm{i}\omega_j$ where $\omega_j \in \mathbb{R}$ are the undamped natural frequencies and $\mathbf{z}_j = \mathbf{x}_j \in \mathbb{R}^N$. In view of the mass normalization relationship in [2.63], $\gamma_j = \frac{1}{2\mathrm{i}\omega_j}$ and equation [5.71] leads to

$$\mathbf{H}(\mathrm{i}\omega) = \sum_{j=1}^{N} \frac{1}{2\mathrm{i}\omega_j} \left[\frac{1}{\mathrm{i}\omega - \mathrm{i}\omega_j} - \frac{1}{\mathrm{i}\omega + \mathrm{i}\omega_j} \right] \mathbf{x}_j \mathbf{x}_j^T = \sum_{j=1}^{N} \frac{\mathbf{x}_j \mathbf{x}_j^T}{\omega_j^2 - \omega^2}. \qquad [5.73]$$

– *Viscously damped systems with non-proportional damping* [LAN 66, VIG 86, GÉR 97]: in this case, $m = 2N$ and $\gamma_j = \frac{1}{\mathbf{z}_j^T[2s_j\mathbf{M}+\mathbf{C}]\mathbf{z}_j}$. These reduce expression [5.71] to

$$\mathbf{H}(\mathrm{i}\omega) = \sum_{j=1}^{N}\left[\frac{\gamma_j\mathbf{z}_j\mathbf{z}_j^T}{\mathrm{i}\omega - s_j} + \frac{\gamma_j^*\mathbf{z}_j^*\mathbf{z}_j^{*^T}}{\mathrm{i}\omega - s_j^*}\right]. \qquad [5.74]$$

5.4. Dynamic response

The steady-state response due to harmonic loads or the response due to broadband random excitation can be obtained directly from the expression of the transfer function matrix in equation [5.71]. In this section, we consider the system response due to transient loads and initial conditions in the time and frequency domains.

Taking the Laplace transform of the equation of motion [5.1] and considering the initial conditions in [5.2], we have

$$s^2\mathbf{M}\bar{\mathbf{q}}(s) - s\mathbf{M}\mathbf{q}_0 - \mathbf{M}\dot{\mathbf{q}}_0 + s\,\mathbf{G}(s)\bar{\mathbf{q}}(s) - \mathbf{G}(s)\mathbf{q}_0 + \mathbf{K}\bar{\mathbf{q}}(s) = \bar{\mathbf{f}}(s)$$

$$\text{or} \quad \left[s^2\mathbf{M} + s\,\mathbf{G}(s) + \mathbf{K}\right]\bar{\mathbf{q}}(s) = \bar{\mathbf{f}}(s) + \mathbf{M}\dot{\mathbf{q}}_0 + \left[s\mathbf{M} + \mathbf{G}(s)\right]\mathbf{q}_0. \qquad [5.75]$$

Using the expression for the transfer function derived before, the response vector $\bar{\mathbf{q}}$ may be obtained as

$$\bar{\mathbf{q}}(s) = \sum_{j=1}^{m}\frac{\gamma_j\mathbf{z}_j\mathbf{z}_j^T}{s - s_j}\left\{\bar{\mathbf{f}}(s) + \mathbf{M}\dot{\mathbf{q}}_0 + \left[s\mathbf{M} + \mathbf{G}(s)\right]\mathbf{q}_0\right\}. \qquad [5.76]$$

This can be simplified further to

$$\bar{\mathbf{q}}(\mathrm{i}\omega) = \sum_{j=1}^{m}\frac{\gamma_j A_j(\mathrm{i}\omega)}{\mathrm{i}\omega - s_j}\mathbf{z}_j \qquad [5.77]$$

where the frequency-dependent complex scalar

$$A_j(\mathrm{i}\omega) = \mathbf{z}_j^T\bar{\mathbf{f}}(\mathrm{i}\omega) + \mathbf{z}_j^T\mathbf{M}\dot{\mathbf{q}}_0 + \mathrm{i}\omega\mathbf{z}_j^T\mathbf{M}\mathbf{q}_0 + \mathbf{z}_j^T\mathbf{G}(\mathrm{i}\omega)\mathbf{q}_0. \qquad [5.78]$$

The summation in equation [5.77] may be split into two different parts: the first part would correspond to the $2N$ complex conjugate pairs of elastic modes and the second part would be the contribution of the non-viscous modes.

The response in the time domain due to any forcing function can be obtained using a convolution integral over the *impulse response function*. From the expression of the transfer function in equation [5.71], the impulse response function matrix $\mathbf{h}(t) \in \mathbb{R}^{N\times N}$ may be obtained as

$$\mathbf{h}(t) = \sum_{j=1}^{N} \left[\gamma_j \mathbf{z}_j \mathbf{z}_j^T e^{s_j t} + \gamma_j^* \mathbf{z}_j^* \mathbf{z}_j^{*^T} e^{s_j^* t} \right] + \sum_{j=2N+1}^{m} \gamma_j \mathbf{z}_j \mathbf{z}_j^T e^{s_j t}. \qquad [5.79]$$

The response due to the initial conditions may also be obtained by taking the inverse transform of equation [5.76]. First, simplify equation [5.76] to obtain

$$\bar{\mathbf{q}}(s) = \sum_{j=1}^{m} \gamma_j \left[\frac{\mathbf{z}_j^T \bar{\mathbf{f}}(s) + \mathbf{z}_j^T \mathbf{G}(s) \mathbf{q}_0}{s - s_j} + \frac{\mathbf{z}_j^T \mathbf{M} \dot{\mathbf{q}}_0}{s - s_j} + \left(1 + \frac{s_j}{s - s_j} \right) \mathbf{z}_j^T \mathbf{M} \mathbf{q}_0 \right] \mathbf{z}_j. \qquad [5.80]$$

From the above, we have

$$\mathbf{q}(t) = \mathcal{L}^{-1}[\bar{\mathbf{q}}(s)] = \sum_{j=1}^{N} \left[\gamma_j a_j(t) \mathbf{z}_j + \gamma_j^* a_j^*(t) \mathbf{z}_j^* \right] + \sum_{j=2N+1}^{m} \gamma_j a_j(t) \mathbf{z}_j \qquad [5.81]$$

where the time-dependent scalar coefficients

$$a_j(t) = \int_0^t e^{s_j(t-\tau)} \left\{ \mathbf{z}_j^T \mathbf{f}(\tau) + \mathbf{z}_j^T \mathcal{G}(\tau) \mathbf{q}_0 \right\} \mathrm{d}\tau + e^{s_j t} \left\{ \mathbf{z}_j^T \mathbf{M} \dot{\mathbf{q}}_0 + s_j \mathbf{z}_j^T \mathbf{M} \mathbf{q}_0 \right\}; \quad \forall t > 0. \qquad [5.82]$$

The expression of the system response, either the frequency-domain description in equation [5.77] or the time-domain description in equation [5.81], is similar to the classical modal superposition result for undamped or proportionally damped systems usually obtained using the mode–orthogonality relationships. Thus, the formulation presented here is a generalization of the classical result where the real normal modes are appropriately replaced by the elastic modes and the non-viscous modes. Also note that we have not used any orthogonality relationship – the expression of the transfer function residue in equation [5.70] allows us to express the response in terms of superposition of individual modes even when the equation of motion cannot be decoupled.

5.4.1. *Summary of the method*

Following the procedure outlined so far, we can obtain the eigensolutions and dynamic response of non-viscously damped linear systems. Here, we briefly summarize the steps to be followed in order to obtain the eigenvalues, eigenvectors and dynamic response of non-viscously damped linear systems:

1) Evaluate the natural frequencies ω_j and eigenvectors $\mathbf{x}_j$ of the undamped system from $\mathbf{K}\mathbf{x}_j = \omega_j^2\mathbf{M}\mathbf{x}_j$ for all $j = 1, \cdots, N$. Normalize so that $\mathbf{x}_l^T\mathbf{M}\mathbf{x}_j = \delta_{lj}$.

2) Obtain the eigenvalues by solving the (say mth order) characteristic equation $\det\left[s^2\mathbf{M} + s\mathbf{G}(s) + \mathbf{K}\right] = 0$. For convenience, arrange the eigenvalues as $s_1, s_2, \cdots, s_N, s_1^*, s_2^*, \cdots, s_N^*, s_{2N+1}, s_{2N+2}, \cdots, s_m$.

3) Set up the $\mathbf{G}'(s_j)$ matrix using $G'_{kl}(s_j) = \mathbf{x}_k^T\mathbf{G}(s_j)\mathbf{x}_l$. Calculate the matrices $\mathbf{R}_j$ using

$$R_{j_{k_1 l_1}} = \frac{-s_j G'_{kl}(s_j)\left(1 - \delta_{k_1 l_1}\right)}{\omega_k^2 + s_j^2 + s_j G'_{kk}(s_j)}. \qquad [5.83]$$

To keep the dimension of $\mathbf{R}_u{}^{(j)}$ to $(N-1)\times(N-1)$, the subscript k_1 is expressed as $k_1 = k - \mathcal{U}_{(k,j)}$. Here, the function $\mathcal{U}_{(k,j)}$ is defined as

$$\mathcal{U}_{(k,j)} = \begin{cases} 0 & \text{if} \quad k < j, \\ 1 & \text{if} \quad k > j, \\ \text{not defined} & \text{if} \quad k = j. \end{cases} \qquad [5.84]$$

4) Obtain the vectors $\mathbf{a}_0^{(j)}$ using

$$a_{j0_l} = \frac{-s_j G'_{lj}(s_j)}{\omega_l^2 + s_j^2 + s_j G'_{ll}(s_j)}. \qquad [5.85]$$

5) Select the number of terms, say n, and calculate $\mathbf{a}_{j_k} = \mathbf{R}_j\mathbf{a}_{j_{k-1}}$ for all $k = 1, \cdots, n$. Subsequently, obtain the complex constants $\mathbf{a}^{(j)} = \sum_{k=1}^{n}\mathbf{a}_{j_{k-1}}$.

6) Determine the *elastic modes* as $\mathbf{u}_j = \mathbf{x}_j + \sum_{\substack{l=1\\l\neq j}}^{N} a_{j0_l}\mathbf{x}_l$ for all $j = 1, \cdots, N$.

7) For $2N < j < 2N + p$, obtain the dynamic stiffness matrix $\mathbf{D}(s_j) = s_j^2\mathbf{M} + s_j\mathbf{G}(s_j) + \mathbf{K}$. Partition as $\mathbf{D}(s_j) = \begin{bmatrix} \mathbf{D_{11}}(s_j) & \mathbf{D_{12}}(s_j) \\ \mathbf{D_{21}}(s_j) & \mathbf{D_{22}}(s_j) \end{bmatrix}$ where $\mathbf{D_{11}}(s_j) \in \mathbb{R}$, $\mathbf{D_{12}}(s_j) \in \mathbb{R}^{1\times(N-1)}$, $\mathbf{D_{21}}(s_j) \in \mathbb{R}^{(N-1)\times 1}$ and $\mathbf{D_{22}}{}^{(j)} \in \mathbb{R}^{(N-1)\times(N-1)}$.

8) Calculate $\mathbf{z_2}_j = -\left[\mathbf{D_{22}}(s_j)\right]^{-1}\mathbf{D_{21}}(s_j)$. Now obtain the *non-viscous* modes as $\mathbf{u}_j = \left\{1, \mathbf{z_2}_j\right\}$ for all $j = 2N+1, \cdots m$.

9) Finally, using the eigensolutions, calculate the response either in the frequency domain from equation [5.77] or, alternatively, in the time domain from equation [5.81].

This procedure is general, simple, direct and provides better physical insights as the familiar N-space eigenvectors are only used. The approach also offers reduction in computational effort because it neither uses the state-space formalism nor utilizes additional dissipation coordinates. Applications of the proposed method are illustrated next.

5.5. Numerical examples

5.5.1. *The system*

We consider a three degree-of-freedom system to illustrate the proposed method. Figure 5.1 shows the example taken together with the numerical values considered for mass and stiffness properties. A similar system with viscous damping has been studied in [NEW 89, pp. 148–151]. Damping is associated only with the middle mass and the kernel function corresponding to this damper has the form

$$\mathcal{G}_{22}(t) = c\,g(t) \qquad [5.86]$$

where c is a damping coefficient and $g(t)$ is the damping function. Two different forms of $g(t)$ available in the literature will be considered here. For the first model (exponential)

$$g(t) = \mu e^{-\mu t}; \quad \mu, t \geq 0 \qquad [5.87]$$

and for the second model (double exponential)

$$g(t) = \frac{1}{2}\left(\mu_1 e^{-\mu_1 t} + \mu_2 e^{-\mu_2 t}\right); \quad \mu_1, \mu_2, t \geq 0. \qquad [5.88]$$

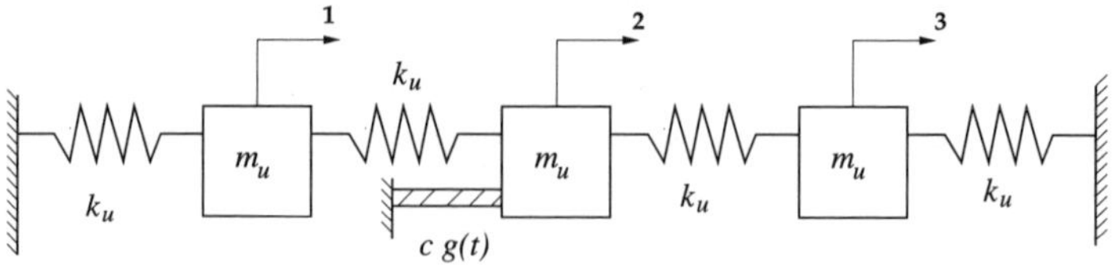

Figure 5.1. *Three degree-of-freedom non-viscously damped system, $m_u = 1$ kg, $k_u = 1$ N/m*

The exponential function in [5.87] is possibly the simplest physically realistic non-viscous damping model. This function, often known as a "relaxation function", was introduced by Biot [BIO 55]. It has been used extensively in the context of viscoelastic systems. In [ADH 14b], Chapter 3, a method will be proposed to identify such damping models using modal testing. The double exponential damping function, known as the Golla-Hughes-McTavish (GHM) model, was introduced in [GOL 85, MCT 93]. Identification of the GHM model has been discussed in [FRI 97]. Both damping functions have been scaled so as to have unit area when integrated to infinity. This makes them directly comparable with the viscous model in which the corresponding damping function would be a unit delta function, $g(t) = \delta(t)$, and the coefficient c would be the usual viscous damping coefficient. The difference between a delta function and $g(t)$ given by equations [5.87] and [5.88] is that at $t = 0$ they start with finite values of μ and $1/2\,(\mu_1 + \mu_2)$, respectively. Thus, the values of μ, μ_1 and μ_2 give a notion of *non-viscousness* – if they are large, the damping behavior will be near viscous, and vice versa.

The mass and stiffness matrices and the damping matrix in the Laplace domain for the problem can be obtained as

$$\mathbf{M} = \begin{bmatrix} m_u & 0 & 0 \\ 0 & m_u & 0 \\ 0 & 0 & m_u \end{bmatrix}, \qquad [5.89]$$

$$\mathbf{K} = \begin{bmatrix} 2k_u & -k_u & 0 \\ -k_u & 2k_u & -k_u \\ 0 & -k_u & 2k_u \end{bmatrix} \qquad [5.90]$$

and

$$\mathbf{G}(s) = \begin{bmatrix} 0 & 0 & 0 \\ 0 & cG(s) & 0 \\ 0 & 0 & 0 \end{bmatrix}. \qquad [5.91]$$

Here, $G(s)$ is the Laplace transform of $g(t)$. Next, the eigensolutions and the dynamic response of the system are discussed for the two functional forms of $g(t)$.

5.5.2. *Example 1: exponential damping*

5.5.2.1. *Eigensolutions*

We assume $c = 0.3$, as considered in [NEW 89, p. 149] for the equivalent viscously damped system. From equation [5.87], we obtain

$$G(s) = \frac{1}{s+\mu}. \qquad [5.92]$$

Using this expression, the characteristic equation can be simplified as

$$m_u^3 s^7 + m_u^3 \mu\, s^6 + \left(2\, m_u^2 k_u + m_u \left(\mu\, c m_u + 4\, m_u k_u\right)\right) s^5 + 6\, k_u m_u^2 \mu\, s^4$$
$$+ \left(2\, k_u \left(\mu\, c m_u + 4\, m_u k_u\right) + m_u \left(2\, \mu\, c k_u + 2\, {k_u}^2\right)\right) s^3 + 10\, {k_u}^2 m_u \mu\, s^2 \qquad [5.93]$$
$$+2\, k_u \left(2\, \mu\, c k_u + 2\, {k_u}^2\right) s + 4\, {k_u}^3 \mu = 0.$$

The order of the above polynomial, $m = 7$. Since the system has three degrees-of-freedom, there are three elastic modes corresponding to the three modes of vibration. The number of the non-viscous modes, $p = m - 2N = 1$.

It is of interest to us to understand the effect of "non-viscousness" on the eigensolutions. Figure 5.2 shows the locus of the third eigenvalue, that is s_3, plotted as a function of μ. It is interesting to observe that the locus is much more sensitive in the region of lower values of μ (i.e. when damping is significantly non-viscous) compared to that in the region of higher values. The eigenvalue of the corresponding viscously damped system is also plotted (marked by *) in the same figure. Note that the non-viscous damping mechanism approaches the viscous damping when $\mu >\approx 50.0$. Similar behavior has also been observed (results not shown here) for the locus of s_1 also. The second mode, in which the middle mass remains stationary, is not affected by damping.

The eigenvectors of the system, i.e. the three elastic modes (together with their complex conjugates) and one non-viscous mode, can be obtained in a straightforward manner by following the steps outlined in the previous section. We select two representative values of μ – one when μ is large (i.e. near viscous case) and the other when μ is small. The undamped eigenvalues and eigenvectors are obtained as

$$\{\omega_1, \omega_2, \omega_3\} = \{0.7654, 1.4142, 1.8478\} \qquad [5.94]$$

and

$$[\mathbf{x}_1, \mathbf{x}_2, \mathbf{x}_3] = \begin{bmatrix} 0.5 & 0.7071 & -0.5 \\ 0.7071 & 0.0 & 0.7071 \\ 0.5 & -0.7071 & -0.5 \end{bmatrix}. \quad [5.95]$$

Using these results, when $\mu = 50.0$, for the elastic modes, we have

$$\{s_1, s_2, s_3\} = \{-0.0757 + 0.7659\text{i}, 1.4142\text{i}, -0.0751 + 1.8416\text{i}\} \quad [5.96]$$

and

$$[\mathbf{z}_1, \mathbf{z}_2, \mathbf{z}_3] = \begin{bmatrix} 0.4983 + 0.0204\text{i} & 0.7071 & -0.5002 + 0.0491\text{i} \\ 0.7095 - 0.0289\text{i} & 0.0 & 0.7069 + 0.0694\text{i} \\ 0.4983 + 0.0204\text{i} & -0.7071 & -0.5002 + 0.0491\text{i} \end{bmatrix}. \quad [5.97]$$

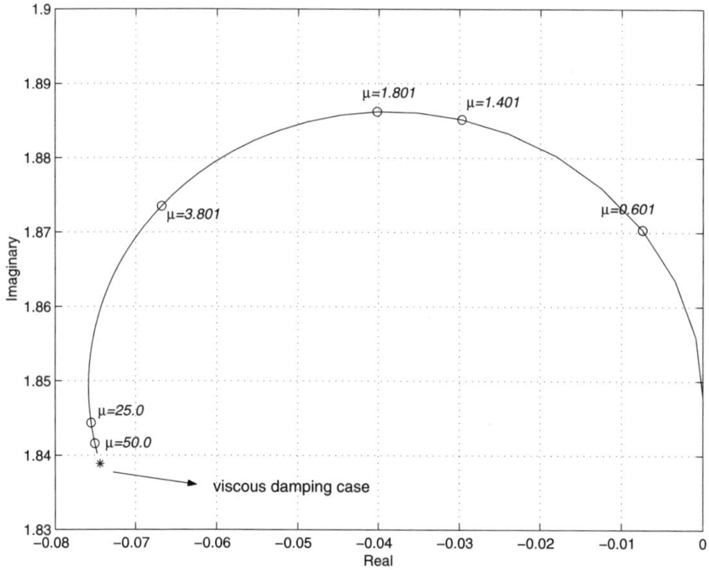

Figure 5.2. *Root-locus diagram showing the locus of the third eigenvalue (s_3) as a function of μ*

The above calculation is performed by retaining five terms in the series [5.32]. It may be verified that, because μ is large (about 27 times the maximum natural frequency), the results obtained are close to the viscously damped case, see [NEW 89, p. 149]. For the one non-viscous mode, we obtain

$$s_7 = -49.6984 \quad \text{and} \quad \mathbf{z}_7 = \begin{Bmatrix} 1.0 \\ 2.4719 \times 10^3 \\ 1.0 \end{Bmatrix}. \qquad [5.98]$$

Because s_7 is purely real and negative, this mode is non-oscillatory (overcritically damped) and stable.

When $\mu = 0.5$, the damping is significantly non-viscous. For this case, performing similar calculations for the elastic modes, we have

$$\{s_1, s_2, s_3\} = \{-0.0207 + 0.8\mathrm{i}, 1.4142\mathrm{i}, -0.0053 + 1.8671\mathrm{i}\} \qquad [5.99]$$

and

$$[\mathbf{z}_1, \mathbf{z}_2, \mathbf{z}_3] = \begin{bmatrix} 0.4983 + 0.0204\mathrm{i} & 0.7071 & -0.5002 + 0.0491\mathrm{i} \\ 0.6787 + 0.0112\mathrm{i} & 0.0 & 0.7442 - 0.0630\mathrm{i} \\ 0.4983 + 0.0204\mathrm{i} & -0.7071 & -0.5002 + 0.0491\mathrm{i} \end{bmatrix}. \qquad [5.100]$$

These values are not significantly different from those obtained for $\mu = 50.0$ in [5.97]. For this problem, the elastic modes are not very sensitive to the damping mechanism. However, we emphasize that this fact *cannot* be generalized to all systems. For the non-viscous mode, we have

$$s_7 = -0.4480 \quad \text{and} \quad \mathbf{z}_7 = \begin{Bmatrix} 1.0 \\ 2.2007 \\ 1.0 \end{Bmatrix}. \qquad [5.101]$$

These values are, however, quite different from those obtained for $\mu = 50.0$ in [5.98]. It is difficult to physically visualize the nature of the non-viscous modes in general. These modes are intrinsic to the dampers and we do not have sufficient generalized coordinates to represent them properly. Nevertheless, they yield non-zero residues in the system transfer functions and thus contribute to the global dynamic response.

5.5.2.2. *Dynamic response analysis*

The problem of stationary random vibration analysis of the system is considered here. Suppose the system is subjected to a band-limited Gaussian *white noise* at the third degree-of-freedom. We are interested in the resulting displacement of the system at the third degree-of-freedom (i.e. $\mathbf{z}_3$). The power spectral density (PSD) of the response (see [NIG 83] for details) can be given by

$$S_{uu}(\mathrm{i}\omega) = |H_{33}(\mathrm{i}\omega)|^2 S_{ff}(\mathrm{i}\omega) \qquad [5.102]$$

where

$$S_{ff}(\mathrm{i}\omega) = \begin{cases} 1 & \text{if} \quad 0 < \omega \leq 2.5\,\text{rad/s} \\ 0 & \text{otherwise} \end{cases}. \qquad [5.103]$$

In Figure 5.3, the PSD of $\mathbf{z}_3$, that is $|H_{33}(\mathrm{i}\omega)|^2$, is plotted for the cases when $\mu = 50.0$ and $\mu = 0.5$. These results are obtained by direct application of equation [5.71]. From the figure, observe that the damping is less for the case when $\mu = 0.5$ than when $\mu = 50.0$. Also note the (horizontal) shift in the position of the natural frequencies. These features may also be observed in the root-locus diagram as shown in Figure 5.2. To understand the effect of "non-viscosity", in the same diagram, we have plotted the non-viscous term (the second term) appearing in equation [5.71] for both values of μ. For this problem, the non-viscous part is quite small and becomes smaller at higher frequencies. Observe that when $\mu = 0.5$, that is when damping is significantly non-viscous, the value of the non-viscous part of the response is more than that when $\mu = 50.0$. This plot also clearly demonstrates that the non-viscous part of the response is *not* oscillatory in nature.

5.5.3. *Example 2: GHM damping*

Taking the Laplace transform of equation [5.88], we obtain

$$G(s) = \frac{(\mu_1 + \mu_2)/2\, s + \mu_1\mu_2}{s^2 + (\mu_1 + \mu_2)\, s + \mu_1\mu_2}. \qquad [5.104]$$

Using this equation, together with the expressions of the system matrices given by equations [5.89]–[5.91], it can be shown that the order of the characteristic polynomial, $m = 8$. Thus, the number of the non-viscous modes, $p = m - 2N = 2$. In this section, we focus our attention on the numerical accuracy of the formulation developed in this chapter.

Regarding the numerical values of the damping parameters, we assume $c = 0.5$, $\mu_1 = 1$ and $\mu_2 = 3$. Small values of μ_1 and μ_2 indicate that the damping mechanism is strongly non-viscous. By solving the characteristic equation, exact eigenvalues corresponding to the three elastic modes can be obtained as

$$\{s_1, s_2, s_3\} = \{-0.0994 + 0.8180\mathrm{i}, 1.4142\mathrm{i}, -0.0687 + 1.9025\mathrm{i}\} \qquad [5.105]$$

and their complex conjugate pairs. Eigenvalues corresponding to the two non-viscous modes are found to be

$$\{s_7, s_8\} = \{-2.7901, -0.8738\}. \tag{5.106}$$

Eigenvalues corresponding to the elastic modes can also be obtained approximately by equation [5.42] in section 5.2.3. Recall that only the undamped eigensolutions are required in order to apply this equation. Approximate eigenvalues using equation [5.42] are calculated as

$$\{s_1, s_2, s_3\}_{\text{approx}} = \{-0.0981 - 0.8105\text{i}, 1.4142\text{i}, -0.0595 - 1.9018\text{i}\}. \tag{5.107}$$

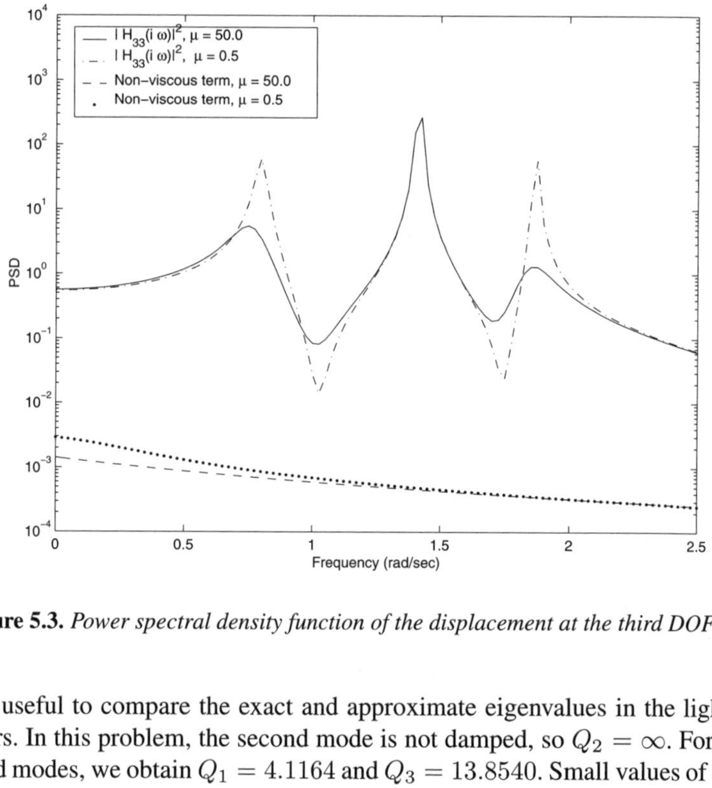

Figure 5.3. *Power spectral density function of the displacement at the third DOF ($\mathbf{z}_3$)*

It is useful to compare the exact and approximate eigenvalues in the light of the Q-factors. In this problem, the second mode is not damped, so $Q_2 = \infty$. For the first and third modes, we obtain $Q_1 = 4.1164$ and $Q_3 = 13.8540$. Small values of Q-factor indicate that these modes, are quite heavily damped. Comparing equations [5.105] and [5.107], it may be observed that the approximate values are quite close to the exact values even when damping is reasonably high.

To check the numerical accuracy of the eigenvectors, first, the exact values are calculated by the matrix inversion method. For the elastic modes, we obtain

$$[\mathbf{z}_1, \mathbf{z}_2, \mathbf{z}_3] = \begin{bmatrix} 0.5114 + 0.0299\mathrm{i} & 0.7071 & -0.4639 + 0.0403\mathrm{i} \\ 0.6905 - 0.0431\mathrm{i} & 0.0 & 0.7596 + 0.0562\mathrm{i} \\ 0.5114 + 0.0299\mathrm{i} & -0.7071 & -0.4639 + 0.0403\mathrm{i} \end{bmatrix} \qquad [5.108]$$

and their complex conjugates. For the two non-viscous modes, we have

$$[\mathbf{z}_7, \mathbf{z}_8] = \begin{bmatrix} 1.0000 & 1.0000 \\ 9.7847 & 2.7636 \\ 1.0000 & 1.0000 \end{bmatrix}. \qquad [5.109]$$

Approximate eigenvectors corresponding to the elastic modes, calculated by equation [5.44], are obtained as

$$[\mathbf{z}_1, \mathbf{z}_2, \mathbf{z}_3]_{\text{approx}} = \begin{bmatrix} 0.5114 + 0.0299\mathrm{i} & 0.7071 & -0.4639 + 0.0403\mathrm{i} \\ 0.6910 - 0.0422\mathrm{i} & 0.0 & 0.7582 + 0.0569\mathrm{i} \\ 0.5114 + 0.0299\mathrm{i} & -0.7071 & -0.4639 + 0.0403\mathrm{i} \end{bmatrix}. \qquad [5.110]$$

The above values are equivalent to performing the calculation by retaining only one term in the series [5.32]. Also recall that the approximate values are obtained from the undamped eigensolutions only. Comparing equation [5.108] and [5.110], it is clear that the results obtained from the approximate method match the exact solutions to an excellent accuracy.

As a final check on the formulation developed in this chapter, we compare the transfer function obtained from equation [5.71] with the exact transfer function calculated by inversion of the dynamic stiffness matrix. Figure 5.4 shows such a comparison for $H_{33}(\mathrm{i}\omega)$. Approximate natural frequencies and modes given by [5.107] and [5.110] are used and also the non-viscous term in equation [5.71] is neglected in order to calculate the approximate transfer function. Thus, in turn, the approximate transfer function in Figure 5.4 is obtained only by proper "post-processing" of the undamped eigensolutions. From this figure, it may be observed that, except in a few places, the approximate transfer function is reasonably close to the exact one. This analysis demonstrates the usefulness of the proposed method.

5.6. Eigenrelations of non-viscously damped systems

In the previous sections, the eigenvalues, eigenvectors and transfer functions associated with multiple-degrees-of-freedom non-viscously damped systems have been discussed. A method was outlined to obtain the eigenvectors and dynamic response of the system. Although the method is analogous to classical modal analysis, unlike the classical modes, little is understood about qualitative properties of the modes of non-viscously damped systems. The purpose of this chapter is to develop some basic relationships satisfied by the eigensolutions and the system matrices of [5.12]. Specifically, we have focused our attention on the normalization and orthogonality relationship of the eigenvectors. Nature of the eigensolutions is briefly reviewed in section 5.6.1. In section 5.6.2, normalization of the eigenvectors is discussed. Orthogonality properties of the eigenvectors are established in section 5.6.3. Relationships between the eigensolutions and the damping function matrix are explored in section 5.6.4. In section 5.6.5, the system matrices are expressed in terms of the eigensolutions. As a special case, eigenrelations for viscously damped systems are derived in section 5.6.6. The theoretical results derived here are illustrated by numerical examples in section 5.6.7.

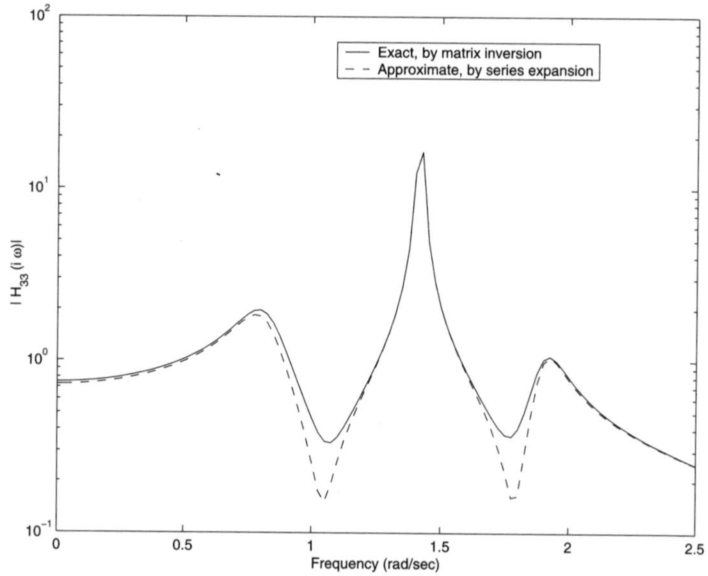

Figure 5.4. *Transfer function* $H_{33}(\mathrm{i}\omega)$

5.6.1. *Nature of the eigensolutions*

The eigenvalue problem associated with equation [5.1] can be defined as

$$\left[s_j^2\mathbf{M} + s_j\,\mathbf{G}(s_j) + \mathbf{K}\right]\mathbf{z}_j = \mathbf{0} \quad \text{or} \quad \mathbf{D}(s_j)\mathbf{z}_j = \mathbf{0},\ \forall\, k = 1, \cdots, m \qquad [5.111]$$

where the *dynamic stiffness matrix*

$$\mathbf{D}(s) = s^2\mathbf{M} + s\,\mathbf{G}(s) + \mathbf{K} \in \mathbb{C}^{N \times N}. \quad [5.112]$$

Here, $\mathbf{z}_j$ is the jth eigenvector and s_j is the jth eigenvalue. In general, the number of eigenvalues, $m = 2N + p; p \geq 0$. It is assumed that *all* m eigenvalues are distinct. It should be noted that eigenvalue problems of this kind are not similar to the eigenvalue problems that arise in the context of Lambda matrices [LAN 66] because $\mathbf{G}(s)$ is not a constant matrix. We consider the damping to be "non-proportional", that is the mass and stiffness matrices as well as the matrix of the kernel functions cannot be simultaneously diagonalized by any linear transformation. It is assumed that $|G_{jk}(s)| < \infty$ when $s \to \infty$. This, in turn, implies that the elements of $\mathbf{G}(s)$ are at the most of order $1/s$ in s or constant, as in the case of viscous damping. Construct the diagonal matrix containing the eigenvalues as

$$\mathbf{S} = \operatorname{diag}\left[s_1, s_2, \cdots, s_m\right] \in \mathbb{C}^{m \times m} \quad [5.113]$$

and the matrix containing the eigenvectors (the modal matrix) as

$$\mathbf{Z} = \left[\mathbf{z}_1, \mathbf{z}_2, \cdots, \mathbf{z}_m\right] \in \mathbb{C}^{N \times m}. \quad [5.114]$$

Next, we consider the normalization relationship of the eigenvectors.

5.6.2. *Normalization of the eigenvectors*

Premultiplying equation [5.111] by $\mathbf{z}_k^T$, applying equation [5.111] for k-th set and postmultiplying by $\mathbf{z}_j$ and subtracting one from the other, we obtain

$$\mathbf{z}_k^T\left[\left(s_j^2 - s_k^2\right)\mathbf{M} + s_j\mathbf{G}(s_j) - s_k\mathbf{G}(s_k)\right]\mathbf{z}_j = 0. \quad [5.115]$$

Since s_j and s_k are distinct for different j and k, the above equation can be divided by $(s_j - s_k)$ to obtain

$$\mathbf{z}_k^T\left[(s_j + s_k)\mathbf{M} + \frac{s_j\mathbf{G}(s_j) - s_k\mathbf{G}(s_k)}{s_j - s_k}\right]\mathbf{z}_j = 0,\ \forall j, k; j \neq k. \quad [5.116]$$

This equation may be regarded as the orthogonality relationship of the eigenvectors. It is easy to verify that in the undamped limit equation [5.116] degenerates to the familiar mass-orthogonality relationship of the undamped eigenvectors. However, this orthogonality relationship is not very useful because it is expressed in terms of the natural frequencies. A frequency-independent orthogonality relationship of the eigenvectors will be derived later in this chapter. Assuming $\delta_s = s_j - s_k$, rewrite equation [5.116] as

$$\mathbf{z}_k^T \left[(\delta_s + 2s_k)\,\mathbf{M} + \frac{(s_k + \delta_s)\,\mathbf{G}(s_k + \delta_s) - s_k \mathbf{G}(s_k)}{\delta_s} \right] \mathbf{z}_j = 0. \qquad [5.117]$$

Consider the case when $s_j \to s_k$, that is $\delta_s \to 0$. For this limiting case, equation [5.117] reads

$$\mathbf{z}_k^T \left[2s_k \mathbf{M} + \frac{\partial\,[s\mathbf{G}(s)]}{\partial s}|_{s_k} \right] \mathbf{z}_k = \gamma_k \qquad [5.118]$$

$$\text{or} \quad \mathbf{z}_k^T \left[2s_k \mathbf{M} + \mathbf{G}(s_k) + s_k \frac{\partial \mathbf{G}(s_k)}{\partial s} \right] \mathbf{z}_k = \gamma_k, \quad \forall k = 1, \cdots, m \qquad [5.119]$$

for some non-zero $\gamma_k \in \mathbb{C}$. Equation [5.119] is the normalization relationship for the eigenvectors of the non-viscously damped system [5.1]. From the expression of the dynamic stiffness matrix in [5.112], the normalization condition in equation [5.119] can also be expressed as

$$\mathbf{z}_k^T \frac{\partial \mathbf{D}(s_k)}{\partial s} \mathbf{z}_k = \gamma_k, \quad \forall k = 1, \cdots, m. \qquad [5.120]$$

Equation [5.119], and consequently equation [5.120], can be regarded as the generalization of the "mass normalization" relationship used in structural dynamics. In the undamped limit when $\mathbf{G}(s)$ is a null matrix, equation [5.119] reduces to the familiar mass normalization relationship for the undamped eigenvectors. For viscously damped systems (see section 5.6.6 for details), a relationship analogous to [5.119] was obtained, for example, in [SES 94] using a state-space approach and in [FAW 76, FAW 77a] using second-order equations of motion. We define the *normalization matrix*, $\boldsymbol{\Theta}$, as

$$\boldsymbol{\Theta} = \text{diag}\,[\gamma_1, \gamma_2, \cdots, \gamma_m] \in \mathbb{C}^{m \times m}. \qquad [5.121]$$

Numerical values of γ_k can be selected in various ways:

– Choose $\gamma_k = 2s_k, \forall k$, that is $\boldsymbol{\Theta} = 2\mathbf{S}$. This reduces to $\mathbf{z}_k^T \mathbf{M} \mathbf{z}_k = 1, \forall k$ when the damping is zero. This is consistent with the unity modal mass convention, often used in experimental modal analysis and finite element methods.

– Choose $\gamma_k = 1 + 0\mathrm{i}, \forall k$, that is $\boldsymbol{\Theta} = \mathbf{I}_m$. Theoretical analysis becomes easiest with this normalization. However, as pointed out in [FAW 77a, VIG 86] in the context of viscously damped systems, this normalization is inconsistent with undamped or classically damped modal theories.

5.6.3. *Orthogonality of the eigenvectors*

The orthogonality relationship of the eigenvectors given by equation [5.116] is not very useful because it is expressed in terms of the eigenvalues of the system. In this section, we will develop an orthogonality relationship which is independent of the eigenvalues. Expressions equivalent to the orthogonality relationships of the undamped eigenvectors with respect to the mass and stiffness matrices will also be established. To derive these results, first recall the expression of the transfer function matrix derived in section 5.3. In the pole-residue form, the inverse of the dynamic stiffness matrix (transfer function matrix) can be expressed as

$$\mathbf{D}^{-1}(s) = \frac{\text{adj}\,[\mathbf{D}(s)]}{\det\,[\mathbf{D}(s)]} = \sum_{j=1}^{m} \frac{\mathbf{R}_j}{s - s_j}. \qquad [5.122]$$

Here, $\mathbf{R}_j$, the residue of $\mathbf{D}^{-1}(s)$ at the pole s_j obtained from equation [5.70], is

$$\mathbf{R}_j = \frac{\mathbf{z}_j \mathbf{z}_j^T}{\mathbf{z}_j^T \frac{\partial \mathbf{D}(s_j)}{\partial s_j} \mathbf{z}_j} \in \mathbb{C}^{N \times N}. \qquad [5.123]$$

Interestingly, observe that denominator of left-hand side of the above equation is exactly the same as the normalization condition given by equation [5.120]. Now, using equations [5.120] and [5.123], we finally obtain the residues as

$$\mathbf{R}_j = \frac{\mathbf{z}_j \mathbf{z}_j^T}{\gamma_j}. \qquad [5.124]$$

For undamped systems, i.e. when $\mathbf{G}(s) = \mathbf{O}_N, \forall\, s$, the eigenvectors satisfy familiar orthogonality relationship over the mass and stiffness matrices as given by equations [2.63] and [2.64]. For viscously damped systems (with non-proportional damping), equivalent relationships may be obtained by converting the equation of motion into the state-space form, see [SES 94]. The eigenproblem in the state-space form is essentially similar to the undamped eigenproblem except that the size of the problem becomes doubled, and the eigensolutions become complex. Thus, from the analysis point of view, the state-space approach offers a significant advantage for viscously damped systems. Unfortunately, for non-viscously damped systems, no advantage can be gained by adopting the state-space formalism as at least one of the system matrices will not be a constant matrix. For this reason, and also realizing that the state-space eigenvectors are not physically appealing, we kept the analysis in the second-order form. One of our main results regarding the orthogonality of the eigenvectors is the following:

THEOREM 5.2.– The modal matrix of a non-viscously damped system, $\mathbf{Z} \in \mathbb{C}^{N\times m}$, satisfies the orthogonality relationship $\mathbf{Z}\mathbf{\Theta}^{-1}\mathbf{Z}^T = \mathbf{O}_N$.

PROOF.– From equations [5.122] and [5.124], we obtain

$$\frac{\operatorname{adj}[\mathbf{D}(s)]}{\det[\mathbf{D}(s)]} = \sum_{j=1}^{m} \frac{1}{s-s_j}\frac{\mathbf{z}_j\mathbf{z}_j^T}{\gamma_j}. \qquad [5.125]$$

Multiplying both sides of the above equation by s and taking limit as $s \to \infty$, we obtain

$$\lim_{s\to\infty} s\frac{\operatorname{adj}[\mathbf{D}(s)]}{\det[\mathbf{D}(s)]} = \lim_{s\to\infty}\sum_{j=1}^{m} \frac{s}{s-s_j}\frac{\mathbf{z}_j\mathbf{z}_j^T}{\gamma_j} = \sum_{j=1}^{m}\frac{\mathbf{z}_j\mathbf{z}_j^T}{\gamma_j}. \qquad [5.126]$$

It is easy to observe that the order of the elements of $\operatorname{adj}[\mathbf{D}(s)]$ is at the most $(m-2)$ in s. Since the order of the determinant, $\det[\mathbf{D}(s)]$, is m, after taking the limit, every element of the left-hand side of equation [5.126] reduces to zero. Thus, in the limit, the left-hand side of equation [5.126] approaches an $N \times N$ null matrix. Finally, writing equation [5.126] in the matrix form, we obtain

$$\mathbf{Z}\mathbf{\Theta}^{-1}\mathbf{Z}^T = \mathbf{O}_N \qquad [5.127]$$

and the theorem is proved.

The result of this theorem is quite general and it does not depend on the nature of the system property matrices. The only requirement of this theorem is that the system must have $m \geq 2N$ distinct eigenvalues. Clearly, the undamped systems as well as the viscously damped systems are also covered as special cases. In the context of viscously damped systems (see section 5.6.6 for details), a similar result has been derived in [FAW 76] by considering the normalization matrix $\mathbf{\Theta}$ as the identity matrix. Later, Fawzy [FAW 77a] generalized this result for the case when $\mathbf{\Theta}$ is not an identity matrix. The result obtained in theorem 5.2 can be viewed as a further generalization of these results to the non-viscously damped systems. Next, we consider the relationship between the eigensolutions and the mass matrix.

THEOREM 5.3.– The modal matrix of a non-viscously damped system, $\mathbf{Z} \in \mathbb{C}^{N\times m}$, satisfies the relationship $\mathbf{Z}\mathbf{\Theta}^{-1}\mathbf{S}\mathbf{Z}^T = \mathbf{M}^{-1}$.

PROOF.– First consider the function $s\mathbf{D}^{-1}(s)$. Following the approach outlined in section 5.3.2 and using the residue theorem, we obtain

$$s\mathbf{D}^{-1}(s) = \sum_{j=1}^{m} \frac{\mathbf{Q}_j}{s - s_j}. \qquad [5.128]$$

Here, the residues $\mathbf{Q}_j$ can be obtained as

$$\mathbf{Q}_j \stackrel{\text{def}}{=} \lim_{s \to s_j} (s - s_j) \left[s\mathbf{D}^{-1}(s)\right] = s_j \frac{\mathbf{z}_j \mathbf{z}_j^T}{\gamma_j}. \qquad [5.129]$$

Using the expression of the dynamic stiffness matrix in equation [5.112], we can deduce

$$\lim_{s \to \infty} \frac{\mathbf{D}(s)}{s^2} = \lim_{s \to \infty} \left[\mathbf{M} + \frac{\mathbf{G}(s)}{s} + \frac{\mathbf{K}}{s^2}\right] = \mathbf{M}. \qquad [5.130]$$

Taking the inverse of the above equation results

$$\lim_{s \to \infty} \left[s^2 \mathbf{D}^{-1}(s)\right] = \mathbf{M}^{-1}. \qquad [5.131]$$

Now, multiplying equation [5.128] by s and taking limit as $s \to \infty$, we obtain

$$\lim_{s \to \infty} \left[s^2 \mathbf{D}^{-1}(s)\right] = \lim_{s \to \infty} \sum_{j=1}^{m} \frac{s}{s - s_j} \frac{s_j \mathbf{z}_j \mathbf{z}_j^T}{\gamma_j} = \sum_{j=1}^{m} \frac{s_j \mathbf{z}_j \mathbf{z}_j^T}{\gamma_j}. \qquad [5.132]$$

Casting the right-hand side of the above equation in the matrix form and equating it with [5.131] results

$$\mathbf{Z}\boldsymbol{\Theta}^{-1}\mathbf{S}\mathbf{Z}^T = \mathbf{M}^{-1} \qquad [5.133]$$

and the theorem is proved.

Since $\boldsymbol{\Theta}$ and $\mathbf{S}$ are diagonal matrices, they commute in product. For this reason, the above result can also be expressed as $\mathbf{Z}\mathbf{S}\boldsymbol{\Theta}^{-1}\mathbf{Z}^T = \mathbf{M}^{-1}$. For viscously damped systems, similar result has been derived in [FAW 76] by considering the normalization matrix $\boldsymbol{\Theta}$ as the identity matrix. It might be thought that by taking the inverse of equation [5.133] and rearranging, the conventional mass–orthogonality relationship

$$\mathbf{Z}^T\mathbf{M}\mathbf{Z} = \mathbf{S}^{-1}\boldsymbol{\Theta} \qquad [5.134]$$

could be obtained. We emphasize that the representation of equation [5.133] in the form of equation [5.134] is *not* always possible. To show this, premultiply equation [5.134] by $\mathbf{Z}\boldsymbol{\Theta}^{-1}$ to obtain

$$\mathbf{Z}\boldsymbol{\Theta}^{-1}\mathbf{Z}^T\mathbf{M}\mathbf{Z} = \mathbf{Z}\boldsymbol{\Theta}^{-1}\mathbf{S}^{-1}\boldsymbol{\Theta} = \mathbf{Z}\mathbf{S}^{-1}. \qquad [5.135]$$

Due to theorem 5.2, the left-hand side of equation [5.135] is a null matrix, while its right-hand side is not. Thus, [5.134] cannot be a valid equation. However, for a special case, when the system is undamped, the modal matrix $\mathbf{Z}$ can be expressed by a square matrix and equation [5.133] can be represented by the classical mass–orthogonality relationship in [5.134]. Thus, theorem 5.3 provides the result equivalent to the classical mass–orthogonality relationship for general cases.

Like the mass–orthogonality relationship of the eigenvectors, the orthogonality relationship with respect to the stiffness matrix can also be obtained. Assuming that $\mathbf{K}^{-1}$ exists, we have the following:

THEOREM 5.4.– The modal matrix of a non-viscously damped system, $\mathbf{Z} \in \mathbb{C}^{N\times m}$, satisfies the relationship $\mathbf{Z}\boldsymbol{\Theta}^{-1}\mathbf{S}^{-1}\mathbf{Z}^T = -\mathbf{K}^{-1}$.

PROOF.– Using the expression of the dynamic stiffness matrix in equation [5.112], we can easily deduce

$$\lim_{s\to 0} \mathbf{D}(s) = \mathbf{K}. \qquad [5.136]$$

Taking the inverse of the above equation results

$$\lim_{s\to 0} \mathbf{D}^{-1}(s) = \mathbf{K}^{-1}. \qquad [5.137]$$

From equations [5.122] and [5.124], we obtain

$$\mathbf{D}^{-1}(s) = \sum_{j=1}^{m} \frac{1}{s - s_j}\frac{\mathbf{z}_j\mathbf{z}_j^T}{\gamma_j}. \qquad [5.138]$$

Taking the limit as $s \to 0$ in equation [5.138], we obtain

$$\lim_{s\to 0} \mathbf{D}^{-1}(s) = \sum_{j=1}^{m} \frac{1}{-s_j}\frac{\mathbf{z}_j\mathbf{z}_j^T}{\gamma_j}. \qquad [5.139]$$

Casting the right-hand side of the previous equation in the matrix form and equating it with [5.137] results in

$$\mathbf{Z}\mathbf{\Theta}^{-1}\mathbf{S}^{-1}\mathbf{Z}^{T} = -\mathbf{K}^{-1} \qquad [5.140]$$

and the theorem is proved.

5.6.4. *Relationships between the eigensolutions and damping*

In this section, some direct relationships have been established between the mass and stiffness matrices and the eigensolutions. In this section, we try to establish the relationships between the damping matrix and the eigensolutions. A major difficulty in this regard is that, unlike the mass and stiffness matrices, the damping matrix, $\mathbf{G}(s)$, is a function of s. To simplify the problem, we consider only two limiting cases: (1) when $s \to \infty$ and (2) when $s \to 0$. Suppose

$$\lim_{s\to\infty} \mathbf{G}(s) = \mathbf{G}_\infty \in \mathbb{R}^{N\times N} \qquad [5.141]$$

$$\text{and} \quad \lim_{s\to 0} \mathbf{G}(s) = \mathbf{G}_0 \in \mathbb{R}^{N\times N} \qquad [5.142]$$

where $\|\mathbf{G}_\infty\|$, $\|\mathbf{G}_0\| < \infty$.

5.6.4.1. *Relationships in terms of* $\mathbf{M}^{-1}$

Casting equation [5.138] into the matrix form, we obtain

$$\mathbf{D}^{-1}(s) = \mathbf{Z}\mathbf{\Theta}^{-1}\left(s\mathbf{I}_m - \mathbf{S}\right)^{-1}\mathbf{Z}^{T}. \qquad [5.143]$$

The previous equation can be expanded as

$$\begin{aligned}\mathbf{D}^{-1}(s) &= \tfrac{1}{s}\mathbf{Z}\mathbf{\Theta}^{-1}\left(\mathbf{I}_m - \tfrac{\mathbf{S}}{s}\right)^{-1}\mathbf{Z}^{T}\\ &= \tfrac{1}{s}\left(\mathbf{Z}\mathbf{\Theta}^{-1}\mathbf{Z}^{T}\right) + \tfrac{1}{s^2}\left(\mathbf{Z}\mathbf{\Theta}^{-1}\mathbf{S}\mathbf{Z}^{T}\right)\\ &\quad + \tfrac{1}{s^3}\left(\mathbf{Z}\mathbf{\Theta}^{-1}\mathbf{S}^2\mathbf{Z}^{T}\right) + \tfrac{1}{s^4}\left(\mathbf{Z}\mathbf{\Theta}^{-1}\mathbf{S}^3\mathbf{Z}^{T}\right) + \cdots\end{aligned} \qquad [5.144]$$

Now, rewrite the expression of the dynamic stiffness matrix in equation [5.112] as

$$\mathbf{D}(s) = s^2\mathbf{M}\left[\mathbf{I}_N + \frac{\mathbf{M}^{-1}}{s}\left(\mathbf{G}(s) + \frac{\mathbf{K}}{s}\right)\right]. \qquad [5.145]$$

Taking the inverse of this equation and expanding the right-hand side, we obtain

$$\mathbf{D}^{-1}(s) = \left[\mathbf{I}_N - \frac{\mathbf{M}^{-1}}{s}\left(\mathbf{G}(s) + \frac{\mathbf{K}}{s}\right) + \left\{\frac{\mathbf{M}^{-1}}{s}\left(\mathbf{G}(s) + \frac{\mathbf{K}}{s}\right)\right\}^2 - \cdots\right]\frac{\mathbf{M}^{-1}}{s^2}. \quad [5.146]$$

Equation [5.146] can be further simplified to obtain

$$\mathbf{D}^{-1}(s) = \frac{\mathbf{M}^{-1}}{s^2} + \frac{1}{s^3}\left(-\mathbf{M}^{-1}\mathbf{G}(s)\mathbf{M}^{-1}\right) + \frac{1}{s^4}\left(\mathbf{M}^{-1}\left[\mathbf{G}(s)\mathbf{M}^{-1}\mathbf{G}(s) - \mathbf{K}\right]\mathbf{M}^{-1}\right) + \cdots \quad [5.147]$$

Comparing equations [5.144] and [5.147], it is clear that their right-hand sides are equal. Theorems 5.1 and 5.2 can be alternatively proved by multiplying these equations by s and s^2, respectively, and taking the limit as $s \to \infty$. Observe that the coefficients associated with the corresponding (negative) powers of s in the series expressions [5.144] and [5.147] cannot be equated because $\mathbf{G}(s)$ is also a function of s. However, in the limit when $s \to \infty$, the variation of $\mathbf{G}(s)$ becomes negligible as by equation [5.141] it approaches to $\mathbf{G}_\infty$. Considering the second term of the right-hand side of equation [5.147], equating it with the corresponding term of equation [5.144] and taking the limit as $s \to \infty$, we obtain

$$\mathbf{Z}\boldsymbol{\Theta}^{-1}\mathbf{S}^2\mathbf{Z}^T = -\mathbf{M}^{-1}\mathbf{G}_\infty\mathbf{M}^{-1}. \quad [5.148]$$

It must be noted that this procedure cannot be extended to further lower order terms as all of them would be effected by the functional variation of $\mathbf{G}(s)$ from previous terms.

Were the system viscously damped $\mathbf{G}(s)$ would be a constant matrix and equating the coefficients associated with different powers of s, we could obtain several relationships between the eigensolutions and the system matrices. Considering the first few terms in the series expressions [5.144] and [5.147], some such relationships are reported in section 5.6.6.

5.6.4.2. *Relationships in terms of* $\mathbf{K}^{-1}$

We rewrite equation [5.143] as

$$\mathbf{D}^{-1}(s) = -\mathbf{Z}\boldsymbol{\Theta}^{-1}\mathbf{S}^{-1}\left(\mathbf{I}_m - s\mathbf{S}^{-1}\right)^{-1}\mathbf{Z}^T. \quad [5.149]$$

Expanding equation [5.149], we obtain

$$\mathbf{D}^{-1}(s) = -\mathbf{Z}\boldsymbol{\Theta}^{-1}\mathbf{S}^{-1}\mathbf{Z}^T - s\left(\mathbf{Z}\boldsymbol{\Theta}^{-1}\mathbf{S}^{-2}\mathbf{Z}^T\right) - s^2\left(\mathbf{Z}\boldsymbol{\Theta}^{-1}\mathbf{S}^{-3}\mathbf{Z}^T\right) - s^3\left(\mathbf{Z}\boldsymbol{\Theta}^{-1}\mathbf{S}^{-4}\mathbf{Z}^T\right) - \cdots \quad [5.150]$$

The expression of the dynamic stiffness matrix in equation [5.112] can be rearranged as

$$\mathbf{D}(s) = \mathbf{K}\left[\mathbf{I}_N + s\left(s\mathbf{K}^{-1}\mathbf{M} + \mathbf{K}^{-1}\mathbf{G}(s)\right)\right]. \qquad [5.151]$$

Taking the inverse of equation [5.151] and expanding the right-hand side, we obtain

$$\begin{aligned}\mathbf{D}^{-1}(s) = & \left[\mathbf{I}_N - s\left(s\mathbf{K}^{-1}\mathbf{M} + \mathbf{K}^{-1}\mathbf{G}(s)\right)\right. \\ & \left. + \left\{s\left(s\mathbf{K}^{-1}\mathbf{M} + \mathbf{K}^{-1}\mathbf{G}(s)\right)\right\}^2 - \cdots\right]\mathbf{K}^{-1}.\end{aligned} \qquad [5.152]$$

The previous equation can be further simplified to obtain

$$\begin{aligned}\mathbf{D}^{-1}(s) = & \ \mathbf{K}^{-1} + s\left(-\mathbf{K}^{-1}\mathbf{G}(s)\mathbf{K}^{-1}\right) \\ & + s^2\left(\mathbf{K}^{-1}\left[\mathbf{G}(s)\mathbf{K}^{-1}\mathbf{G}(s) - \mathbf{M}\right]\mathbf{K}^{-1}\right) + \cdots\end{aligned} \qquad [5.153]$$

Comparing the right-hand side of equations [5.150] and [5.153], theorem 5.4 can be proved alternatively by taking the limit as $s \to 0$. Considering the second term of the right-hand side of equation [5.153], equating it with the second term of equation [5.150] and taking the limit as $s \to 0$, we obtain

$$\mathbf{Z}\boldsymbol{\Theta}^{-1}\mathbf{S}^{-2}\mathbf{Z}^T = \mathbf{K}^{-1}\mathbf{G}_0\mathbf{K}^{-1}. \qquad [5.154]$$

Again, note that this approach cannot be extended to the higher order terms as all of them would be effected by the functional variation of $\mathbf{G}(s)$ from previous terms.

5.6.5. *System matrices in terms of the eigensolutions*

Theorems 5.3 and 5.4 and equations [5.148] and [5.154] allow us to represent the system property matrices explicitly in terms of the eigensolutions. This might be useful in system identification problems where the eigensolutions of a structure can be measured from experiments. Using the eigensolutions, we define two matrices

$$\mathbf{P}_1 = \mathbf{Z}\boldsymbol{\Theta}^{-1}\mathbf{S}\mathbf{Z}^T \qquad [5.155]$$

and $$\mathbf{P}_2 = \mathbf{Z}\boldsymbol{\Theta}^{-1}\mathbf{S}^{-1}\mathbf{Z}^T. \qquad [5.156]$$

Using these equations, from equation [5.133], we obtain the mass matrix as

$$\mathbf{M} = \mathbf{P}_1^{-1}. \qquad [5.157]$$

Similarly, from equation [5.140], the stiffness matrix can be obtained as

$$\mathbf{K} = -\mathbf{P}_2^{-1}. \qquad [5.158]$$

The damping matrix in the Laplace domain, $\mathbf{G}(s)$, can be obtained only at the two limiting values when $s \to \infty$ and $s \to 0$. From equations [5.148] and [5.154], we obtain

$$\mathbf{G}_\infty = -\mathbf{P}_1^{-1}\left[\mathbf{Z}\boldsymbol{\Theta}^{-1}\mathbf{S}^2\mathbf{Z}^T\right]\mathbf{P}_1^{-1} \qquad [5.159]$$

$$\text{and} \quad \mathbf{G}_0 = \mathbf{P}_2^{-1}\left[\mathbf{Z}\boldsymbol{\Theta}^{-1}\mathbf{S}^{-2}\mathbf{Z}^T\right]\mathbf{P}_2^{-1}. \qquad [5.160]$$

These results, however, do not give any indication regarding the functional behavior of $\mathbf{G}(s)$ between these two extreme values.

5.6.6. *Eigenrelations for viscously damped systems*

Viscously damped systems arise as a special case of the more general non-viscously damped systems when the damping matrix become a constant matrix, that is $\mathbf{G}(s) = \mathbf{C} \in \mathbb{R}^{N\times N}, \forall\, s$. Here, several relationships satisfied by the eigensolutions and the system matrices will be derived for this special case.

For viscously damped systems, the order of the characteristic polynomial $m = 2N$, and consequently the modal matrix $\mathbf{Z} \in \mathbb{C}^{N\times 2N}$ and the diagonal matrices $\mathbf{S}, \boldsymbol{\Theta} \in \mathbb{C}^{2N\times 2N}$. From equation [5.119], the normalization relationship reads

$$\mathbf{z}_k^T\left[2s_k\mathbf{M} + \mathbf{C}\right]\mathbf{z}_k = \gamma_k, \quad \forall k = 1, \cdots, 2N. \qquad [5.161]$$

Now, consider the series expansion of $\mathbf{D}^{-1}(s)$ given by equations [5.144] and [5.147]. Equating the coefficients of $1/s$, we obtain the mode–orthogonality relationship

$$\mathbf{Z}\boldsymbol{\Theta}^{-1}\mathbf{Z}^T = \mathbf{O}_N. \qquad [5.162]$$

This relationship was also derived in [FAW 77a]. Now, equating the coefficients of $1/s^2, \cdots, 1/s^5$ in the right-hand sides of equations [5.144] and [5.147], several relationships involving the eigensolutions and $\mathbf{M}^{-1}$, $\mathbf{C}$ and $\mathbf{K}$ may be obtained

$$\mathbf{Z}\boldsymbol{\Theta}^{-1}\mathbf{S}\mathbf{Z}^T = \mathbf{M}^{-1} \qquad [5.163]$$

$$\mathbf{Z}\boldsymbol{\Theta}^{-1}\mathbf{S}^2\mathbf{Z}^T = -\mathbf{M}^{-1}\mathbf{C}\mathbf{M}^{-1} \qquad [5.164]$$

$$\mathbf{Z}\boldsymbol{\Theta}^{-1}\mathbf{S}^3\mathbf{Z}^T = \mathbf{M}^{-1}\left[\mathbf{C}\mathbf{M}^{-1}\mathbf{C} - \mathbf{K}\right]\mathbf{M}^{-1} \qquad [5.165]$$

and

$$\mathbf{Z}\boldsymbol{\Theta}^{-1}\mathbf{S}^{4}\mathbf{Z}^{T} = \mathbf{M}^{-1}\left[\mathbf{K}\mathbf{M}^{-1}\mathbf{C} + \mathbf{C}\mathbf{M}^{-1}\mathbf{K} - \mathbf{C}\mathbf{M}^{-1}\mathbf{C}\mathbf{M}^{-1}\mathbf{C}\right]\mathbf{M}^{-1}. \quad [5.166]$$

This procedure can be extended to obtain further higher order terms involving **S**.

Similarly, equating the coefficients of $s^0, \cdots, s^3$ in the right-hand sides of equations [5.150] and [5.153], several relationships involving the eigensolutions and $\mathbf{K}^{-1}$, **C** and **M** may be obtained

$$\mathbf{Z}\boldsymbol{\Theta}^{-1}\mathbf{S}^{-1}\mathbf{Z}^{T} = -\mathbf{K}^{-1} \quad [5.167]$$

$$\mathbf{Z}\boldsymbol{\Theta}^{-1}\mathbf{S}^{-2}\mathbf{Z}^{T} = \mathbf{K}^{-1}\mathbf{C}\mathbf{K}^{-1} \quad [5.168]$$

$$\mathbf{Z}\boldsymbol{\Theta}^{-1}\mathbf{S}^{-3}\mathbf{Z}^{T} = \mathbf{K}^{-1}\left[\mathbf{M} - \mathbf{C}\mathbf{K}^{-1}\mathbf{C}\right]\mathbf{K}^{-1} \quad [5.169]$$

and

$$\mathbf{Z}\boldsymbol{\Theta}^{-1}\mathbf{S}^{-4}\mathbf{Z}^{T} = \mathbf{K}^{-1}\left[\mathbf{C}\mathbf{K}^{-1}\mathbf{C}\mathbf{K}^{-1}\mathbf{C} - \mathbf{M}\mathbf{K}^{-1}\mathbf{C} + \mathbf{C}\mathbf{K}^{-1}\mathbf{M}\right]\mathbf{K}^{-1}. \quad [5.170]$$

This procedure can be extended to obtain further lower order terms involving **S**. Using a different approach, and considering the normalization matrix $\boldsymbol{\Theta}$ as the identity matrix, Fawzy and Bishop [FAW 76] obtained expressions similar to equations [5.163]–[5.165] and [5.167]–[5.169]. Thus, the relationships derived here extend their results to generally normalized eigenvectors.

5.6.7. *Numerical examples*

5.6.7.1. *A three degree-of-freedom system*

A three degree-of-freedom system, similar to what was considered in section 5.5.1, is used to illustrate the results derived in this section. The mass and stiffness matrices are assumed to be

$$\mathbf{M} = \begin{bmatrix} 3 & 0 & 0 \\ 0 & 3 & 0 \\ 0 & 0 & 3 \end{bmatrix} \quad [5.171]$$

$$\text{and} \quad \mathbf{K} = \begin{bmatrix} 4 & -2 & 0 \\ -2 & 4 & -2 \\ 0 & -2 & 4 \end{bmatrix}. \quad [5.172]$$

Numerical values for these matrices are taken by assuming $m_u = 3$ and $k_u = 2$ in the example considered in section 5.5.1. The matrix of the damping functions is assumed to be of the form

$$\boldsymbol{\mathcal{G}}(t) = \begin{bmatrix} 0 & 0 & 0 \\ 0 & 1.5g(t) & 0 \\ 0 & 0 & 0 \end{bmatrix} \qquad [5.173]$$

where

$$g(t) = \delta(t) + \left(\mu_1 e^{-\mu_1 t} + \mu_2 e^{-\mu_2 t}\right); \quad \mu_1, \mu_2 > 0. \qquad [5.174]$$

The damping matrix in the Laplace domain, $\mathbf{G}(s)$, can be obtained by taking the Laplace transform of equation [5.173]. The Laplace transform of $g(t)$ given by [5.174] can be obtained as

$$G(s) = 1 + \frac{(\mu_1 + \mu_2)\, s + 2\mu_1\mu_2}{s^2 + (\mu_1 + \mu_2)\, s + \mu_1\mu_2}. \qquad [5.175]$$

This damping model is a linear combination of the viscous model and the GHM model discussed in Chapter 4. Regarding the numerical values of the damping parameters, we assume $\mu_1 = 1.5$ and $\mu_2 = 0.1$.

5.6.7.2. *Eigenvalues and eigenvectors*

Using equation [5.175], together with the expressions of the system matrices given by equations [5.171]–[5.173], it can be shown that the order of the characteristic polynomial, $m = 8$. It has been mentioned that (for lightly damped systems), among the m eigenvalues, $2N = 6$ appear in complex conjugate pairs (elastic modes) and the rest $p = m - 2N = 2$ eigenvalues become purely real (non-viscous modes).

Solving the characteristic equation, the diagonal matrix containing the eigenvalues can be expressed as

$$\mathbf{S} = \mathrm{diag}\left[\mathbf{s}_e, \mathbf{s}_e^*, \mathbf{s}_{nv}\right] \in \mathbb{C}^{8\times 8}. \qquad [5.176]$$

Here, the eigenvalues corresponding to the three elastic modes are

$$\mathbf{s}_e = \{-0.2632 + 0.7363\mathrm{i}, 1.1547\mathrm{i}, -0.2392 + 1.5177\mathrm{i}, \}. \qquad [5.177]$$

The eigenvalues corresponding to the two non-viscous modes are found to be

$$\mathbf{s}_{nv} = \{-1.0029, -0.0921\}. \qquad [5.178]$$

Since these eigenvalues are purely real and negative, it implies that the non-viscous modes are stable and non-oscillatory in nature (i.e. overcritically damped).

The eigenvectors can be obtained by applying the procedure outlined in Chapter 4. The matrix of eigenvectors can be expressed as

$$\mathbf{Z} = [\mathbf{Z}_e, \mathbf{Z}_e^*, \mathbf{Z}_{nv}] \in \mathbb{C}^{3\times 8}. \quad [5.179]$$

Here, $\mathbf{Z}_e$, the matrix of eigenvectors corresponding to the three elastic modes, is

$$\mathbf{Z}_e = \begin{bmatrix} 1 & 1 & 1 \\ 1.2908 - 0.5814\mathrm{i} & 0 & -1.3690 - 1.0893\mathrm{i} \\ 1 & -1 & 1 \end{bmatrix}. \quad [5.180]$$

The matrix of eigenvectors corresponding to the two non-viscous modes is calculated as

$$\mathbf{Z}_{nv} = \begin{bmatrix} 1 & 1 \\ 3.5088 & 2.0127 \\ 1 & 1 \end{bmatrix}. \quad [5.181]$$

Using the eigenvectors given by equations [5.180]–[5.181], the normalization matrix $\boldsymbol{\Theta}$ can be obtained from equation [5.119] as

$$\boldsymbol{\Theta} = \mathrm{diag}\left[\boldsymbol{\theta}_e, \boldsymbol{\theta}_e^*, \boldsymbol{\theta}_{nv}\right]. \quad [5.182]$$

Here

$$\boldsymbol{\theta}_e = \{0.0223 + 0.1180\mathrm{i}, 0.1386\mathrm{i}, -0.2756 + 0.2359\mathrm{i}\} \quad [5.183]$$

and

$$\boldsymbol{\theta}_{nv} = \{1.0075, 9.8738\}. \quad [5.184]$$

5.6.7.3. *Orthogonality relationships*

Using $\mathbf{Z}$ and $\boldsymbol{\Theta}$, we can easily verify that the mode–orthogonality relationship given by theorem 5.2 is satisfied, that is,

$$\mathbf{Z}\boldsymbol{\Theta}^{-1}\mathbf{Z}^T = \mathbf{O}_3. \quad [5.185]$$

Now, in the line of theorem 5.3, we calculate

$$\mathbf{Z}\mathbf{\Theta}^{-1}\mathbf{S}\mathbf{Z}^T = \begin{bmatrix} 0.3333 & 0 & 0 \\ 0 & 0.3333 & 0 \\ 0 & 0 & 0.3333 \end{bmatrix} = \mathbf{M}^{-1}. \qquad [5.186]$$

Similarly, following theorem 5.4, we obtain

$$-\mathbf{Z}\mathbf{\Theta}^{-1}\mathbf{S}^{-1}\mathbf{Z}^T = \begin{bmatrix} 0.375 & 0.250 & 0.125 \\ 0.250 & 0.500 & 0.250 \\ 0.125 & 0.250 & 0.375 \end{bmatrix} = \mathbf{K}^{-1}. \qquad [5.187]$$

5.6.7.4. *Relationships with the damping matrix*

Taking the Laplace transform of equation [5.173] and considering the limiting cases as $s \to \infty$ and $s \to 0$, we obtain

$$\mathbf{G}_\infty = \begin{bmatrix} 0 & 0 & 0 \\ 0 & 1.5 & 0 \\ 0 & 0 & 0 \end{bmatrix} \qquad [5.188]$$

$$\text{and} \quad \mathbf{G}_0 = \begin{bmatrix} 0 & 0 & 0 \\ 0 & 4.5 & 0 \\ 0 & 0 & 0 \end{bmatrix}. \qquad [5.189]$$

Note that the matrices $\mathbf{P}_1$ and $\mathbf{P}_2$, defined in equations [5.155] and [5.156], respectively, can be directly obtained from [5.186] and [5.187]. Using these matrices, the truth of equations [5.159] and [5.160], which relate $\mathbf{G}_\infty$ and $\mathbf{G}_0$ to the eigensolutions, can be verified. Thus, the damping matrix, $\mathbf{G}(s)$, can be reconstructed from the eigensolutions for the cases when $s \to \infty$ and $s \to 0$.

5.7. Rayleigh quotient for non-viscously damped systems

Rayleigh quotient for viscously damped damped systems was introduced in section 2.6. Both proportional and non-proportionally damped systems were considered. In this section, we extend the idea of the Rayleigh quotient to non-viscously damped systems. Since the Rayleigh quotient has been widely applied in the analysis of many vibrating systems and their associated linear algebraic eigenvalue problems, we aim to understand its role in the context of systems with general damping models.

Suppose $\lambda_j, \mathbf{z}_j$ is jth complex natural frequency and complex mode shape. In the context of the notations used in section 5.2.3, $s_j = \mathrm{i}\lambda_j$. Using this, from equation

[5.42], approximate expression for the complex natural frequencies can be obtained as

$$\lambda_j \approx \pm\omega_j + \mathrm{i}G'_{jj}(\pm\omega_j)/2 \tag{5.190}$$

where $G'_{kl}(\omega_j) = \mathbf{x}_k^T\mathbf{G}(\omega_j)\mathbf{x}_l$ is the frequency-dependent damping matrix expressed in normal coordinates. Since the inverse Fourier transform of $\mathbf{G}(\omega)$ must be real, it must satisfy the condition $\mathbf{G}(-\omega) = \mathbf{G}(\omega)^*$, where $(\bullet)^*$ denotes complex conjugation. It follows that the eigenvalues of the generally damped system appear in pairs λ and $-\lambda^*$ (unless λ is purely imaginary). The first-order approximate expression for the complex eigenvectors can be obtained as a special case of equation [5.44] as

$$\mathbf{z}_j \approx \mathbf{x}_j + \mathrm{i}\sum_{k\neq n}\beta_{kn}\mathbf{x}_k, \quad \text{where} \quad \beta_{kj} = \frac{\omega_j G'_{kj}(\omega_j)}{(\omega_j^2-\omega_k^2)}. \tag{5.191}$$

Note that the eigenvectors also appear in complex conjugate pairs. Since, in general, $G'_{kj}(\omega_j)$ will be complex, in contrast to the viscously damped case, the real part of natural frequencies and mode shapes do not coincide with the undamped ones. Adequacy of the first-order theory for systems with small damping has been investigated in section 5.5.

Since the complex eigenvectors $\mathbf{z}_i, i = 1\dots N$ form the complete basis of an N-dimensional complex vector space, any arbitrary complex vector ψ can be expressed as

$$\psi = \sum_i c_i\mathbf{z}_i. \tag{5.192}$$

Consider a vector close to $\mathbf{z}_r$ that can be written as

$$\psi = \mathbf{z}_r + \sum_{i\neq r}\epsilon_i\mathbf{z}_i, \quad |\, \epsilon_i \,| = \left|\frac{c_i}{c_r}\right| << 1. \tag{5.193}$$

Replacing the matrix $\mathbf{G}(\omega_r)$ with $\mathbf{M}$ and noting the orthogonality properties given in equation [2.259], we obtain

$$\mathbf{z}_r^H\mathbf{M}\mathbf{z}_r = 1 + O(|\beta|^2) \tag{5.194}$$

$$\mathbf{z}_i^H\mathbf{M}\mathbf{z}_r = O(|\beta|^2). \tag{5.195}$$

Similarly, with $\mathbf{K}$ we obtain

$$\mathbf{z}_r^H \mathbf{K} \mathbf{z}_r = \omega_r^2 + O(|\beta|^2) \tag{5.196}$$

$$\mathbf{z}_i^H \mathbf{K} \mathbf{z}_r = O(|\beta|^2). \tag{5.197}$$

Consider the first Rayleigh quotient

$$R_1(\psi) = \frac{\psi^H \mathbf{K} \psi}{\psi^H \mathbf{M} \psi} = \frac{\mathbf{z}_r^H \mathbf{K} \mathbf{z}_r + 2\sum_{i \neq r} \Re(\epsilon_i) \mathbf{z}_i^H \mathbf{K} \mathbf{z}_r + O(|\epsilon|^2)}{\mathbf{z}_r^H \mathbf{M} \mathbf{z}_r + 2\sum_{i \neq r} \Re(\epsilon_i) \mathbf{z}_i^H \mathbf{M} \mathbf{z}_r + O(|\epsilon|^2)}. \tag{5.198}$$

Substituting [5.194–5.197] in the above equation, we obtain

$$R_1(\psi) = \frac{\omega_r^2 + O(|\beta|^2) + O(\epsilon^2)}{1 + O(|\beta|^2) + O(\epsilon^2)} \approx \omega_r^2 (1 - O(\epsilon^2)) \tag{5.199}$$

which proves the stationarity of $R_1(\psi)$.

The second and third Rayleigh quotients involving the damping term need to be carefully defined. The difference between the viscous and the non-viscous case is that the (effective) damping matrix for the non-viscous case is complex valued and a function of frequency. Therefore, in order to define a meaningful Rayleigh quotient we need to select a value of frequency. If we are interested in studying stationary behavior of rth mode, then it is logical to select the frequency value as ω_r. We define the real-valued Rayleigh quotient for a non-viscously damped systems as

$$R_2(\psi, \omega_r) = \frac{\left|\psi^H \mathbf{G}(\omega_r) \psi\right|}{\psi^H \mathbf{M} \psi}. \tag{5.200}$$

For a viscously damped system $\mathbf{G}(\omega_r) = \mathbf{C}, \forall\, r$ and because $\mathbf{C}$ is a real matrix, equation [5.200] reduces to equation [2.276] as a special case. Therefore, equation [5.200] can be viewed a generalization of the Rayleigh quotient defined in equation [2.276].

Substituting equation [5.191] into the above equation leads to

$$R_2(\psi) = \frac{\psi^H \mathbf{G}(\omega_r) \psi}{\psi^H \mathbf{M} \psi} = \frac{\left|\mathbf{z}_r^H \mathbf{G}(\omega_r) \mathbf{z}_r + 2\sum_{i \neq r} \Re(\epsilon_i) \mathbf{z}_i^H \mathbf{G}(\omega_r) \mathbf{z}_r + O(|\epsilon|^2)\right|}{\mathbf{z}_r^H \mathbf{M} \mathbf{z}_r + 2\sum_{i \neq r} \Re(\epsilon_i) \mathbf{z}_i^H \mathbf{M} \mathbf{z}_r + O(|\epsilon|^2)}. \tag{5.201}$$

The first term in numerator can be expressed as

$$\begin{aligned}
\mathbf{z}_r^H \mathbf{G}(\omega_r)\mathbf{z}_r &= [\mathbf{x}_r^T - \mathrm{i}\sum_{k\neq r} \beta_{kr}^* \mathbf{x}_k^T]\mathbf{G}(\omega_r)[\mathbf{x}_r + \mathrm{i}\sum_{k\neq r} \beta_{kr}\mathbf{x}_k] \\
&= \mathbf{x}_r^T\mathbf{G}(\omega_r)\mathbf{x}_r - \mathrm{i}\sum_{k\neq r}(\beta_{kr}^* - \beta_{kr})\left[\mathbf{x}_k^T\mathbf{G}(\omega_r)\mathbf{x}_r\right. \\
&\quad \left. - \mathbf{x}_r^T\mathbf{G}(\omega_r)\mathbf{x}_k\right] + O(|\beta|^2) \\
&= G'_{rr} + 2\mathrm{i}\sum_{k\neq r}\Im(\beta_{kr})\left[\mathbf{x}_k^T\mathbf{G}(\omega_r)\mathbf{x}_r - \mathbf{x}_r^T\mathbf{G}(\omega_r)\mathbf{x}_k\right] + O(|\beta|^2) \\
&= G'_{rr} + O(\Im(\beta)) + O(|\beta|^2) \qquad [5.202]
\end{aligned}$$

where $G'_{rr} = \mathbf{x}_r^T\mathbf{G}(\omega_r)\mathbf{x}_r$. Note that $\mathbf{G}(\omega_r)$ is assumed to be symmetric in simplifying the above equation. From the second term in the numerator of equation [5.201], we have

$$\begin{aligned}
\mathbf{z}_i^H \mathbf{G}(\omega_r)\mathbf{z}_r &= [\mathbf{x}_i^T - \mathrm{i}\sum_{k\neq i} \beta_{ki}^* \mathbf{x}_k^T]\mathbf{G}(\omega_r)[\mathbf{x}_r + \mathrm{i}\sum_{k\neq r} \beta_{kr}\mathbf{x}_k] \\
&= \mathbf{x}_i^T\mathbf{G}(\omega_r)\mathbf{x}_r - \mathrm{i}\sum_{k\neq i}\beta_{ki}^*\mathbf{x}_k^T\mathbf{G}(\omega_r)\mathbf{x}_r + \mathrm{i}\sum_{k\neq r}\beta_{kr}\mathbf{x}_i^T\mathbf{G}(\omega_r)\mathbf{x}_k \\
&\quad + O(|\beta|^2) \\
&= G'_{ir} + O(\beta) + O(|\beta|^2) \qquad [5.203]
\end{aligned}$$

where $G'_{ir} = \mathbf{x}_i^T\mathbf{G}(\omega_r)\mathbf{x}_r$.

Substituting equations [5.194]–[5.197], [5.202] and [5.203], into equation [5.201], we obtain

$$\begin{aligned}
R_2(\psi) &= \frac{\left|G'_{rr} + 2\sum_{i\neq r}\Re(\epsilon_i)G'_{ir} + O(\epsilon)O(\beta) + O(\epsilon^2) + O(\Im(\beta)) + O(|\beta|^2)\right|}{1 + O(\epsilon^2) + O(|\beta|^2)} \\
&< \frac{|G'_{rr}| + 2\sum_{i\neq r}\Re(\epsilon_i)|G'_{ir}| + O(\epsilon)O(\beta) + O(\epsilon^2) + O(\Im(\beta)) + O(|\beta|^2)}{1 + O(\epsilon^2) + O(|\beta|^2)}.
\end{aligned} \qquad [5.204]$$

The last line in the above equation follows from the triangle inequality. The terms involving $O(\Im(\beta))$ are smaller than $O(|\beta|)$ terms. Moreover, for lightly non-viscous systems, the terms involving $O(\Im(\beta))$ are expected to be smaller than $O(\Re(\beta))$ terms, as can be seen in section 5.2.3. As a result, we expect to have the inequality

$$O(\Im(\beta)) < O(\Re(\beta)) < O(|\beta|). \qquad [5.205]$$

From equation [5.204], it can be seen that first-order changes in ψ lead to first-order changes in $R_2(\psi)$. However, if the complex modal damping matrix is diagonally dominant, i.e.

$$\frac{|G'_{ir}|}{|G'_{rr}|} << 1 \qquad [5.206]$$

then first-order changes in ψ lead to second-order changes in $R_2(\psi)$. In this case, stationarity of the Rayleigh quotient is obtained.

Returning to the third quotient, the equation corresponding to equation [5.201] in the case of $R_3(\psi)$ is

$$\begin{aligned} R_3(\psi) &= \frac{|\psi^H \mathbf{G}(\omega_r)\psi|}{\psi^H \mathbf{K}\psi} \\ &< \frac{|G'_{rr}| + 2\sum_{i\neq r} \Re(\epsilon_i)|G'_{ir}| + O(\epsilon)O(\beta) + O(\epsilon^2) + O(\Im(\beta)) + O(|\beta|^2)}{\omega_r^2 + O(\epsilon^2) + O(|\beta|^2)}. \end{aligned} \qquad [5.207]$$

The above quotient is not stationary. However, when the modal damping matrix is diagonally dominant in accordance with equation [2.281], stationarity of $R_3(\psi)$ holds.

5.8. Summary

The problem of dynamic analysis of non-viscously damped multiple-degrees-of-freedom linear systems has been considered. The assumed non-viscous damping model is such that the damping forces depend on the past history of motion via convolution integrals over some kernel functions. It has been assumed that, in general, the mass and stiffness matrices as well as the matrix of the kernel functions cannot be simultaneously diagonalized by any linear transformation. The analysis is, however, restricted to systems with non-repetitive eigenvalues and non-singular mass matrices. The concept of proportional damping was extended to non-viscously damped systems. Conditions for existence of proportional damping in non-viscously damped systems were derived.

For the general non-proportionally damped case, the system eigenvalues were obtained by solving the characteristic equation. It turns out that, unlike viscously damped systems, the order of the characteristic equation for an N degrees-of-freedom system is more than $2N$. As a result, the number of modes becomes more than $2N$ and they were divided into two types: (1) elastic modes and (2) non-viscous modes. It is assumed that the elastic modes appear in complex conjugate pairs, that is they are subcritically damped. The elastic modes, which consist of N eigenvectors together with their complex conjugate pairs, correspond to N modes of vibration of the structural system. These N eigenvectors were expressed as a complex linear combination of the (real) eigenvectors of the corresponding undamped system. The vectors of these complex constants were further determined from a series obtained by the Neumann expansion method. On the basis of this analysis, some approximate formulas for the eigenvalues and eigenvectors were suggested and their accuracy were verified using numerical examples. The non-viscous modes, which occur due to the non-viscous damping mechanism, are real, over-critically damped and non-oscillatory in nature. These modes were obtained by inversion of a partition of the dynamic stiffness matrix evaluated at the corresponding eigenvalues.

The transfer function matrix of the system was derived in terms of the eigenvalues and eigenvectors of the second-order system. Exact closed-form expressions of the response due to arbitrary forcing functions and initial conditions were obtained. The response can be expressed as a sum of two parts, one that arises in usual viscously damped systems and the other that occurs due to non-viscous damping mechanisms. Through an example, it was shown that the non-viscous part of the response is purely dissipative and non-oscillatory in nature. The method developed here is analogous to classical modal analysis where undamped natural frequencies and modes have to be appropriately replaced by elastic modes and non-viscous modes of the non-conservative system. The method presented offers a reduction in computational effort because neither the first-order formalisms nor the additional dissipation coordinates are used. Moreover, this approach also provides better physical insight as familiar N-space eigenvectors are utilized.

We have also developed several eigenrelations for non-viscously damped multiple-degrees-of-freedom linear dynamic systems. It has been assumed that, in general, the mass and stiffness matrices as well as the matrix of the kernel functions cannot be simultaneously diagonalized by any linear transformation. The analysis is, however, restricted to systems with non-repetitive eigenvalues and non-singular mass matrices. Relationships regarding the normalization and the orthogonality of the (complex) eigenvectors have been established (theorem 5.2). Expressions equivalent to the orthogonality of the undamped modes over the mass and stiffness matrices have been proposed (theorems 5.3 and 5.4). It was shown that the classical relationships can be obtained as special cases of these general results. On the basis of these results, we have shown that the mass and stiffness matrices can be uniquely expressed in terms of the eigensolutions. The damping matrix, $\mathbf{G}(s)$, cannot be reconstructed using this approach because it is not a constant matrix. However, we have provided expressions that relate the damping matrix to the eigensolutions for the cases when $s \to \infty$ and $s \to 0$. Whenever applicable, viscously damped counterparts of the newly developed results were provided.

The concept of the Rayleigh quotient was investigated for general non-viscously damped systems. Three new Rayleigh quotients are proposed and their stationary properties are investigated. The Rayleigh quotient involving mass and stiffness matrix is still stationary while the remaining two involving the frequency-dependent damping matrix are not. Stationarity in this case is subject to (1) the diagonal dominance of the absolute value of the frequency-dependent complex modal damping matrix and (2) light non-viscous damping. For an arbitrarily chosen non-viscous damping function, the stationarity property does not hold true.

This chapter provided quantitative and qualitative description of the dynamics of general non-viscously damped systems. The eigenvalues in the configuration space was obtained by solving the characteristic equation directly. This can be a computationally challenging problem for large systems. To address this, in Chapter 6, we propose some reduced computational approaches for generally damped linear dynamic systems.

Chapter 6

Reduced Computational Methods for Damped Systems

Dynamics of viscously damped systems and general non-viscously damped systems have been discussed in the previous chapters. It was shown that even when the system is not proportionally damped, the dynamic response can be expressed as superposition of complex modes. Therefore, the determination of the eigensolutions of the system is of crucial importance. Some perturbation-based methods have been proposed for the calculation of the eigenvalues and eigenvectors so far. In this chapter, we explore this problem in further detail. The aim is to develop computational methods for eigensolutions, which are computationally efficient and, at the same time, accurate. These two demands are often conflicting in nature. We develop some iterative methods that can address these issues to some extent. Once the eigensolutions can be obtained using the approach proposed in this chapter, they can be used in the expressions of the dynamic response derived in the previous chapters.

In section 6.1, we consider reduced computational methods for general non-proportionally damped systems with viscous damping. An iterative approach is presented here. In section 6.2, approximate methods for single-degree-of-freedom (SDOF) non-viscously damped systems are discussed. These ideas are extended to multiple-degrees-of-freedom (MDOF) non-viscously damped systems in section 6.3. In section 6.4, a different approach to reduce computational cost is pursued. Here, an equivalent second-order form is presented for general non-viscously damped systems. Finally, section 6.5 summarizes the studies undertaken in this chapter.

6.1. General non-proportionally damped systems with viscous damping

The equation of motion of an N degree-of-freedom (DOF) linear viscously damped system can be expressed by coupled differential equations as

$$\mathbf{M}\ddot{\mathbf{q}}(t) + \mathbf{C}\dot{\mathbf{q}}(t) + \mathbf{K}\mathbf{q}(t) = \mathbf{f}(t). \tag{6.1}$$

Here, $\mathbf{q}(t) \in \mathbb{R}^N$ is the displacement vector, $\mathbf{f}(t) \in \mathbb{R}^N$ is the forcing vector, $\mathbf{M}, \mathbf{K}, \mathbf{C} \in \mathbb{R}^{N\times N}$ are, respectively, the mass matrix, stiffness and the viscous damping matrix. In general, $\mathbf{M}$ is a positive definite symmetric matrix, and $\mathbf{C}$ and $\mathbf{K}$ are non-negative definite symmetric matrices. The natural frequencies ($\omega_j \in \mathbb{R}$) and the mode shapes ($\mathbf{x}_j \in \mathbb{R}^N$) of the corresponding undamped system can be obtained [MEI 97] by solving the matrix eigenvalue problem (given in [2.55])

$$\mathbf{K}\mathbf{x}_j = \omega_j^2 \mathbf{M}\mathbf{x}_j, \quad \forall j = 1, 2, \ldots, N. \tag{6.2}$$

The undamped eigenvectors satisfy an orthogonality relationship over the mass and stiffness matrices (see equations [2.63] and [2.64]), that is

$$\mathbf{x}_k^T \mathbf{M}\mathbf{x}_j = \delta_{kj} \tag{6.3}$$

$$\text{and} \quad \mathbf{x}_k^T \mathbf{K}\mathbf{x}_j = \omega_j^2 \delta_{kj}, \quad \forall\, k, j = 1, 2, \ldots, N \tag{6.4}$$

where δ_{kj} is the Kroneker delta function. We construct the modal matrix

$$\mathbf{X} = [\mathbf{x}_1, \mathbf{x}_2, \ldots, \mathbf{x}_N] \in \mathbb{R}^{N\times N}. \tag{6.5}$$

For proportionally damped systems (see section 2.4), the modal matrix can be used to diagonalize system [6.1] and the dynamic response can be obtained in term of the classical normal modes. The concern of this section is, when this condition is not met, the most likely case for many practical applications. In particular, due to the recent developments in actively controlled structures and the increasing use of composite and smart materials, the need to consider general non-proportionally damped linear dynamic systems is greater than ever before.

For non-proportionally damped systems, the modal damping matrix

$$\mathbf{C}' = \mathbf{X}^T \mathbf{C}\mathbf{X} \tag{6.6}$$

is not a diagonal matrix. Such problems can be solved using a spectral approach similar to the undamped or proportionally damped system by transforming equation [6.1] into a state-space form as shown in section 2.5.1.1. The state-space approach is not only computationally more expensive, it also lacks the physical insight provided by the classical normal mode based approach. Therefore, many authors have developed approximate methods in the original space. Rayleigh [RAY 77] proposed a perturbation method that forms the basis of many contemporary approximation methods [ADH 99a, ELB 09, ADH 99b]. It is now known that either the frequency separation between the normal modes [HAS 76], often known as "Hasselsman's criteria", or some form of diagonal dominance [SHA 88, MOR 09, ADH 04a, MOR 08b] in the modal damping matrix $\mathbf{C}'$ is sufficient for neglecting modal coupling. Udwadia [UDW 09] proved that for systems with non-repeated eigenvalues, the best approximation of a diagonal modal damping matrix is simply to consider the diagonal of the $\mathbf{C}'$ matrix. More discussion of this can be found in section 1.2.

The eigenvalue problem corresponding to system [6.1] can be expressed as

$$\left[s_j^2\mathbf{M} + s_j\mathbf{C} + \mathbf{K}\right]\mathbf{z}_j = \mathbf{0}, \quad \forall j = 1, 2, \ldots, 2N \qquad [6.7]$$

where $s_j \in \mathbb{C}$ are the eigenvalues and $\mathbf{z}_j \in \mathbb{C}^N$ are the eigenvectors. Comprehensive details on this type of quadratic eigenvalue problem can be found in Tisseur and Meerbergen [TIS 01]. Since $\mathbf{M}, \mathbf{K}$ and $\mathbf{C}$ are all real matrices, the eigensolutions are either real or they appear in the complex conjugate pairs. In this section, we consider complex conjugate eigensolutions only, as for stable systems such eigenvalues are of great practical importance. Using the eigensolutions, the frequency response function (FRF) can be obtained (see, for example [ADH 99a, TIS 01], equation [2.239]) as

$$\mathbf{H}(\mathrm{i}\omega) = \sum_{j=1}^{N}\left[\frac{\gamma_j\mathbf{z}_j\mathbf{z}_j^T}{\mathrm{i}\omega - s_j} + \frac{\gamma_j^*\mathbf{z}_j^*\mathbf{z}_j^{*T}}{\mathrm{i}\omega - s_j^*}\right] \quad \text{where } \gamma_j = \frac{1}{\mathbf{z}_j^T\left[2s_j\mathbf{M} + \mathbf{C}\right]\mathbf{z}_j}. \qquad [6.8]$$

Here, $(\bullet)^*$ denotes complex conjugation and $(\bullet)^T$ denotes matrix transposition. This equation shows that if the complex eigensolutions s_j and $\mathbf{z}_j$ can be obtained efficiently, the dynamic response can be obtained exactly using equation [6.8]. In this section, an iterative approach is developed to obtain the complex eigensolutions of non-proportionally damped systems from the undamped eigensolutions.

6.1.1. *Iterative approach for the eigensolutions*

Ibrahimbegovic and Wilson [IBR 89] have developed a procedure for analyzing non-proportionally damped systems using a subspace with a vector basis generated from the mass and stiffness matrices. Their approach avoids the use of complex eigensolutions. In the time domain, an iterative approach for solving the coupled equations was developed by Udwadia and Esfandiari [UDW 90] based on updating the forcing term appropriately. In the method proposed here, we obtain the complex modes and complex frequencies in an iterative manner.

For distinct undamped eigenvalues (ω_l^2), $\mathbf{x}_l$, $\forall\, l = 1, \ldots, N$, form a complete set of vectors. For this reason, $\mathbf{z}_j$ can be expanded as a complex linear combination of $\mathbf{x}_l$. Thus, an expansion of the form

$$\mathbf{z}_j = \sum_{l=1}^{N} \alpha_l^{(j)} \mathbf{x}_l \qquad [6.9]$$

may be considered. Without any loss of generality, we can assume that $\alpha_j^{(j)} = 1$ (normalization), which leaves us to determine $\alpha_l^{(j)}, \forall l \neq j$. Substituting the expansion of $\mathbf{z}_j$ into the eigenvalue equation [6.7], we obtain the approximation error for the jth mode as

$$\varepsilon_j = \sum_{l=1}^{N} s_j^2 \alpha_l^{(j)} \mathbf{M}\mathbf{x}_l + s_j \alpha_l^{(j)} \mathbf{C}\mathbf{x}_l + \alpha_l^{(j)} \mathbf{K}\mathbf{x}_l. \qquad [6.10]$$

We use a Galerkin approach to minimize this error by viewing the expansion [6.9] as a projection in the basis functions $\mathbf{x}_l \in \mathbb{R}^N, \forall l = 1, 2, \ldots N$. Therefore, we make the error orthogonal to the basis functions, that is

$$\varepsilon_j \perp \mathbf{x}_l \quad \text{or} \quad \mathbf{x}_k^T \varepsilon_j = 0 \quad \forall\, k = 1, 2, \ldots, N. \qquad [6.11]$$

Using the orthogonality property of the undamped eigenvectors described by [6.3] and [6.4], we obtain

$$s_j^2 \alpha_k^{(j)} + s_j \sum_{l=1}^{N} \alpha_l^{(j)} C'_{kl} + \omega_k^2 \alpha_k^{(j)} = 0, \quad \forall k = 1, \ldots, N \qquad [6.12]$$

where $C'_{kl} = \mathbf{x}_k^T \mathbf{C} \mathbf{x}_l$ are the elements of the modal damping matrix $\mathbf{C}'$ defined in equation [6.6]. The jth equation of this set obtained by setting $k = j$ can be written as

$$\left(s_j^2 + s_j C'_{jj} + \omega_j^2\right) \alpha_j^{(j)} + s_j \sum_{l \neq j}^{N} \alpha_l^{(j)} C'_{jl} = 0. \qquad [6.13]$$

Recalling that $\alpha_j^{(j)} = 1$ and $\mathbf{C}'$ is a symmetric matrix, this equation can be rewritten as

$$s_j^2 + s_j \underbrace{\left(C'_{jj} + \sum_{l \neq j}^{N} \alpha_l^{(j)} C'_{lj} \right)}_{\theta_j} + \omega_j^2 = 0 \qquad [6.14]$$

where

$$\theta_j = C'_{jj} + \mathbf{b}_j^T \mathbf{a}_j \qquad [6.15]$$

$$\mathbf{b}_j = \left\{ C'_{1j}, C'_{2j}, \ldots, ^{\{j\text{th term deleted}\}}, \ldots, C'_{nj} \right\}^T \in \mathbb{R}^{(N-1)} \qquad [6.16]$$

$$\text{and} \quad \mathbf{a}_j = \left\{ \alpha_1^{(j)}, \alpha_2^{(j)}, \ldots, ^{\{j\text{th term deleted}\}}, \ldots, \alpha_N^{(j)} \right\}^T \in \mathbb{C}^{(N-1)}. \qquad [6.17]$$

The vector $\mathbf{a}_j$ is unknown and can be obtained by excluding the $j = k$ case in equation [6.12]. Excluding this case, we have

$$s_j^2 \alpha_k^{(j)} + s_j \left(C'_{kj} + \alpha_k^{(j)} C'_{kk} + \sum_{l \neq k \neq j}^{N} \alpha_l^{(j)} C'_{kl} \right) + \omega_k^2 \alpha_k^{(j)} = 0,$$

$$\text{or} \quad \left(s_j^2 + \omega_k^2 + C'_{kk} \right) \alpha_k^{(j)} + s_j \sum_{l \neq k \neq j}^{N} C'_{kl} \alpha_l^{(j)} = -s_j C'_{kj}, \quad \forall k = 1, \ldots, N; \neq j. \qquad [6.18]$$

These equations can be combined into a matrix form as

$$\left[\mathbf{P}_j - \mathbf{Q}_j \right] \mathbf{a}_j = \mathbf{b}_j. \qquad [6.19]$$

In the above equation, the vectors $\mathbf{a}_j$ and $\mathbf{b}_j$ have been defined before. The matrices $\mathbf{P}_j$ and $\mathbf{Q}_j$ are defined as

$$\mathbf{P}_j = \operatorname{diag} \left[\frac{s_j^2 + s_j C'_{11} + \omega_1^2}{-s_j}, \ldots, ^{\{j-\text{th term deleted}\}}, \ldots, \frac{s_j^2 + s_j C'_{NN} + \omega_N^2}{-s_j} \right] \in \mathbb{C}^{(N-1)\times(N-1)}, \qquad [6.20]$$

and

$$\mathbf{Q}_j = \begin{bmatrix} 0 & C'_{12} & \cdots & \{j\text{th term deleted}\} & \cdots & C'_{1N} \\ C'_{21} & 0 & \vdots & \vdots & \vdots & C'_{2N} \\ \vdots & \vdots & \vdots & \{j\text{th term deleted}\} & \vdots & \vdots \\ \vdots & \vdots & \vdots & \vdots & \vdots & \vdots \\ C'_{N1} & C'_{N2} & \cdots & \{j\text{th term deleted}\} & \cdots & 0 \end{bmatrix} \in \mathbb{R}^{(N-1)\times(N-1)}. \quad [6.21]$$

From equation [6.19], $\mathbf{a}_j$ should be obtained by solving the set of linear equations. Because $\mathbf{P}_j$ is a diagonal matrix, one way to do this is by using the Neumann expansion, method [ADH 99a]. Using the Neumann expansion we have

$$\mathbf{a}_j = \left[\mathbf{I}_{N-1} - \mathbf{P}_j^{-1}\mathbf{Q}_j\right]^{-1} \left\{\mathbf{P}_j^{-1}\mathbf{b}_j\right\} = \left[\mathbf{I}_{N-1} + \mathbf{R}_j + \mathbf{R}_j^2 + \mathbf{R}_j^3 + \ldots\right] \mathbf{a}_{j_0} \quad [6.22]$$

where $\mathbf{I}_{N-1}$ is a $(N-1) \times (N-1)$ identity matrix,

$$\mathbf{R}_j = \mathbf{P}_j^{-1}\mathbf{Q}_j \in \mathbb{C}^{(N-1)\times(N-1)} \quad \text{and} \quad \mathbf{a}_{j_0} = \mathbf{P}_j^{-1}\mathbf{b}_j \in \mathbb{C}^{(N-1)}. \quad [6.23]$$

Because $\mathbf{P}_j$ is a diagonal matrix, its inversion can be carried out analytically and subsequently the closed-form expressions of the elements of $\mathbf{a}_j$ can be obtained. Keeping one term in the series [6.22], the first-order expression of the elements of $\mathbf{a}_j$ can be obtained as

$$\mathbf{a}_j \equiv \left\{\alpha_k^{(j)}\right\}_{\forall k \neq j} = \frac{s_j C'_{kj}}{\omega_k^2 + s_j^2 + s_j C'_{kk}}. \quad [6.24]$$

Similarly, the second-order expression of the elements of $\mathbf{a}_j$ can be obtained as

$$\mathbf{a}_j \equiv \left\{\alpha_k^{(j)}\right\}_{\forall k \neq j} = \frac{s_j C'_{kj}}{\omega_k^2 + s_j^2 + s_j C'_{kk}} + \sum_{\substack{l=1 \\ l \neq j \neq k}}^{N} \frac{s_j^2 C'_{kl} C'_{lj}}{\left(\omega_k^2 + s_j^2 + s_j C'_{kk}\right)\left(\omega_l^2 + s_j^2 + s_j C'_{ll}\right)}. \quad [6.25]$$

The vector $\mathbf{a}_j$ obtained using this way can be substituted back in the expression of the eigenvalues in [6.14], which in turn can be solved for s_j as

$$s_j = -\left(\theta_j \pm \mathrm{i}\sqrt{4\omega_j^2 - \theta_j^2}\right)/2. \quad [6.26]$$

However, the vector $\mathbf{a}_j$ is also a function of s_j. As a result, θ_j becomes a function of s_j. This forms the basis of the iterative approach, as from equation [6.26] we can write

$$s_j^{(r+1)} = -\theta_j(s_j^{(r)})/2 \pm \mathrm{i}\sqrt{\omega_j^2 - \theta_j^2(s_j^{(r)})/4}; \quad r = 0, 1, 2, \ldots \qquad [6.27]$$

For every iteration step, the vector $\mathbf{a}_j$ gets updated based on new values of s_j using equation [6.24] or [6.25] depending on the order of terms retained in the series [6.22]. The iteration can be started with the equivalent proportional damping assumption [UDW 09], namely

$$s_j^{(0)} = -C'_{jj}/2 \pm \mathrm{i}\sqrt{\omega_j^2 - C'^2{}_{jj}/4}. \qquad [6.28]$$

The iteration can be stopped when the successive values of s_j or $\mathbf{a}_j$ do not change significantly. Once the final values of $\alpha_k^{(j)}, \forall k$ are obtained, the jth complex mode $\mathbf{z}_j$ can be obtained from the series [6.9].

The necessary and sufficient conditions for the convergence of the proposed method are difficult to obtain. Below we give a sufficient condition.

PROPOSITION 6.1.– A sufficient condition for the convergence of the proposed iterative method is that $\mathbf{C}'$ is a diagonally dominant matrix.

PROOF.– During the iteration process, the value of s_j changes for different iteration steps. We aim to derive the condition for the convergence of series [6.22] for an arbitrary value of s_j is. This will guarantee the convergence of the iterative method, no matter what the value of s_j is. The complex matrix power series [6.22] converges if, and only if, for all the eigenvalues $\sigma_l^{(j)}$ of the matrix $\mathbf{R}_j$, the inequality $|\sigma_l^{(j)}| < 1$ holds. Although this condition is both necessary and sufficient, checking convergence for all $j = 1, \cdots, N$ is not feasible for every iteration step. So we look for a sufficient condition that is relatively easy to check and that ensures convergence for all $j = 1, \cdots, N$.

For an arbitrary rth iteration, let us denote the matrix $\mathbf{R}_j$ defined in equation [6.23] as $\mathbf{R}_j^{(r)}$. Suppose the value of s_j for the rth iteration step is $s_j^{(r)}$. The klth element of the matrix $\mathbf{R}_j^{(r)}$ can be obtained as

$$R^{(r)}{}_{jkl} = \frac{-s_j^{(r)} C'_{kl}\,(1 - \delta_{kl})}{\omega_k^2 + s_j^{(r)^2} + s_j^{(r)} C'_{kk}}, \quad \forall k, l \neq j. \qquad [6.29]$$

Since a matrix norm is always greater than or equal to its maximum eigenvalue, it follows from the inequality $|\sigma_l^{(j)}| < 1$ that the convergence of the series is guaranteed if $\| \mathbf{R}_j^{(r)} \| < 1$. Writing the sum of absolute values of entries of $\mathbf{R}_j^{(r)}$ results in the following inequality as the required sufficient condition for the convergence

$$\sum_{\substack{k=1 \\ k \neq j}}^{N} \sum_{\substack{l=1 \\ l \neq j}}^{N} \left| \frac{s_j^{(r)} C'_{kl}}{\omega_k^2 + s_j^{(r)^2} + s_j^{(r)} C'_{kk}} \right| (1 - \delta_{lk}) < 1. \qquad [6.30]$$

Dividing both the numerator and denominator by $s_j^{(r)}$, the above inequality can be written as

$$\sum_{\substack{k=1 \\ k \neq j}}^{N} \sum_{\substack{l=1 \\ l \neq i \neq k}}^{N} \frac{|C'_{kl}|}{\left|1/s_j^{(r)} \left(\omega_k^2 + s_j^{(r)^2}\right) + C'_{kk}\right|} < 1. \qquad [6.31]$$

Taking the maximum for all $k \neq j$, this condition can further be represented as

$$\max_{k \neq j} \frac{\sum_{\substack{l=1 \\ l \neq j,k}}^{N} |C'_{kl}|}{\left|1/s_j^{(r)} \left(\omega_k^2 + s_j^{(r)^2}\right) + C'_{kk}\right|} < 1 \qquad [6.32]$$

It is clear that [6.32] always holds if

$$\sum_{\substack{l=1 \\ l \neq j \neq k}}^{N} |C'_{kl}| < |C'_{kk}|, \quad \forall k \neq j \qquad [6.33]$$

which in turn implies that, for all $j = 1 \cdots N$, the inequality $\| \mathbf{R}_j^{(r)} \| < 1$ holds if $\mathbf{C}'$ is a diagonally dominant matrix. It is important to note that the diagonal dominance of $\mathbf{C}'$ is only a sufficient condition and the lack of it does not necessarily prevent convergence of the proposed iterative method.

6.1.2. *Summary of the algorithm*

In this section, we propose a simple iterative algorithm to implement the idea developed in section 6.1.1. We select a tolerance between the difference of the successive values of s_j, denoted by ϵ_m. A small value, say $\epsilon_m = 0.001$, can be selected for numerical calculations. Considering that the undamped eigensolutions (ω_j and $\mathbf{x}_j$) and the modal damping matrix $\mathbf{C}'$ are known, the complex eigensolutions (s_j and $\mathbf{z}_j$) can be obtained using the following iterative algorithm:

for $j = 1, 2, \ldots N$ **do**
Initialize $\epsilon = 100$, $r = 0$
$s_j^{(r)} = -C'_{jj}/2 \pm \mathrm{i}\sqrt{\omega_j^2 - C'^2{}_{jj}/4}$
$\mathbf{b}_j = \left\{C'_{1j}, C'_{2j}, \ldots, ^{\{j\text{th term deleted}\}}, \ldots, C'_{nj}\right\}^T$
while $\epsilon > \epsilon_m$ **do**
$\mathbf{a}_j(s_j^{(r)}) \equiv \left\{\alpha_k^{(j)}\right\}_{\forall k \neq j} = \frac{s_j^{(r)} C'_{kj}}{\omega_k^2 + s_j^{(r)2} + s_j^{(r)} C'_{kk}}$
$\theta_j = C'_{jj} + \mathbf{b}_j^T \mathbf{a}_j(s_j^{(r)})$
$s_j^{(r+1)} = -\theta_j(s_j^{(r)})/2 \pm \mathrm{i}\sqrt{\omega_j^2 - \theta_j^2(s_j^{(r)})/4}$
$\epsilon = |s_j^{(r+1)} - s_j^{(r)}|/|s_j^{(r)}|$
$r = r + 1$
end while
$\mathbf{z}_j = \sum_{k=1}^N \alpha_k^{(j)} \mathbf{x}_k$
end for

The algorithm is outlined for the first-order expression of $\alpha_k^{(j)}$ given by equation [6.24]. However, the extension to the second-order or higher order expressions is straightforward. We simply need to change the expression of $\mathbf{a}_j(s_j^{(r)})$ in this algorithm. If the higher order terms are used, then a lower number of steps in the iteration are needed. Once the complex eigensolutions s_j and $\mathbf{z}_j$ are obtained using this method for all j, the dynamic response such as the FRF can be obtained exactly using equation [6.8].

The computational complexity to solve algebraic eigenvalue problems scales cubically with the dimension [WIL 88]. Therefore, an estimation of the order of calculations needed to solve the eigenvalue problem of a matrix of size N is $\mathcal{O}(N^3)$ for large N. This is the computational time for an undamped system of dimension N. Suppose the computational cost for the iteration of each eigensolution pair (s_j and $\mathbf{z}_j$) is proportional to c_I. The value of c_I will be higher if a greater number of iterations are used. The total cost of the iteration would be in the order Nc_I. Adding these two, the order of calculations needed to approximate the eigensolutions with the proposed method is $\mathcal{O}(N^3 + C_I N)$. For a general non-proportionally damped system, the size of the state-space matrix is $2N$. Therefore, the computational time is in the order $\mathcal{O}((2N)^3) = \mathcal{O}(8N^3)$. It is important to note that the order of computation related to the iteration scales linearly with N, as opposed to cubically in the case of, direct state-space method. Therefore, when N becomes large, as expected for practical problems, the proposed method has a clear advantage as $\mathcal{O}(8N^3) > \mathcal{O}(N^3 + C_I N)$. The idealized computational efficiency can be obtained as

$$\frac{8N^3}{N^3 + C_I N} \to 8 \text{ (when } N \to \infty). \qquad [6.34]$$

Next, we illustrate this new method using a numerical example.

6.1.3. *Numerical example*

We consider a three DOF system with the mass, stiffness and damping matrices given by

$$\mathbf{M} = \begin{bmatrix} 3 & 0 & 0 \\ 0 & 3 & 0 \\ 0 & 0 & 3 \end{bmatrix}, \quad \mathbf{K} = \begin{bmatrix} 4 & -2 & 0 \\ -2 & 4 & -2 \\ 0 & -2 & 4 \end{bmatrix} \quad \text{and} \quad \mathbf{C} = \begin{bmatrix} 0 & 0 & 0 \\ 0 & 1.75 & -1.75 \\ 0 & -1.75 & 1.75 \end{bmatrix}. \qquad [6.35]$$

The undamped eigenvalues and eigenvectors are obtained as

$$\{\omega_1, \omega_2, \omega_3\} = \{0.6249, 1.1547, 1.5087\}$$

$$\text{and} \qquad [\mathbf{x}_1, \mathbf{x}_2, \mathbf{x}_3] = \frac{1}{2\sqrt{3}} \begin{bmatrix} 1 & -\sqrt{2} & -1 \\ \sqrt{2} & 0 & \sqrt{2} \\ 1 & \sqrt{2} & -1 \end{bmatrix}. \qquad [6.36]$$

Eigenvalue number	State-space (exact)	First-order perturbation	Proposed iterative method
1	−0.0103+0.6298i	−0.0125+0.6249i(0.8552)	−0.0114+0.6306i, (0.2018), 2 iterations
2	−0.0478+1.2407i	−0.1458+1.1547i(10.5023)	−0.0468+1.2379i, (0.2428), 6 iterations
3	−0.5252+1.2890i	−0.4250+1.5087i(17.3472)	−0.4314+1.3687i, (8.8436), 8 iterations

Table 6.1. *The complex eigenvalues of the system obtained using the proposed method is compared with the exact state-space method and the first-order perturbation method. The numbers in the parenthesis represent the percentage error*

The last two undamped eigenvalues are very close and therefore we would expect significant modal coupling. The eigenvalues obtained using the proposed iterative method are compared with the exact state-space method and the first-order perturbation method in Table 6.1. We have used the first-order expression of $\alpha_k^{(j)}$ given by equation [6.24] and considered the error tolerance to be $\epsilon_m = 0.001$. For the three eigenvalues, two, six and eight iterations are used, respectively. The first-order perturbation results are obtained from equation [6.28]. The percentage errors are calculated with respect to the exact state-space results. Due to the high modal coupling, the first-order perturbation results corresponding to modes 2 and 3 have considerable errors (10.5023% and 17.3472%, respectively). Using the proposed iterative method, errors corresponding to these two modes reduce to 0.2428% and 8.8436% using six and eight iterations, respectively. The computational time (obtained using Matlab's `tic` and `toc` command) for the state-space method is 0.8912×10^{-2} s. The computational times for the first-order perturbation and the proposed iterative method are, respectively, 0.2367×10^{-2} s and 0.2774×10^{-2} s. This implies that for this problem, the proposed approach is about 3.2 times faster compared to the state-space method. As discussed before, theoretically the maximum computational efficiency can be up to eight times. However, this can only be realized for very large systems. In Figure 6.1, two typical FRFs of the system calculated from equation [6.8] using the three methods are shown. The modal damping factors for the three modes are, respectively, 0.0164, 0.0386 and 0.4074. Due to the high damping in the third mode, the resonance peak is not visible. As expected, the first-order perturbation performs poorly around the second and third mode due to the significant modal coupling. From these results, the relative accuracy of the proposed iterative method can be observed. Next, we consider non-viscously damped systems.

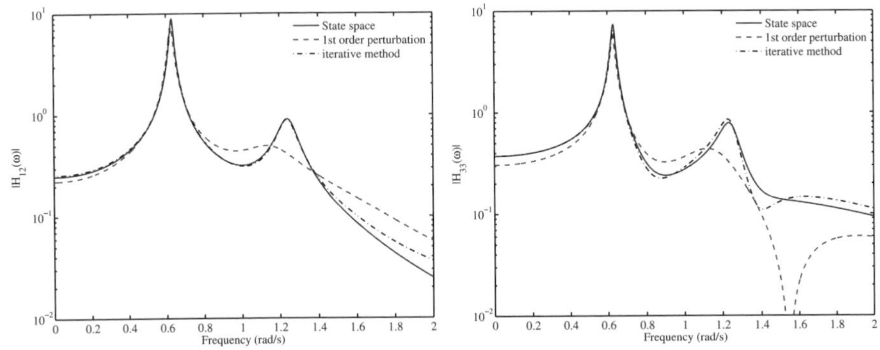

(a) Cross FRF $H_{12}(\omega)$ (b) Driving-point FRF $H_{33}(\omega)$

Figure 6.1. *A cross FRF and a driving-point FRF of the system calculated using the exact state-space method, the first-order perturbation method and the proposed iterative method*

6.2. Single-degree-of-freedom non-viscously damped systems

A key feature of non-viscously damped systems is the incorporation of the time history of the state-variables in the equation of motion. Several different models are available for non-viscously damped systems. We use the Biot model [BIO 58] that allows us to incorporate a wide range of functions in the frequency domain by means of summation of simple "pole residue forms". Several authors have considered this model due to its simplicity and generality (see, for example, [MUR 97b, MUR 98a, ZHA 07a]). The equation of motion of an SDOF linear non-viscously damped system can be expressed by

$$m\ddot{q}(t) + \int_0^t \mathcal{G}(t-\tau)\,\dot{q}(\tau)\,\mathrm{d}\tau + kq(t) = f(t). \qquad [6.37]$$

Here, $q(t)$ is the displacement, $f(t)$ is the forcing, m is the mass, k is the elastic stiffness and $\mathcal{G}(t)$ is the non-viscously damped stiffness kernel function. The kernel function $\mathcal{G}(t)$, or functions similar to it, is known by many names such as retardation functions, heredity functions, after-effect functions or relaxation functions in the context of different subjects. Equation [6.37] is very general and for any engineering applications some specific form of $\mathcal{G}(t)$ has to be assumed. A wide variety of mathematical expressions could be used for the kernel functions $\mathcal{G}(t)$ as long as the rate of energy dissipation is non-negative (see Table 1.1). Here, we will use a non-viscously damped material model for which the kernel function has the form

$$\mathcal{G}(t) = c_0\delta(t) + \sum_{k=1}^{n} a_k e^{-b_k t} c_k \qquad [6.38]$$

or in the Laplace domain

$$G(s) = c_0 + \sum_{k=1}^{n} \frac{a_k}{s+b_k} c_k. \qquad [6.39]$$

The constants a_k, b_k, c_0 and c_k are the non-viscously damped material parameters and n denotes the number of terms in the series. This model is similar to the exponential model discussed in Chapter 4. In fact, it can be seen that in the special case when $a_k = b_k = \mu_k \mathrm{m}$, we regain the model discussed in Chapter 4. Here, we use the more general model given in [6.38]. Some other non-viscously damped modeling approaches, such as the Golla-Hughes-McTavish (GHM) approach [GOL 85, MCT 93], use the following expression for the kernel function

$$sG(s) = G^{\infty}\left[1 + \sum_k \alpha_k \frac{s^2 + 2\xi_k\omega_k s}{s^2 + 2\xi_k\omega_k s + \omega_k^2}\right]. \qquad [6.40]$$

For the fractional derivative approach proposed by Bagley and Torvik [BAG 83], the kernel function takes the form

$$G(s) = \frac{E_1 s^{\alpha} - E_0 b s^{\beta}}{1 + b s^{\beta}} \quad (0 < \alpha, \beta < 1). \qquad [6.41]$$

For the anelastic displacement field (ADF) approach [LES 90, LES 95], we have

$$G(s) = 1 + \sum_{k=1}^{n} \frac{\Delta_k s}{s + \beta_k}. \quad [6.42]$$

Although these models are physically different, it will be seen later that they can be treated in an unified way for the calculation of some of the eigenvalues of the system.

Equation [6.37], together with the kernel in equation [6.38], represents an integro-differential equation. Several authors have proposed [BAG 83, GOL 85, MCT 93, LES 90, LES 95, MUR 97b, MUR 98a, WAG 03, ADH 03a] state-space approach based on the internal variables for this type of equation. The main reasons for seeking an alternative to the state-space approach include the following:

– Although exact in nature, the state-space approach for non-viscously damped systems is computationally very intensive for real-life MDOF systems due to the huge number of internal variables.

– The physical insights offered by methods in the original space (e.g. the modal analysis) are lost in a state-space-based approach.

Regarding the first point, an approximate method in the space of the original problem was discussed in section 5.2. A direct time-domain approach to obtain the solution of equation [6.37] was proposed in section 4.7. This method is computationally efficient and accurate but does not provide much physical insight. Based on a variational principle, Qian and Hansen [QIA 95] derived a substructure synthesis method where the non-viscously damped system eigensolution is obtained from the undamped system eigensolution. Daya and Potier-Ferry [DAY 01] proposed an asymptotic numerical method for the calculation of natural frequencies and loss factors of non-viscously damped systems. Further discussion on this topic can be found in section 1.3.

The calculation of the eigenvalues by solving the nonlinear eigenvalue problem corresponding to the equation of motion [6.37] is the main topic of this section. The objective is to derive the approximations in such a way so that they can be easily extended to MDOF systems where the computational efficiency can make a real difference. We have derived closed-form approximate expressions of the eigenvalues of the system for three mathematically different cases based on the number of kernel functions. First, SDOF systems are considered and then the results are extended to MDOF systems. The approximations utilize Taylor series expansion in the complex domain and are based on certain simplifying physical assumptions. The validity of the assumptions and the accuracy of the results are verified by numerical calculations.

6.2.1. *Nonlinear eigenvalue problem for non-viscously damped systems*

The equation of motion [6.37] can be expressed in the Laplace domain as

$$\begin{aligned} &s^2 m\bar{q}(s) + s\, G(s)\bar{q}(s) + k\bar{q}(s) = \bar{f}(s) \\ &\text{or} \quad d(s)\bar{q}(s) = \bar{f}(s). \end{aligned} \tag{6.43}$$

Here, the dynamic stiffness

$$d(s) = s^2 m + s\, G(s) + k \tag{6.44}$$

and $\bar{q}(s)$, $\bar{f}(s)$ and $G(s)$ are, respectively, the Laplace transforms of displacement, forcing and non-viscously damped kernel function. The eigenvalues of the system can be obtained by solving the characteristic equation

$$\begin{aligned} &s_j^2 m + s_j G(s_j) + k = 0 \\ \text{or} \quad &s_j^2 m + s_j \left(c_0 + \textstyle\sum_{k=1}^{n} \frac{a_k}{s_j + b_k} c_k \right) + k = 0, \quad j = 1, \cdots, m \end{aligned} \tag{6.45}$$

where m is the order of the characteristic polynomial. Following the approach outlined in section 4.5.2, the dynamic response, that is the solution of equation [6.37], can be expressed in terms of the eigenvalues as

$$\begin{aligned} q(t) &= \sum_{j=1}^{m} \gamma_j \left[\int_0^t \left\{ e^{s_j(t-\tau)} f(\tau) + \mathcal{G}(\tau) u_0 \right\} d\tau + e^{s_j t} \left\{ m\dot{q}_0 + s_j m q_0 \right\} \right], \\ &\forall\, t > 0. \end{aligned} \tag{6.46}$$

Here, u_0 and $\dot{u}_0$ are, respectively, the initial displacement and velocity and the constants γ_j can be expressed as

$$\gamma_j = \frac{1}{\frac{\partial d(s)}{\partial s}\big|_{s=s_j}}. \tag{6.47}$$

For systems with only the elastic stiffness term, the order of the characteristic polynomial is $m = 2$. For non-viscously damped systems, in general, m is more than two. From equation [6.45], we can observe that in general $m = 2 + n$. This is the key difference between a non-viscously damped system and an elastic system where the number of eigenvalues is exactly two. The two conjugated complex eigenvalues correspond to the oscillatory motions of the system and are called *elastic mode* or *vibration mode*. The modes corresponding to the "additional" n eigenvalues are called *damping modes* or *overdamped modes*. For stable passive systems, the damping modes are overcritically damped (i.e. negative real eigenvalues) and not oscillatory in nature. An extensive discussion on the nature of the eigenvalues, when $n = 1$, was undertaken up in Chapter 3. Here, both the complex conjugate modes and the damping modes will be derived. In the following sections, closed-form approximate expressions of eigenvalues are derived for the elastic modes and damping modes.

6.2.2. *Complex conjugate eigenvalues*

The main motivation behind the proposed approximations is that the approximate eigenvalues can be "constructed" from the eigenvalues of the underlying elastic system. The eigenvalues of the underlying damped elastic system can in turn be expressed in terms of the undamped eigenvalues. Combining these together, we can therefore obtain the eigenvalues of non-viscously damped systems by simple "post-processing" of the eigenvalues of the underlying undamped systems only. The eigenvalues (appearing in a complex conjugate pair) of the underlying damped elastic system [MEI 67] are given by

$$s_0 = -\zeta_n\omega_n \pm \mathrm{i}\omega_n\sqrt{1-\zeta_n^2} \approx -\zeta_n\omega_n \pm \mathrm{i}\omega_n \qquad [6.48]$$

where the undamped natural frequency is

$$\omega_n = \sqrt{k/m} \qquad [6.49]$$

and ζ_n is the viscous damping factor. Viscously damped elastic system can be considered as a special case of equation [6.43] when the function $G(s)$ is replaced by $G(s \to 0)$. Therefore, for the purpose of numerical approximations, we can obtain an equivalent viscous damping coefficient

$$c = \lim_{s\to 0} G(s) = c_0 + \sum_{k=1}^{n} a_k\, c_k/b_k. \qquad [6.50]$$

From this expression, the equivalent viscous damping factor ζ_n can be obtained as

$$\zeta_n = c/2\sqrt{km} = \left(c_0 + \sum_{k=1}^{n} a_k\, c_k/b_k\right)/2\sqrt{km}. \qquad [6.51]$$

For a truly damped elastic system, the solution given by equation [6.48] would have been the exact solution of the characteristic equation [6.45]. Since, in general, this is not the case, the difference between the elastic solution and the true solution of the characteristic equation [6.45] essentially arises due to the "varying" nature of the function $G(s)$. The approximate solution obtained here is based on keeping this fact in mind.

The central idea is that the actual solution of the characteristic equation [6.45] can be obtained by expanding the solution in a Taylor series around s_0. The error arising in the resulting solution would then depend on the "degree of variability" of the function $G(s)$. We assume that the true solution of equation [6.45] can be expressed as

$$s = s_0 + \delta \qquad [6.52]$$

where δ is a small quantity. Substituting this into the characteristic equation, we have

$$(s_0 + \delta)^2 m + (s_0 + \delta) G(s_0 + \delta) + k = 0. \qquad [6.53]$$

Expanding $G(s_0 + \delta)$ in a Taylor series in δ around s_0, we have

$$(s_0 + \delta)^2 m \; + \; (s_0 + \delta) \left(G(s_0) + \delta \frac{\partial G(s_0)}{\partial s} + \cdots \right) + k = 0, \quad \text{or}$$

$$\left(m + \frac{\partial G(s_0)}{\partial s} \right) \delta^2 \; + \; \left(s_0 \left(2m + \frac{\partial G(s_0)}{\partial s} \right) + G(s_0) \right) \delta$$
$$+ \; s_0 \left(s_0 m + G(s_0) \right) + k \approx 0. \qquad [6.54]$$

Keeping only the first-order terms in δ, we have

$$\delta^{(1)} \approx - \frac{s_0 (s_0 m + G(s_0)) + k}{s_0 (2m + \frac{\partial G}{\partial s}(s_0)) + G(s_0)}. \qquad [6.55]$$

The superscript 1 in equation [6.55] is used to denote that this is a first-order approximation.

We can improve the accuracy by retaining higher order terms in δ. Retaining up to second-order terms in δ in the Taylor expansion of equation [6.54], we have

$$\delta^{(2)} = \frac{-B - \sqrt{B^2 - 4AC}}{2A} \qquad [6.56]$$

where

$$A = m + \frac{1}{2} \frac{\partial^2 G(s_0)}{\partial s^2} s_0 + \frac{\partial G(s_0)}{\partial s}, \quad B = 2m s_0 + s_0 \frac{\partial G(s_0)}{\partial s} + G(s_0) \qquad [6.57]$$

$$\text{and} \quad C = s_0^2 m + s_0 G(s_0) + k. \qquad [6.58]$$

In the above expressions, $\frac{\partial G(s_0)}{\partial s}$ and $\frac{\partial^2 G(s_0)}{\partial s^2}$ are, respectively, the first- and second-order derivative of $G(s)$ evaluated at $s = s_0$. When the kernel function is given by equation [6.39], we have

$$\begin{aligned} \frac{\partial G(s_0)}{\partial s} &= -\sum_{k=1}^{n} \frac{a_k}{\left(s_0 + b_k\right)^2} c_k \\ \frac{\partial^2 G(s_0)}{\partial s^2} &= \sum_{k=1}^{n} \frac{2a_k}{\left(s_0 + b_k\right)^3} c_k. \end{aligned} \qquad [6.59]$$

Our numerical works show that retaining terms higher than the second-order results in considerably complex expressions and the accuracy gained is not very significant. As a result, we have not perused this approach in the rest of this section. On the basis of our numerical works, we recommend the second-order expression in equation [6.56] as it gives an excellent accuracy and additional computation cost is just marginally higher compared to the first-order approximation. The expression of the approximate eigenvalue derived here shows that the complex conjugate eigenvalues of a general non-viscously damped system can be obtained by post-processing of the undamped eigenvalue ω_n and equivalent viscous damping factor ζ_n. This approximation of the complex conjugate eigenvalues is valid for any kernel function. It could therefore be used with non-viscously damped models given by equations [6.40]–[6.42]. On the contrary, the real eigenvalues are specific to a particular kernel function. In section 6.2.3, the approximation to the real eigenvalues is derived for the Biot model.

6.2.3. *Real eigenvalues*

While the complex conjugate eigenvalues can be expected to be close to the eigenvalues of the equivalent elastic damped system, no such analogy can be made for the real solution as the equivalent elastic system does not have one. This makes the calculation of the real eigenvalues more challenging. The nature of the real eigenvalues is expected to be specific to the corresponding kernel functions. For this reason, the methods proposed here are only valid for the Biot model. For the convenience of analytical developments, the following three cases are considered separately:

– SDOF system with single exponential kernel ($n = 1$);

– SDOF system with double exponential kernels ($n = 2$);

– SDOF system with multiple exponential kernels ($n > 2$).

6.2.3.1. *Systems with single exponential kernel*

For this case, when $n = 1$, the eigenvalue equation can be simplified from equation [6.45] as

$$s^2 m + sG(s) + k = 0 \quad \text{where} \quad G(s) = c_0 + \frac{a_1}{s + b_1} c_1. \qquad [6.60]$$

Equation [6.60] is a third-order polynomial in s and it can be solved exactly in closed form. A more detailed study on the properties of the exact solutions has been carried out in Chapter 3. Here, we propose an approximate solution so that the approximation method can be extended to more complex cases.

We first multiply the characteristic equation [6.60] by $(s + b_1)$ and rewrite

$$(s^2 m + sc_0 + k)(b_1 + s) + sa_1 c_1 = 0. \qquad [6.61]$$

To obtain an initial estimation, the "small non-viscously damping" approximation is introduced so that $sa_1 c_1 \approx 0$. Since $(s^2 m + sc_0 + k) \neq 0$, as we are considering the real solution only, the first guess is obtained as

$$b_1 + s_0 = 0 \quad \text{or} \quad s_0 = -b_1. \qquad [6.62]$$

We take the first approximation of the real root as

$$s = s_0 + \Delta = -b_1 + \Delta \qquad [6.63]$$

where Δ is a small quantity. Equation [6.63] indicates that the real eigenvalues are expected to be close to the relaxation parameters. This is one of the most crucial observations made here. Substituting s from [6.63] into the characteristic equation, we obtain

$$\left((-b_1 + \Delta)^2 m + (-b_1 + \Delta)c_0 + k\right)\Delta + (-b_1 + \Delta)a_1 c_1 = 0. \qquad [6.64]$$

Expanding this equation results

$$m\Delta^3 + (-b_1 m + c_0)\Delta^2 + \left({b_1}^2 m - b_1 c_0 + k + a_1 c_1\right)\Delta - a_1 b_1 c_1 = 0. \qquad [6.65]$$

After neglecting all the terms associated with Δ^n for $n > 1$, this equation results in

$$\Delta \approx \frac{a_1 b_1 c_1}{{b_1}^2 m - b_1 c_0 + k + a_1 c_1}. \qquad [6.66]$$

Therefore, the approximate real solution is given by

$$s = s_0 + \Delta \approx -b_1 + \frac{a_1 b_1 c_1}{{b_1}^2 m - b_1 c_0 + k + a_1 c_1}. \qquad [6.67]$$

The third-order characteristic polynomial in equation [6.65] can be solved exactly. However, recall that the main reason for this approximation is not to solve this specific problem, but to solve MDOF systems with multiple kernels using the insight obtained from this simple problem. In section 6.2.3.2, this approximation technique is extended to SDOF systems with two kernels.

6.2.3.2. *Systems with two exponential kernels*

This section is the next logical step toward solving the general multiple kernel case. For this case, the kernel function $G(s)$ takes the form

$$G(s) = c_0 + \frac{a_1}{s + b_1} c_1 + \frac{a_2}{s + b_2} c_2. \qquad [6.68]$$

The characteristic function with this type of kernel function can be expressed as

$$s^2 m + s \left(c_0 + \frac{a_1}{s + b_1} c_1 + \frac{a_2}{s + b_2} c_2 \right) + k = 0. \qquad [6.69]$$

This is a polynomial in s of order four and therefore has four roots. The two complex conjugate roots can be obtained using the expressions derived before. In this section, only the real roots are considered. Multiplying equation [6.69] by $(s + b_1)(s + b_2)$, we have

$$(s^2 m + s c_0 + k)(s + b_1)(s + b_2) + s\left(a_1 c_1 (s + b_2) + a_2 c_2 (s + b_1)\right) = 0. \qquad [6.70]$$

Similar to the previous case, the first solution is approximated as

$$s_1 = -b_1 + \Delta_1. \qquad [6.71]$$

Substituting this into equation [6.70] and neglecting the higher order terms Δ_1^n for $n > 1$, we obtain

$$\Delta_1 \approx \frac{(b_1 - b_2) a_1 b_1 c_1}{(b_2 - b_1)\left(b_1^2 m - b_1 c_0 + k + a_1 c_1\right) - b_1 \left(a_1 c_1 + a_2 c_2\right)}. \qquad [6.72]$$

Using this, the approximation to the first real solution is given by

$$s_1 \approx -b_1 + \frac{(b_1 - b_2)a_1 b_1 c_1}{(b_2 - b_1)\left(b_1^2 m - b_1 c_0 + k + a_1 c_1\right) - b_1\left(a_1 c_1 + a_2 c_2\right)}. \qquad [6.73]$$

The second solution can be obtained from the same expression by swapping the subscripts 1 and 2.

6.2.3.3. *The general case: systems with multiple exponential kernels*

For this case, the kernel function $G(s)$ is given by equation [6.39]. The characteristic equation with this type of kernel function can be expressed by [6.45]. This is a polynomial in s of order $(n + 2)$ and therefore has $(n + 2)$ roots. In this section, we derive approximate solutions with the view of generalizing them to MDOF systems.

There are n number of pure real roots corresponding to n terms in the series in equation [6.39]. Multiplying the characteristic equation [6.45] by the product $\prod_j^n (s + b_j)$, we have

$$(s^2 m + s c_0 + k) \prod_{j=1}^{n} (s + b_j) + s \sum_{l=1}^{n} \Big(a_l c_l \prod_{\substack{j=1 \\ j \neq l}}^{n} (b_j + s) \Big) = 0. \qquad [6.74]$$

Like the previous case, we use the approximation

$$s_k = -b_k + \Delta_k, \quad k = 1, 2, \ldots n. \qquad [6.75]$$

Substituting this into the characteristic equation results

$$\left((-b_k + \Delta_k)^2 m + (\Delta_k - b_k) c_0 + k\right) \Delta_k \prod_{\substack{j=1 \\ j \neq k}}^{n} (\Delta_k + b_j - b_k)$$

$$+(\Delta_k - b_k) \left(a_k c_k \prod_{\substack{j=1 \\ j \neq k}}^{n} (\Delta_k + b_j - b_k) \sum_{l=1}^{n} a_l c_l \Delta_k \prod_{\substack{j=1 \\ j \neq l \\ j \neq k}}^{n} (\Delta_k + b_j - b_k) \right) = 0. \quad [6.76]$$

Expanding further and retaining only the first-order terms in Δ_k, after some simplifications, we have

$$\Delta_k \approx \frac{b_k a_k c_k p_1}{\left[b_k^2 m - b_k c_0 + k\right] p_1 + \left[-b_k(p_2 + p_3) + a_k c_0 p_1\right]} \tag{6.77}$$

with

$$\begin{aligned} p_1 &= \prod_{\substack{j=1\\ j\neq k}}^{n} (b_j - b_k), \quad p_2 = a_k c_k \sum_{\substack{j=1\\ j\neq k}}^{n} \prod_{\substack{r=1\\ r\neq j\\ r\neq k}}^{n} (b_r - b_k) \\ \text{and } p_3 &= \sum_{\substack{j=1\\ j\neq k}}^{n} a_j c_j \prod_{\substack{r=1\\ r\neq j,k}}^{n} (b_r - b_k). \end{aligned} \tag{6.78}$$

It can be verified that the general expression [6.77] reduces to equation [6.72] derived in section 6.2.3.2 when $n = 2$. Equations [6.75], [6.77] and [6.78] completely define the n real eigenvalues of an SDOF non-viscously damped system with a general kernel function. In section 6.2.4, the accuracy of these approximate expressions are verified by numerical examples.

6.2.4. *Numerical examples*

To understand the accuracy provided by the approximate expressions derived in the previous sections, we consider three representative numerical examples covering the three cases. The approximate results obtained from these expressions are compared with the exact solutions obtained from the state-space method. For the calculation of percentage error, the state-space results are considered as benchmarks.

6.2.4.1. *An SDOF system with a single kernel*

An SDOF system with one kernel is studied to investigate the accuracy of the approximate eigenvalues derived in sections 6.2.2 and 6.2.3.1. The accuracy of the proposed approximation is presented in Table 6.2.

For the numerical calculations, we consider $m = 1$ kg, $k = 2$ N/m and $c_0 = 0$. It is considered that $a_1 c_1 / b_1 = 2\zeta_n \omega_n$, $b_1 = 1.5$ and the damping factor constant $\zeta_n = 0.1$. The complex conjugate eigenvalues turn out to be almost exact while there is a small error in the real eigenvalue.

b_1	Exact solution (state space)	Proposed approximate solution	Percentage error
1.5	-1.4500	-1.4517	0.12
Complex conjugate solution	-0.0250 ± 1.4382i	-0.0250 ± 1.4382i	-0.00 $\pm$ 0.00i

Table 6.2. *The exact and approximate eigenvalues of an SDOF system with a single kernel*

6.2.4.2. *An SDOF system with two kernels*

We consider an SDOF system with two exponential kernels to investigate the accuracy of the approximate eigenvalues derived in sections 6.2.2 and 6.2.3.2. For the numerical calculations, we consider $m = 1$ kg, $k = 2$ N/m and $c_0 = 0$. It is assumed that all $a_k c_k / b_k$ are of the same value so that $a_k c_k / b_k = 2\zeta_n \omega_n, \forall k = 1, 2,$ $b_1 = 1$, $b_2 = 5$ and the damping factor constant $\zeta_n = 0.1$. The accuracy of the proposed approximation is presented in Table 6.3. The complex conjugate eigenvalues turn out to be almost exact while there are small errors in the two real eigenvalues. Error corresponding to the larger value of b_k is smaller than error corresponding to the smaller value of b_k.

$b_k, k = 1, 2$	Exact solution (state space)	Proposed approximate solution	Percentage error
1.0	-0.9531	-0.9539	0.0868
5.0	-4.9739	-4.9743	0.0084
Complex conjugate solution	-0.0365 ± 1.4519i	-0.0365 ± 1.4519i	0.00 $\pm$0.00i

Table 6.3. *The exact and approximate eigenvalues of an SDOF system with two kernels*

6.2.4.3. *The general case: an SDOF system with multiple kernels*

We consider an SDOF system with eight kernels to investigate the accuracy of the approximate eigenvalues derived in sections 6.2.2 and 6.2.3.3. For the numerical calculations, we consider $m = 1$ kg, $k = 2$ N/m and $c_0 = 0$. It is assumed that all $a_k c_k / b_k$ are of the same value so that $a_k c_k / b_k = 2\zeta_n \omega_n, \forall k = 1, 2, \ldots, 8$. The damping factor constant $\zeta_n = 0.1$ and the values of b_k for $k = 1, 2, \ldots, 8$ are selected as 1.4973, 1.5231, 1.7454, 1.7657, 1.9317, 1.9442, 1.9558 and 2.0677. The approximate eigenvalues obtained using the proposed method are compared with the results obtained from exact state-space solution in Table 6.4. The complex conjugate eigenvalues are obtained very accurately using the proposed approximation. The real eigenvalues are not as accurate as the complex conjugate eigenvalues. However, recall that the motion corresponding to the real eigenvalues is purely dissipative in nature and therefore does not significantly affect the dynamic response of the system.

$b_k, k = 1, 2, \ldots, 8$	Exact solution (state space)	Proposed approximate solution	Percentage error
1.4973	-1.3025	-1.4003	7.5121
1.5231	-1.5105	-1.4825	1.8499
1.7454	-1.6586	-1.6581	0.0291
1.7657	-1.7560	-1.7286	1.5607
1.9317	-1.8668	-1.7315	7.2454
1.9442	-1.9375	-1.8981	2.0334
1.9558	-1.9513	-1.9313	1.0219
2.0677	-2.0511	-2.0302	1.0184
Complex conjugate solution	$-0.1984 \pm 1.5920i$	$-0.1984 \pm 1.5915i$	$0.0335 \pm 0.0056i$

Table 6.4. *The exact and approximate eigenvalues of an SDOF system with eight kernels*

6.3. Multiple-degrees-of-freedom non-viscously damped systems

In this section, the results obtained for the SDOF system are extended to general MDOF systems. The equation of motion of an N DOF linear non-viscously damped system can be expressed by extending the corresponding SDOF equation [6.37] as

$$\mathbf{M}\ddot{\mathbf{q}}(t) + \int_0^t \mathcal{G}(t-\tau)\,\dot{\mathbf{q}}(\tau)\,\mathrm{d}\tau + \mathbf{K}\mathbf{q}(t) = \mathbf{f}(t). \qquad [6.79]$$

Here, $\mathbf{q}(t) \in \mathbb{R}^N$ is the displacement vector, $\mathbf{f}(t) \in \mathbb{R}^N$ is the forcing vector, $\mathbf{M} \in \mathbb{R}^{N\times N}$ is the mass matrix, $\mathbf{K} \in \mathbb{R}^{N\times N}$ is the elastic stiffness matrix and $\mathcal{G}(t)$ is the matrix of non-viscously damped kernel functions. In the Laplace domain, the equation of motion can be expressed as

$$\mathbf{D}(s)\bar{\mathbf{q}}(s) = \mathbf{0} \qquad [6.80]$$

where the dynamic stiffness matrix $\mathbf{D}(s)$ is given by

$$\mathbf{D}(s) = s^2\mathbf{M} + s\mathbf{G}(s) + \mathbf{K} \in \mathbb{C}^{N\times N} \qquad [6.81]$$

with

$$\mathbf{G}(s) = \mathbf{C}_0 + \sum_{k=1}^{n} \frac{a_k}{s + b_k} \mathbf{C}_k \in \mathbb{C}^{N \times N}. \qquad [6.82]$$

Equation [6.82] is the extension of the corresponding scalar equation [6.39] to the matrix case. Again note that in the special case when $a_k = b_k = \mu_k$m, we regain the model discussed in Chapter 4. The eigenvalues of the system s_j can be obtained by solving the characteristic polynomial

$$\det\left[\mathbf{D}(s_j)\right] = 0, \quad j = 1, 2, \ldots m. \qquad [6.83]$$

We assume that the order of the characteristic polynomial is m so that there are m eigenvalues of the system. In this chapter, we consider that all the eigenvalues are distinct. For undamped elastic systems or elastic systems with viscous damping, the order of the characteristic polynomial is $m = 2N$. For non-viscously damped systems, in general, m is more than $2N$, that is $m = 2N + p$; $p \geq 0$. If all of the $\mathbf{C}_k$ matrices are of full rank, then $p = nN$. This shows that although the system has N degrees-of-freedom, the number of eigenvalues is more than $2N$. This is a major difference between a non-viscously damped system and an elastic system where the number of eigenvalues is exactly $2N$, including any multiplicities. Like the SDOF case, N complex conjugate pair of eigenvalues corresponds to *elastic modes* or *vibration modes*. These modes are related to the N modes of vibration of the structural system. Physically, the assumption of "$2N$ complex conjugate pairs of eigenvalues" implies that all the elastic modes are oscillatory in nature, that is they are subcritically damped. The additional nN eigenvalues correspond to *nonviscous modes* or *overdamped modes*. Here, the complex eigenvalues are derived for a matrix with general kernel functions. For the real eigenvalues, the Biot model with two cases, namely when $n = 1$ and when $n > 1$, are considered for analytical convenience.

6.3.1. *Complex conjugate eigenvalues*

The aim of this section is to obtain the complex conjugate eigenvalues using the undamped elastic eigenvalues. The undamped elastic eigenvalue problem of an MDOF system is given by

$$\mathbf{K}\mathbf{x}_j = \omega_j^2 \mathbf{M}\mathbf{x}_j, \quad j = 1, 2, \ldots, N \qquad [6.84]$$

where ω_j^2 and $\mathbf{x}_j$ are the eigenvalues and mass-normalized eigenvectors of the system. We define the matrices

$$\boldsymbol{\Omega} = \text{diag}\left[\omega_1, \omega_2, \ldots, \omega_N\right] \tag{6.85}$$

$$\text{and} \quad \mathbf{X} = \left[\mathbf{x}_1, \mathbf{x}_2, \ldots, \mathbf{x}_N\right]. \tag{6.86}$$

There are various efficient methods available to obtain the undamped eigensolution (see, for example, [WIL 88]) and many of them have been integrated within the commercially available general purpose finite element software. Using these, the characteristic equation [6.83] can be transformed into the modal coordinates as

$$\det\left[s_j^2 \mathbf{I}_N + s_j \mathbf{G}'(s_j) + \boldsymbol{\Omega}^2\right] = 0, \quad j = 1, 2, \ldots, m \tag{6.87}$$

where $\mathbf{I}_N$ is an N dimensional identity matrix. The non-viscously damped kernel function matrix in the modal coordinates $\mathbf{G}'(s)$ is defined as

$$\begin{aligned} \mathbf{G}'(s) &= \mathbf{X}^T \mathbf{G}(s) \mathbf{X} = (\mathbf{X}^T \mathbf{C}_0 \mathbf{X}) + \textstyle\sum_{k=1}^{n} \frac{a_k}{s+b_k} (\mathbf{X}^T \mathbf{C}_k \mathbf{X}) \\ &= \mathbf{C}_0' + \textstyle\sum_{k=1}^{n} \frac{a_k}{s+b_k} \mathbf{C}_k' \end{aligned} \tag{6.88}$$

where the matrices

$$\mathbf{C}_0' = \mathbf{X}^T \mathbf{C}_0 \mathbf{X} \quad \text{and} \quad \mathbf{C}_k' = \mathbf{X}^T \mathbf{C}_k \mathbf{X}, \quad k = 1, 2, \ldots, n. \tag{6.89}$$

We consider that the system has small non-proportionality so that the off-diagonal entries of the $\mathbf{G}'$ matrix are small compared to the diagonal entries, that is $G'_{kl}(s_j) \leq G'_{kk}(s_j), \forall s_j, k \neq l$. This approximation is often used in the dynamic analysis of damped systems [CAU 65, SHA 88, SHA 90, ADH 01a]. Considering the jth set of equation [6.87], we obtain

$$s_j^2 + s_j G'_{jj}(s_j) + \omega_j^2 \approx 0. \tag{6.90}$$

This equation is similar to equation [6.45] and can be solved in a similar way. The modal damping factor ζ_j can be defined as

$$\zeta_j = \frac{\lim_{s \to 0} G'_{jj}(s)}{2\omega_j} = \left(C'_{0_{jj}} + \sum_{k=1}^{n} \frac{a_k}{b_k} C'_{k_{jj}} \right) / 2\omega_j \tag{6.91}$$

where the matrices $\mathbf{C}_0'$ and $\mathbf{C}_k'$ are defined in equation [6.89]. The complete solution can be written as

$$s_j = s_{0_j} + \delta_j \tag{6.92}$$

where

$$s_{0_j} = -\zeta_j \omega_j \pm \mathrm{i}\omega_j \sqrt{1-\zeta_j^2}, \qquad [6.93]$$

$$\delta_j \approx \frac{-B_j - \sqrt{B_j^2 - 4A_jC_j}}{2A_j}, \qquad [6.94]$$

$$A_j = 1 + \frac{1}{2}\frac{\partial^2 G'_{jj}\left(s_{0_j}\right)}{\partial s^2} s_{0_j} + \frac{\partial G'_{jj}\left(s_{0_j}\right)}{\partial s}, \qquad [6.95]$$

$$B_j = 2s_{0_j} + s_{0_j}\frac{\partial G'_{jj}\left(s_{0_j}\right)}{\partial s} + G'_{jj}\left(s_{0_j}\right) \qquad [6.96]$$

$$\text{and} \quad C_j = s_{0_j}^2 + s_{0_j} G'_{jj}\left(s_{0_j}\right) + \omega_j^2. \qquad [6.97]$$

In the above expressions, $\frac{\partial G'_{jj}(s_{0_j})}{\partial s}$ and $\frac{\partial^2 G'_{jj}(s_{0_j})}{\partial s^2}$ are, respectively, the first- and second-order derivatives of $G'_{jj}(s)$ evaluated at $s = s_{0_j}$ and can be obtained following equation [6.59]. Equations [6.91]–[6.97] completely define the complex conjugate eigenvalues of an MDOF non-viscously damped system.

6.3.2. *Real eigenvalues*

The real eigenvalues are obtained using an approach similar to the SDOF system. After neglecting the off-diagonal terms of the $\mathbf{G}'$ matrix, the governing characteristic equation for every mode can be expressed by equation [6.90]. This equation can be solved for the real eigenvalues. For systems with single exponential kernel, there are in total $3N$ number of eigenvalues of which N are real. Following equations [6.63]–[6.67], the real eigenvalues can be expressed as

$$s_j = -b_1 + \Delta_j, \quad \forall j = 1, \ldots, N \qquad [6.98]$$

where

$$\Delta_j \approx \frac{a_1 b_1 C'_{1_{jj}}}{b_1^2 - b_1 C'_{0_{jj}} + \omega_j^2 + a_1 C'_{1_{jj}}}. \qquad [6.99]$$

This equation and equation [6.98] completely define all the N real eigenvalues.

For systems with n kernels, there are, in general, nN number of purely real eigenvalues. The approximate eigenvalues can be obtained by extending equations [6.75] and [6.77] as

$$s_{jk} = -b_k + \Delta_{jk}, \quad \forall j = 1, \ldots, N \quad k = 1, \ldots, n \qquad [6.100]$$

where

$$\Delta_{jk} \approx \frac{b_k a_k C'_{k_{jj}} p_1}{\left[b_k^2 - b_k C'_{0_{jj}} + \omega_j^2\right] p_1 + \left[-b_k(p_2 + p_3) + a_k C'_{0_{jj}} p_1\right]} \quad [6.101]$$

with

$$\begin{aligned} p_1 &= \prod_{\substack{l=1 \\ l \neq k}}^{n} (b_l - b_k), \quad p_2 = a_k C'_{k_{jj}} \textstyle\sum_{\substack{l=1 \\ l \neq k}}^{n} \prod_{\substack{r=1 \\ r \neq l \\ r \neq k}}^{n} (b_r - b_k) \\ \text{and } p_3 &= \textstyle\sum_{\substack{m=1 \\ m \neq k}}^{n} a_m C'_{m_{jj}} \prod_{\substack{r=1 \\ r \neq m,k}}^{n} (b_r - b_k). \end{aligned} \quad [6.102]$$

From equation [6.100], note that there are N real solutions around each b_k. Equations [6.100]–[6.102] completely define *all* the nN number of real eigenvalues.

6.3.3. *Numerical example*

We consider a three DOF system to illustrate the proposed method and verify the accuracy of the approximate expressions. The mass, stiffness and non-viscously damped matrices in the Laplace domain for the problem are considered as:

$$\mathbf{M} = \begin{bmatrix} m & 0 & 0 \\ 0 & m & 0 \\ 0 & 0 & m \end{bmatrix}, \quad \mathbf{K} = \begin{bmatrix} 2k_u & -k_u & 0 \\ -k_u & 2k_u & -k_u \\ 0 & -k_u & 2k_u \end{bmatrix} \quad [6.103]$$

and

$$\mathbf{G}(s) = \mathbf{C} \sum_{k=1}^{6} \frac{a_k}{s + b_k}, \quad \text{where} \quad \mathbf{C} = \begin{bmatrix} 0.30 & -0.15 & -0.05 \\ -0.15 & 0.30 & -0.15 \\ -0.05 & -0.15 & 0.30 \end{bmatrix}. \quad [6.104]$$

For the numerical value, we consider $m = 1$ kg, $k_u = 1$N/m and $a_k = b_k$. It is assumed that $\mathbf{C}_0 = \mathbf{0}$, all the $\mathbf{C}_k$ matrices are the same and the values of b_k for $k = 1, 2, .., 6$ are selected as 1.4565, 1.0185, 1.8214, 1.4447, 1.6154 and 1.7919. The approximate eigenvalues obtained using the proposed method are compared with the results obtained from exact state-space solution in Table 6.5. Like the SDOF examples considered before, the complex conjugate eigenvalues are obtained very accurately using the proposed approximation. The real eigenvalues are not as accurate as the complex conjugate eigenvalues. However, recall that the motion corresponding to the real eigenvalues is purely dissipative in nature and therefore may not significantly contribute to the dynamic response of the system around the resonant frequencies.

Next, we propose a different reduced approach by expressing the equation of motion into an equivalent second-order form.

Exact solution (state-space)	Proposed approximate solution	Percentage error
Real eigenvalues		
−0.9380	−0.9425	0.4797
−1.2995	−1.3044	0.3771
−1.4507	−1.7413	20.0317
−1.5754	−1.8096	14.8661
−1.7405	−1.5761	9.4456
−1.8095	−1.4507	19.8287
−0.6301	−0.6301	0
−1.1276	−1.1276	0
−1.4505	−1.7096	17.8628
−1.5507	−1.8081	16.5990
−1.7096	−1.5507	9.2946
−1.8081	−1.4505	19.7777
−0.6798	−0.6731	0.9856
−1.1295	−1.1289	0.0531
−1.4505	−1.7085	17.7870
−1.5501	−1.8080	16.6376
−1.7086	−1.5501	9.2766
−1.8080	−1.4505	19.7732
Complex conjugate eigenvalues		
$-0.4109 \pm 2.6579i$	$-0.4116 \pm 2.6591i$	$0.1704 \pm 0.0451i$
$-0.4359 \pm 2.0939i$	$-0.4359 \pm 2.0939i$	$0 \pm 0i$
$-0.1674 \pm 0.8523i$	$-0.1649 \pm 0.8528i$	$1.4934 \pm 0.0587i$

Table 6.5. *Exact and approximate eigenvalues of the three DOF system*

6.4. Reduced second-order approach for non-viscously damped systems

Current methods for dynamic analysis of non-viscously damped systems are dominated by state-space-based approaches, as discussed in Chapter 4. Bagley and Torvik [BAG 83] have used an extended state-space approach for linear systems with fractional derivative damping models. They have expressed the extended state-vector in terms of various fractional order differentials of the displacement vector. Golla and Hughes [GOL 85] and McTavish and Hughes [MCT 93] have used an internal variable based approach (GHM method) in the context of non-viscously damped structures. Another approach, known as the ADF method, has been developed by Lesieutre and Mingori [LES 90] and Lesieutre and Bianchini [LES 95]. Like GHM, the ADF method is also an internal variable based non-viscously damped model, but distinguished from GHM in that it is first-order in time, not second-order. Muravyov and Hutton [MUR 97b] and Muravyov [MUR 98a] have proposed an extended state-space method for systems with exponential non-viscously damped kernels associated with the stiffness operator. Muscolino *et al.* [MUS 05] and Palmeri *et al.* [PAL 03, PAL 04] have used a state-space-based approach for the time-domain analysis of non-viscously damped systems subjected to random excitations. In Chapter 4, we showed the symmetric state-space approach for linear systems with the exponential non-viscous damping model.

Methods that are not based on state-space approach are less common. The main reasons for seeking an alternative to the state-space approach are (1) although exact in nature, the state-space approach for linear non-viscously damped systems is relatively computationally intensive for real-life MDOF systems due to the huge number of internal variables and (2) the well-understood mathematical insights offered by methods in the original space (e.g. the modal analysis) are lost in a state-space-based approach. For example, a 30 DOF oscillatory system with a four-term Biot model (see Table 1.1) will, in general, result into a state-space matrix of dimension $30 \times (2 + 4) = 180$. Compared to this, a non-proportionally viscously damped system will result into a state-space matrix of dimension $30 \times 2 = 60$. This is a key motivation to seek for reduced non-state-space-based approaches.

It should be emphasized that some kind of approximation needs to be used for non-state-space approaches. Woodhouse [WOO 98] proposed approximate methods based on small damping assumption. Some modal approximation techniques are discussed in Chapter 5. Based on a variational principle, Qian and Hansen [QIA 95] derived a substructure synthesis method where the non-viscously damped system eigensolution is obtained from the undamped system eigensolution. Several authors [MUL 05, ADH 05] have discussed the nature of eigensolutions of non-viscously damped systems to develop an understanding for approximate analyses. Daya and Potier-Ferry [DAY 01] proposed an asymptotic numerical method for the calculation of natural frequencies and loss-factors of non-viscously damped systems. Using the small non-viscously damping assumption, Segalman [SEG 87] presented an approach to obtain the damping and stiffness matrices for non-viscously damped systems. Bilbao *et al.* [BIL 06] considered proportional damping approximation for structures with non-viscously damped dampers. Friswell *et al.* [FRI 97, FRI 99] proposed reduced-order models for non-viscously damped systems with the GHM model. Approximation methods based on the Taylor series expansion of the kernel function $\mathbf{G}(s)$ in the complex plane was discussed earlier this chapter.

We aim to obtain an equivalent second-order system for a general non-viscously damped system. This is effectively a strategy for the reduction of model order. The main motivations behind seeking this approximation are as follows:

– The physical understandings are very well developed for second-order systems compared to non-viscously damped systems that are in general higher order systems.

– Wide ranging resources, such as computational tools, books and software, are available for second-order systems.

– The computational cost for the analysis of a second-order system is significantly less compared to a non-viscously damped system as there is no need to use additional dissipation coordinates.

The approximation proposed here utilizes the generalized proportional damping proposed in section 2.4.2 and approximate eigenvalues of non-viscously damped systems shown in section 6.2 and section 6.3. First, proportionally damped systems are considered in section 6.4.1 and later the results are extended to the general case in section 6.4.2. The newly derived results are applied to a linear array of non-viscously damped spring-mass system with 25 DOF in section 6.4.3 for numerical illustration.

6.4.1. *Proportionally damped systems*

6.4.1.1. *Approximation of the eigenvalues*

In section 6.2.2, we discussed an approximation method for the eigenvalues of SDOF systems. Here, we consider MDOF systems, however, for a simple case when the kernel function matrix takes the form

$$\mathbf{G}(s) = G(s)\mathbf{C} \qquad [6.105]$$

and the coefficient matrix $\mathbf{C}$ can be simultaneously diagonalized with the mass and stiffness matrices using the undamped eigenvector corresponding to the underlying elastic system. As shown in Chapter 2, the condition for the simultaneous diagonalization can be expressed [CAU 65, ADH 01a] as $\mathbf{M}\mathbf{C}^{-1}\mathbf{K} = \mathbf{K}\mathbf{C}^{-1}\mathbf{M}$ (see section 2.4). The undamped eigenvalue problem is given by

$$\mathbf{K}\mathbf{x}_j = \omega_j^2\mathbf{M}\mathbf{x}_j, \quad j = 1, 2, \ldots, N \qquad [6.106]$$

where ω_j^2 and $\mathbf{x}_j$ are, respectively, the eigenvalues and mass-normalized eigenvectors of the system. We define the matrices

$$\boldsymbol{\Omega} = \text{diag}\left[\omega_1, \omega_2, \ldots, \omega_N\right] \quad \text{and} \quad \mathbf{X} = \left[\mathbf{x}_1, \mathbf{x}_2, \ldots, \mathbf{x}_N\right] \qquad [6.107]$$

so that

$$\mathbf{X}^T\mathbf{K}\mathbf{X} = \boldsymbol{\Omega}^2 \quad \text{and} \quad \mathbf{X}^T\mathbf{M}\mathbf{X} = \mathbf{I}_N \qquad [6.108]$$

where $\mathbf{I}_N$ is an N dimensional identity matrix. Using these, equation [6.80] can be transformed into the modal coordinates as

$$\left[s^2\mathbf{I}_N + s\mathbf{G}'(s) + \mathbf{\Omega}^2\right]\bar{\mathbf{u}}'(s) = \bar{\mathbf{f}}'(s) \qquad [6.109]$$

where $(\bullet)'$ denotes the quantities in the modal coordinates:

$$\mathbf{G}'(s) = \mathbf{X}^T\mathbf{G}(s)\mathbf{X} = G(s)\mathbf{C}', \quad \bar{\mathbf{u}}(s) = \mathbf{X}\bar{\mathbf{u}}'(s) \quad \text{and} \quad \bar{\mathbf{f}}'(s) = \mathbf{X}^T\bar{\mathbf{f}}(s). \qquad [6.110]$$

Here, $\mathbf{C}' = \mathbf{X}^T\mathbf{C}\mathbf{X}$ is a diagonal matrix. Denoting the eigenvalues of the system as s_j, the jth characteristic equation corresponding to [6.109] can be obtained as

$$s_j^2 + s_j G'_{jj}(s_j) + \omega_j^2 = 0 \qquad [6.111]$$

$$\text{or} \quad d_j(s_j) = 0, \quad \text{where } d_j(s) = s^2 + sG'_{jj}(s) + \omega_j^2. \qquad [6.112]$$

Using the small non-viscous damping assumption [SEG 87], the solution of this equation can be approximated using the methods outlined earlier in this chapter. A detailed discussion on the quantification of non-viscous damping is given later ([ADH 14b], Chapter 4). In simple mathematical terms, small non-viscous damping implies less variation of the function $\mathbf{G}(s)$ with respect to s. Here, based on earlier developments in this chapter, we outline an approach suitable for the derivation of equivalent second-order system. We define the modal damping factor ζ_j as

$$\zeta_j = \frac{\lim_{s\to 0} G'_{jj}(s)}{2\omega_j}. = C'_{jj}/2\omega_j \qquad [6.113]$$

If [6.109] was a truly second-order viscously damped system, its eigenvalues [MEI 67] would have been given by

$$s_{0_j} = -\zeta_j\omega_j \pm \mathrm{i}\omega_j\sqrt{1-\zeta_j^2}, \approx -\zeta_j\omega_j \pm \mathrm{i}\omega_j. \qquad [6.114]$$

Since, in general, this is not the case, the difference between the elastic solution and the true solution of the characteristic equation [6.111] is essentially arising due to the "varying" nature of the function $G(s)$. The approximate solutions obtained here are based on keeping this fact in mind.

The central idea is that the actual solution of the characteristic equation [6.111] can be obtained by expanding the solution in a Taylor series around s_{0_j}. The error arising in the resulting solution would then depend on the "degree of variability" of the function $G(s)$. We assume that the true solution of equation [6.111] can be expressed as

$$s_j = s_{0_j} + \delta_j \tag{6.115}$$

where δ_j is a small quantity. Substituting this into the characteristic equation [6.112], we have

$$d_j\left(s_{0_j} + \delta_j\right) = 0. \tag{6.116}$$

Expanding $d_j\left(s_{0_j} + \delta_j\right)$ in a Taylor series in δ_j around s_{0_j}, we have

$$d_j\left(s_{0_j}\right) + \delta_j \frac{\partial d_j\left(s_{0_j}\right)}{\partial s} + \frac{1}{2}\delta_j^2 \frac{\partial^2 d_j\left(s_{0_j}\right)}{\partial s^2} + \ldots = 0. \tag{6.117}$$

Keeping only the first-order terms in δ_j, we have

$$\delta_j \approx -\left[\frac{\partial d_j\left(s_{0_j}\right)}{\partial s}\right]^{-1} d_j\left(s_{0_j}\right) \tag{6.118}$$

where

$$\frac{\partial d_j\left(s_{0_j}\right)}{\partial s} = s_{0_j}\left(2 + \frac{\partial G_{jj}\left(s_{0_j}\right)}{\partial s}\right) + G_{jj}\left(s_{0_j}\right). \tag{6.119}$$

The δ_j in equation [6.118] is a first-order approximation. Higher order approximations can be obtained by retaining higher order terms in δ_j in equation [6.117]. Thus, the complete approximate solution can be expressed as

$$s_j \approx \widetilde{s}_j = -\zeta_j\omega_j + \mathrm{i}\omega_j - \delta_j, -\zeta_j\omega_j - \mathrm{i}\omega_j - \delta_j^*. \tag{6.120}$$

The expression of the approximate eigenvalue derived here shows that they can be obtained by post-processing of the undamped eigenvalue ω_n and equivalent viscous damping factor ζ_n. This approximation of the complex conjugate eigenvalues is valid for any kernel function. It could therefore be used with any non-viscously damped models presented in Table 1.1.

Once the approximate eigenvalues of the system are known, the key idea proposed here is that $\mathbf{G}'(s)$ in equation [6.109] can be approximated by $\mathbf{G}'\left(\widetilde{s}_j\right)$, that is

$$\mathbf{G}'_{eqv} = \mathbf{G}'\left(\widetilde{s}_j\right), \forall\ j. \tag{6.121}$$

This approximation is based on the physical understanding that damping has a significant effect on the dynamic response only around the natural frequencies. Using this approximation, we have

$$\left[s^2\mathbf{I}_N + s\mathbf{G}'(s) + \mathbf{\Omega}^2\right] \approx \left[s^2\mathbf{I}_N + s\mathbf{G}'_{eqv} + \mathbf{\Omega}^2\right]. \qquad [6.122]$$

Because the matrix $\mathbf{G}'\left(\widetilde{s}_j\right)$ is a diagonal matrix for all j, the dynamic response of the system can be obtained using the usual modal analysis approach. This approximation can be expressed in a more elegant way using the generalized proportional damping approach.

6.4.1.2. *Generalized proportional damping*

The concept of generalized proportional damping was introduced in section 2.4. The generalized proportional damping expresses the damping matrix in terms of a smooth continuous function involving specially arranged mass and stiffness matrices so that the system still possesses classical normal modes. The main result can be expressed as below.

THEOREM 6.1.– Viscously damped linear systems will have classical normal modes if the damping matrix can be represented by $\mathbf{C} = \mathbf{M}f(\mathbf{M}^{-1}\mathbf{K})$ where $f(\bullet)$ is a smooth analytic function in the neighborhood of all the undamped eigenvalues.

This enables us to model the variations in the modal damping factors with respect to the frequency in a simplified manner. We use this observation to obtain an equivalent second-order system. In theorem 2.2, a more general result was proved. Using $\mathbf{C} = \mathbf{M}f(\mathbf{M}^{-1}\mathbf{K})$, it was shown earlier that

$$\mathbf{X}^T\mathbf{C}\mathbf{X} = f\left(\mathbf{\Omega}^2\right) \quad \text{or} \quad 2\zeta_j\omega_j = f\left(\omega_j^2\right), \ \forall j. \qquad [6.123]$$

The reconstruction of a damping matrix using the generalized proportional damping requires the functional variation to be expressed as a function of ω_j^2. Therefore, it is convenient to write the approximate eigenvalues in equation [6.120] as

$$\widetilde{s}_j = h\left(\omega_j^2\right) \qquad [6.124]$$

where $h(\bullet)$ is a smooth function. Applying equation [6.123] to the $\mathbf{G}'\left(\widetilde{s}_j\right)$ and using the expression of $\mathbf{G}(s)$ in equation [6.105], we can write

$$G'_{jj}\left(\widetilde{s}_j\right) = C'_{jj}G\left(\widetilde{s}_j\right) = C'_{jj}G\left(h\left(\omega_j^2\right)\right), \ \forall j. \qquad [6.125]$$

From this equation, the generalized proportional damping can be identified as

$$\mathbf{G}_{eqv} = \mathbf{C}G\left(h(\mathbf{M}^{-1}\mathbf{K})\right). \qquad [6.126]$$

Observe that $G\left(h(\mathbf{M}^{-1}\mathbf{K})\right)$ is now a function of a matrix argument. The scalar functions presented in Table 1.1 need to be adjusted to be used with matrix arguments. For example, for the Biot model, $G(s) = \sum_{k=1}^{n} a_k/(s+b_k)$ needs to be expressed as $G(\mathbf{S}) = \sum_{k=1}^{n} a_k[\mathbf{S} + b_k\mathbf{I}_N]^{-1}$ where $\mathbf{S}$ is an $N \times N$ complex matrix.

The equivalence between equations [6.125] and [6.126] is a key result for the developments proposed here. To prove this, using equation [6.108], we have

$$\mathbf{M} = \mathbf{X}^{-T}\mathbf{X}^{-1}, \quad \mathbf{K} = \mathbf{X}^{-T}\boldsymbol{\Omega}^2\mathbf{X}^{-1} \quad \text{and} \quad \mathbf{M}^{-1}\mathbf{K} = \mathbf{X}\boldsymbol{\Omega}^2\mathbf{X}^{-1}. \qquad [6.127]$$

Because the function $f(\bullet)$ is assumed to be analytic in the neighborhood of all the eigenvalues of $\mathbf{M}^{-1}\mathbf{K}$, it can be expressed in polynomial forms using the Taylor series expansion. Following Bellman [BEL 60] (Chapter 6), we can obtain

$$f(\mathbf{M}^{-1}\mathbf{K}) = \mathbf{X}\, f\left(\boldsymbol{\Omega}^2\right)\mathbf{X}^{-1}. \qquad [6.128]$$

Premultiplying and postmultiplying equation [6.126] by $\mathbf{X}^T$ and $\mathbf{X}$, respectively, we obtain

$$\mathbf{G}'_{eqv} = \mathbf{X}^T\mathbf{G}_{eqv}\mathbf{X} = \mathbf{X}^T\mathbf{C}G\left(h(\mathbf{M}^{-1}\mathbf{K})\right)\mathbf{X}. \qquad [6.129]$$

Using the relationship in equation [6.128] and considering $f(\bullet) = G(h(\bullet))$, this equation can be expressed as

$$\begin{aligned}\mathbf{G}'_{eqv} &= \mathbf{X}^T\mathbf{C}\left[\mathbf{X}G\left(h\left(\boldsymbol{\Omega}^2\right)\right)\mathbf{X}^{-1}\right]\mathbf{X} = \left[\mathbf{X}^T\mathbf{C}\mathbf{X}\right]G\left(h\left(\boldsymbol{\Omega}^2\right)\right)\\ &= \mathbf{C}'G\left(h\left(\boldsymbol{\Omega}^2\right)\right).\end{aligned} \qquad [6.130]$$

The diagonal of the previous equation is equation [6.125], which completes the proof. One fact immediately obvious from this proof is that, on the contrary to what has been assumed before, the matrix $\mathbf{C}$ does not have to be simultaneously diagonalizable with $\mathbf{M}$ and $\mathbf{K}$ for the equivalence between equations [6.125] and [6.126]. The only consequence will be the fact that the eigenvalue equation [6.111] will become approximate for a non-proportional $\mathbf{C}$ matrix. On the basis of this insight, we generalize this approach to a non-proportionally damped system with a general kernel function in section 6.4.2.

6.4.2. *The general case*

In this section, we consider a general form of the $\mathbf{G}(s)$ matrix. For example, generalizing the Biot model to the matrix case, we have

$$\mathbf{G}(s) = \mathbf{C}_0 + \sum_{k=1}^{n} \frac{a_k}{s + b_k} \mathbf{C}_k \qquad [6.131]$$

where $\mathbf{C}_j, j = 0, 1, 2, \ldots$ are $N \times N$ real symmetric matrices. For a proportionally damped system, the validity of the equivalent second-order system depends on the accuracy of two approximations, namely (1) approximate eigenvalues of the system $\widetilde{s}_j$ given by equation [6.120] and (2) $\mathbf{G}(s)$ can be "replaced" by $\mathbf{G}\left(\widetilde{s}_j\right)$. This approximation may likely be valid for lightly non-viscously damped systems. For systems with general non-proportional $\mathbf{G}(s)$, it is further assumed that the system is lightly non-proportionally damped for the purpose of the calculation of the approximate eigenvalues. This implies that $\mathbf{G}'(s)$ is a diagonally dominant matrix. Therefore, when the equation for approximate eigenvalues [6.120] is applied to non-proportionally damped systems, it contains two approximations, namely (1) the system is lightly non-viscously damped and (2) the system is lightly non-proportional. Under these assumptions, we can extend the approximations, derived in the previous section of the general case and have

$$\mathbf{G}_{eqv} = \mathbf{G}\left(h\big(\mathbf{M}^{-1}\mathbf{K}\big)\right). \qquad [6.132]$$

This is the main result of this section. Using this equivalent damping matrix, the equation of motion of a non-viscously damped system can be approximately expressed in the second-order form. Now, we assume $\widetilde{\mathbf{S}} = h\big(\mathbf{M}^{-1}\mathbf{K}\big)$ for notational convenience and simplify the term $h\big(\mathbf{M}^{-1}\mathbf{K}\big)$ for easy practical implementation.

From equation [6.113], note that

$$\omega_j \zeta_j = \lim_{s \to 0} G'_{jj}(s)/2 = G_{0_{jj}}/2 \qquad [6.133]$$

where

$$\mathbf{G}_0 = \lim_{s \to 0} \mathbf{G}(s). \qquad [6.134]$$

As an example, for the Biot model in [6.131], we have $\mathbf{G}_0 = \mathbf{C}_0 + \sum_{k=1}^{n} \frac{a_k}{b_k} \mathbf{C}_k$. Rewriting equation [6.133] in the matrix form, we have

$$\boldsymbol{\Omega}\, \boldsymbol{\zeta} = \frac{1}{2} \mathbf{M}^{-1} \mathbf{G}_0 \qquad [6.135]$$

where $\boldsymbol{\zeta}$ is the diagonal matrix containing the damping factors ζ_j:

$$\boldsymbol{\zeta} = \text{diag}\left[\zeta_1, \zeta_2, \dots, \zeta_N\right]. \qquad [6.136]$$

Using [6.135] and taking the positive sign for illustration, the eigenvalue matrix can be expressed from equation [6.114] as

$$\mathbf{S}_0 = -\mathbf{M}^{-1}\mathbf{G}_0/2 + \mathrm{i}\sqrt{\mathbf{M}^{-1}\mathbf{K}}. \qquad [6.137]$$

From equation [6.118] and using the idea of generalized proportional damping, the correction term δ_j can be expressed in a matrix form as

$$\boldsymbol{\Delta} = -\left[\mathbf{D}_p\left(\mathbf{S}_0\right)\right]^{-1}\mathbf{D}\left(\mathbf{S}_0\right) \qquad [6.138]$$

where

$$\mathbf{D}\left(\mathbf{S}_0\right) = \mathbf{S}_0^2\mathbf{M} + \mathbf{S}_0\mathbf{G}\left(\mathbf{S}_0\right) + \mathbf{K} \qquad [6.139]$$

and

$$\mathbf{D}_p\left(\mathbf{S}_0\right) = \left.\frac{\partial \mathbf{D}(s)}{\partial s}\right|_{s=\mathbf{S}_0} = \mathbf{S}_0\left(2\mathbf{M} + \left.\frac{\partial \mathbf{G}(s)}{\partial s}\right|_{s=\mathbf{S}_0}\right) + \mathbf{G}(\mathbf{S}_0). \qquad [6.140]$$

As an example, for the Biot model in [6.131], we can obtain

$$\frac{\partial \mathbf{G}(s)}{\partial s} = -\sum_{k=1}^{n} \frac{a_k}{\left(s + b_k\right)^2}\mathbf{C}_k \qquad [6.141]$$

so that

$$\mathbf{G}(\mathbf{S}_0) = \mathbf{C}_0 + \sum_{k=1}^{n} a_k\mathbf{C}_k\left(\mathbf{S}_0 + b_k\mathbf{I}_N\right)^{-1} \qquad [6.142]$$

and $$\left.\frac{\partial \mathbf{G}(s)}{\partial s}\right|_{s=\mathbf{S}_0} = -\sum_{k=1}^{n} a_k\mathbf{C}_k\left(\mathbf{S}_0 + b_k\mathbf{I}_N\right)^{-2}. \qquad [6.143]$$

Using these expressions, from equation [6.115], we have

$$\widetilde{\mathbf{S}} = \mathbf{S}_0 + \boldsymbol{\Delta} \qquad [6.144]$$

and

$$\mathbf{G}_{eqv} = \mathbf{G}\left(\widetilde{\mathbf{S}}\right) \in \mathbb{C}^{N \times N}. \qquad [6.145]$$

The results obtained in this section can be conveniently summarized in the following conjecture:

CONJECTURE.– A non-viscously damped system with a kernel function matrix $\mathbf{G}(s) \in \mathbb{C}^{N \times N}$ in the Laplace domain can be approximated by a second-order system with an equivalent damping matrix obtained by simply replacing the argument s with a complex $N \times N$ matrix $\widetilde{\mathbf{S}}$, that is $\mathbf{G}_{eqv} = \mathbf{G}(\widetilde{\mathbf{S}}) \in \mathbb{C}^{N \times N}$, where

$$\begin{aligned} \widetilde{\mathbf{S}} &= \mathbf{S}_0 + \mathbf{\Delta} \in \mathbb{C}^{N \times N} \\ \mathbf{S}_0 &= -\mathbf{M}^{-1}\mathbf{G}_0/2 + \mathrm{i}\sqrt{\mathbf{M}^{-1}\mathbf{K}} \in \mathbb{C}^{N \times N} \\ \mathbf{G}_0 &= \lim_{s \to 0} \mathbf{G}(s) \in \mathbb{R}^{N \times N} \\ \mathbf{\Delta} &= -\left[\mathbf{D}_p\left(\mathbf{S}_0\right)\right]^{-1}\left[\mathbf{S}_0^2\mathbf{M} + \mathbf{S}_0\mathbf{G}\left(\mathbf{S}_0\right) + \mathbf{K}\right] \in \mathbb{C}^{N \times N} \\ \text{and} \quad \mathbf{D}_p\left(\mathbf{S}_0\right) &= \mathbf{S}_0\left(2\mathbf{M} + \left.\frac{\partial \mathbf{G}(s)}{\partial s}\right|_{s=\mathbf{S}_0}\right) + \mathbf{G}(\mathbf{S}_0) \in \mathbb{C}^{N \times N}. \end{aligned} \qquad [6.146]$$

This conjecture completely defines the proposed reduced-order approximation and it is valid for a general $\mathbf{G}(s)$ as presented in Table 1.1. The approximation scheme will remain the same, the only changes will be due to the different functional forms of $\mathbf{G}(s)$. We only need to solve the undamped eigenvalue problem to obtain the equivalent second-order system to use this approximation. As a result of this conjecture, the equation of motion in the Laplace domain can be expressed by

$$\widetilde{\mathbf{D}}(s)\bar{\mathbf{u}}(s) = \bar{\mathbf{f}}(s) \qquad [6.147]$$

where the equivalent second-order dynamic stiffness matrix $\widetilde{\mathbf{D}}(s)$ is given by

$$\widetilde{\mathbf{D}}(s) = s^2\mathbf{M} + s\mathbf{G}_{eqv} + \mathbf{K} \in \mathbb{C}^{N \times N}. \qquad [6.148]$$

The matrix $\mathbf{G}_{eqv}$ is, in general, non-proportional and complex. All conventional approaches for non-proportionally damped systems can be used for equation [6.147]. This approximation eliminates the need to use additional dissipation coordinates and significantly reduces the computational cost. The suggested approximation is simple and involves only elementary matrix operations involving the system matrices. In section 6.4.3, we verify the accuracy of the proposed approximation.

6.4.3. *Numerical examples*

A partly damped linear array of a spring-mass oscillator is considered to illustrate the application of a proposed approximation. The system, together with the numerical values assumed for different parameters, is shown in Figure 6.2. The mass matrix of the system can be given by $\mathbf{M} = m\mathbf{I}$ where $\mathbf{I}$ is the $N \times N$ identity matrix. The stiffness matrix of the system is given by

$$\mathbf{K}_e = k_u \mathbf{J}, \quad \text{where} \quad \mathbf{J} = \begin{bmatrix} 2 & -1 & & & & \\ -1 & 2 & -1 & & & \\ & \ddots & \ddots & \ddots & & \\ & & -1 & 2 & -1 & \\ & & & \ddots & \ddots & -1 \\ & & & & -1 & 2 \end{bmatrix}. \qquad [6.149]$$

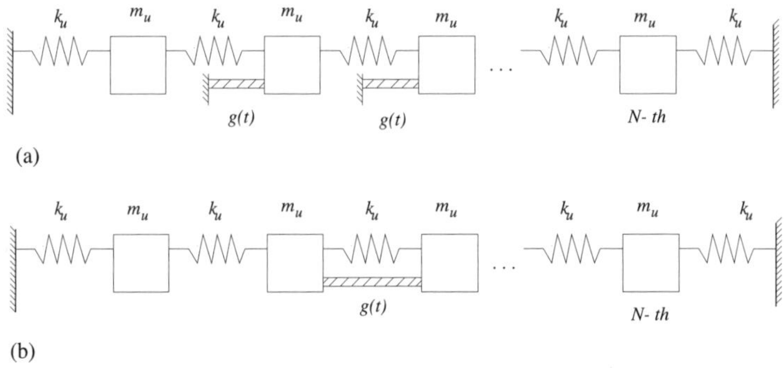

Figure 6.2. *Linear array of N spring-mass oscillators, $N = 25$, $m = 1$ kg, $k_u = 4.0 \times 10^5$ N/m. Non-viscously damped dampers attached between 8th and 17th masses with $c_a = 40.0$ Ns/m and $c_b = 120.0$ Ns/m for the two cases are considered*

The following two cases are considered for the non-viscously damped kernel:

– For the system shown in Figure 6.2(a), certain of the masses have dissipative elements connecting them to the ground. In this case, the non-viscously damped damping force depends only on the absolute motion of the individual masses. For this case, the kernel function matrix has the form of equation [6.105], that is,

$$\mathbf{G}(s) = G(s)\mathbf{C} \qquad [6.150]$$

where the coefficient matrix $\mathbf{C} = c_a\widetilde{\mathbf{I}}$. The matrix $\widetilde{\mathbf{I}}$ has similar form of an $N \times N$ identity matrix except that it has non-zero diagonal entries corresponding to the masses attached with the dampers only.

– For the system shown in Figure 6.2(b), by contrast, dissipative elements are connected between certain adjacent pairs of masses. In this case, the non-viscously damped damping force depends on the relative motion of the two adjacent masses. The kernel function matrix has the form of equation [6.150] with the coefficient matrix $\mathbf{C} = c_b\widetilde{\mathbf{J}}$. The matrix $\widetilde{\mathbf{J}}$ has similar form to the matrix $\mathbf{J}$ in equation [6.149] except that it has non-zero entries corresponding to the masses attached with the non-viscously damped dampers only.

With the $\mathbf{C}$ matrices considered here, it is easy to verify that for both the cases the system is actually non-proportionally damped. For numerical calculations, we have considered a 25 DOF system so that $N = 25$. Values of the mass and stiffness associated with each unit are assumed to be the same with numerical values of $m = 1$ kg and $k = 4.00 \times 10^5$ N/m. The resulting undamped natural frequencies then range from approximately 20–200 Hz. The value $c_a = 40.0$ Ns/m and $c_b = 120.0$ Ns/m has been used for cases (a) and (b), respectively. The function $G(s)$ is assumed to be the Biot model with four terms as

$$G(s) = \sum_{k=1}^{4} \frac{b_k}{s + b_k} \qquad [6.151]$$

with $b_k = \{20.4703, 188.1373, 243.7953, 300.2269\}\ \text{s}^{-1}$. The accuracy of the approximate second-order system proposed here depends of the accuracy of the approximate eigenvalues. As a result, we first look at how the eigenvalues are approximated using the two approximate expressions derived in section 6.4.1.1.

In Figure 6.3, percentage errors in the real sand imaginary parts of the eigenvalues obtained using the two approximate expressions for case (a) have been shown. For the calculation of percentage error, the eigenvalues obtained from the state-space approach involving additional dissipation coordinates, as discussed in Chapter 4, have been used. For this 25 DOF system, the order of the state-space problem turns out be 90. This demonstrates the computational need for the model order reduction of non-viscously damped systems and highlights the justification behind the approximation proposed in this section. The zero-order approximation shown in Figure 6.3 corresponds to the eigenvalues obtained using equation [6.114]. The first-order approximation is obtained using equation [6.120]. As expected, the first-order approximation is more accurate that the zero-order approximation. A driving point and a cross FRF of the system corresponding to case (a) are shown in Figure 6.4. With the exact FRF, three more approximations are shown in this diagram. The FRF with $\mathbf{G}_0$ is obtained from a second-order system with a damping matrix given by equation [6.134]. This is perhaps the simplest possible approximation. However, as can be seen from the graphs, this approximation introduces significant error, which is clearly not suitable for this problem. The other two approximations, namely zeroth- and first-order approximations, produce fairly similar results. The zero-order approximation can be obtained from equation [6.146] by substituting $\mathbf{\Delta} = \mathbf{O}$. In particular, this is a very simple approximation for which $\mathbf{G}_{eqv} = \mathbf{G}\left(-\mathbf{M}^{-1}\mathbf{G}_0/2 + \mathrm{i}\sqrt{\mathbf{M}^{-1}\mathbf{K}}\right) \in \mathbb{C}^{N\times N}$. The first-order approximation can be obtained by using all the equations in [6.146]. The computational time is slightly more than the zero-order approximation as we have to calculate a matrix inversion to obtain the correction term $\mathbf{\Delta}$. From Figure 6.4, it can be observed that both of these approximations turn out to be accurate for this problem. The approximations are less accurate in the lower frequency range, which collaborate with the observation in Figure 6.3 that there are more errors in the lower modes than in the higher modes.

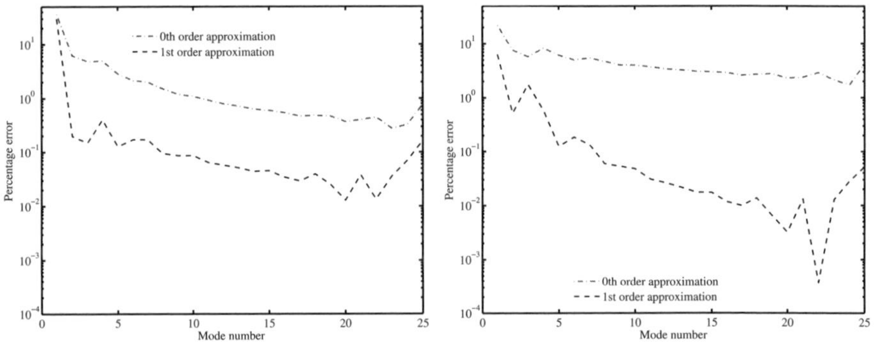

(a) Error in the real parts

(b) Error in the imaginary parts

Figure 6.3. *Percentage error in the eigenvalues obtained using the approximate expressions for case (a)*

Percentage errors in the real sand imaginary parts of the eigenvalues obtained using the two approximate expressions for case (b) are shown in Figure 6.5. As expected, the first-order approximation is more accurate that the zeroth-order approximation. The trend of errors in these diagrams is opposite to the previous case shown in Figure 6.3. Here, errors increase with increasing frequency. As a result, error in using the proposed second-order approximation might be more in the high-frequency regions compared to the low-frequency regions. This can be confirmed from Figure 6.6 where the exact and approximate FRFs are plotted for case (b). Like the previous case, we observe that the zeroth- and first-order approximations are very close and significantly better than that approximation with matrix $\mathbf{G}_0$ only. These numerical results show that the simple approximation developed here can be used to significantly reduce the order of linear MDOF non-viscously damped systems.

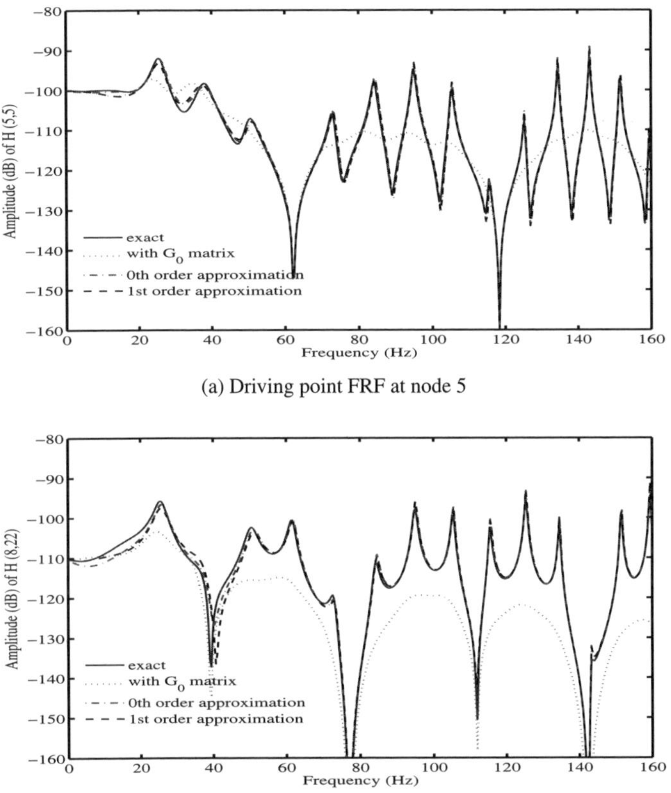

(a) Driving point FRF at node 5

(b) Cross FRF between node 8 and 22

Figure 6.4. *Frequency response functions (FRF) for case (a)*

6.5. Summary

Because of the recent developments in actively controlled structures and the increasing use of composite and smart materials, there is a renewed interest to consider general non-proportionally damped linear dynamic systems. State-space-based methods are normally used to address such problems. Such methods are computationally more expensive and often do not give the physical insight compared to the classical normal mode based method. An iterative approach has been introduced to obtain the complex eigensolutions of a general non-proportionally viscously damped system from the undamped eigensolutions. It is assumed that all the eigenvalues are distinct and appear in complex conjugate pairs. The proposed method exploits a mathematical construction where complex eigenvalues and eigenvectors can be updated from their previous values in an iterative manner. A sufficient condition for the convergence of the proposed iterative method is derived. A simple algorithm is proposed to implement this method. The computational complexity of the proposed method is discussed and compared with the state-space method. The applicability of the proposed method is investigated using an example with a high degree of modal coupling. Acceptable accuracy has been observed. Using the iterative method developed here, it is possible to obtain the eigenvalues, eigenvectors and consequently the dynamic response of non-proportionally viscously damped systems by post-processing of the undamped eigenvalues and eigenvectors, which in turn can be obtained using general purpose finite element software.

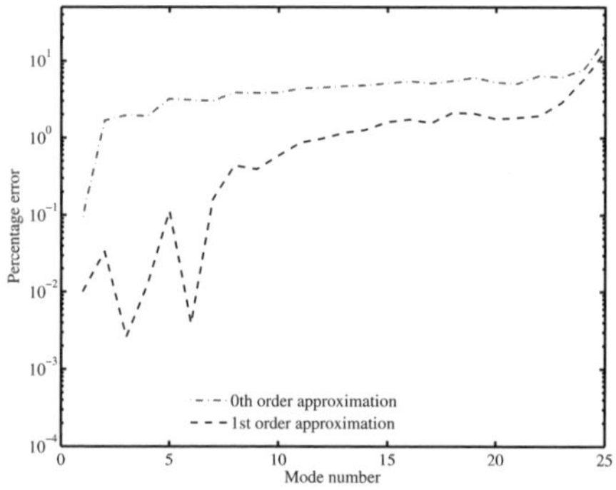

(a) Error in the real parts

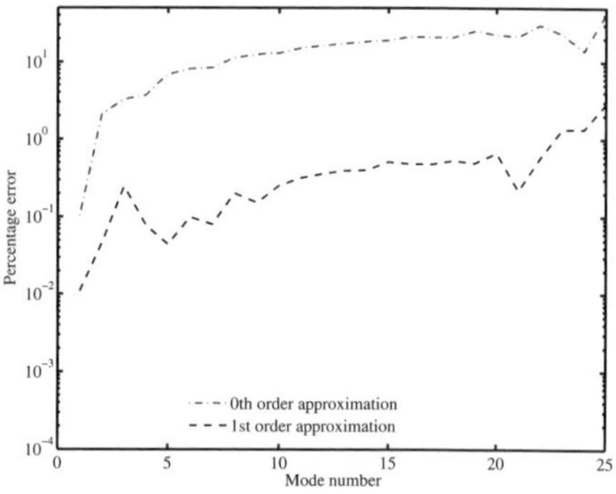

(b) Error in the imaginary parts

Figure 6.5. *Percentage error in the eigenvalues obtained using the approximate expressions for case (b)*

SDOF and MDOF linear non-viscously damped systems are also considered with the same motivation. The calculation of eigenvalues of non-viscously damped systems requires the solution of a nonlinear eigenvalue problem. A non-state-space approach has been discussed for such nonlinear eigenvalue problems. A key motivation behind this is to reduce the additional computational cost necessary for non-viscously damped systems compared to undamped systems. Approximate expressions have been derived for the complex and real eigenvalues with single and multiple exponential kernels. It is assumed that all the eigenvalues are distinct. For the SDOF case, it has been assumed that the non-viscously damped system can be considered as a small perturbation from the underlying elastic system. For the MDOF case, it has been additionally assumed that the system can be approximately diagonalized using the undamped modes. These approximations allow us to obtain the eigenvalues of non-viscously damped systems by simple post-processing of the undamped eigenvalues, which in turn can be obtained using a general purpose finite element software. The complex eigenvalues are close to the eigenvalues of the underlying damped system while the real eigenvalues are close to the relaxation parameters. The expression for the complex eigenvalues is applicable to any general non-viscously damped kernel while the expression for the real eigenvalues is specific to the exponential model only. Although the real solution is applicable to the exponential model only, the technique developed for this purpose may inspire solutions to other types of non-viscously damped models available in literature. The accuracy of the proposed approximations was demonstrated numerically against the exact state-space eigenvalues for a few example problems. The complex conjugate eigenvalues turn out to be more accurate compared to the real eigenvalues. In particular, this is encouraging because complex eigenvalues dominate the dynamic response of linear systems. The method presented offers a reduction in computational effort because neither the state-space formalisms nor the additional internal variables are used. This approach might provide further physical insight and computational advantage for MDOF systems as only familiar undamped natural frequencies and damping factors are required to obtain the eigenvalues of the system.

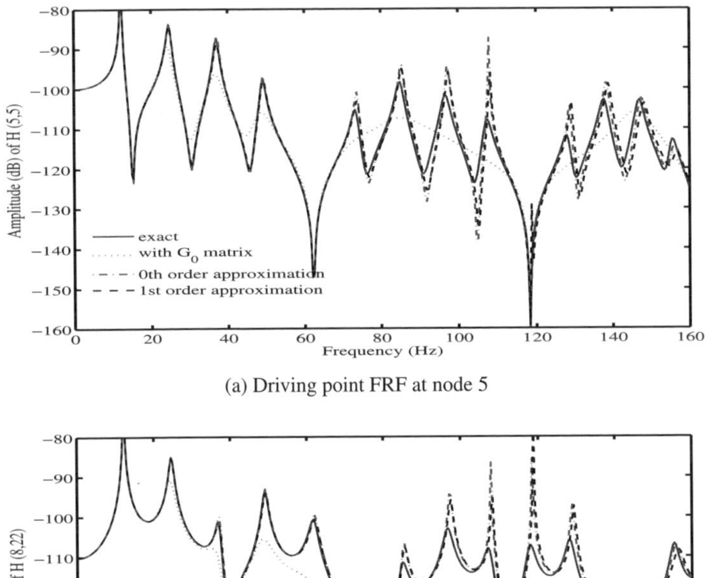

(a) Driving point FRF at node 5

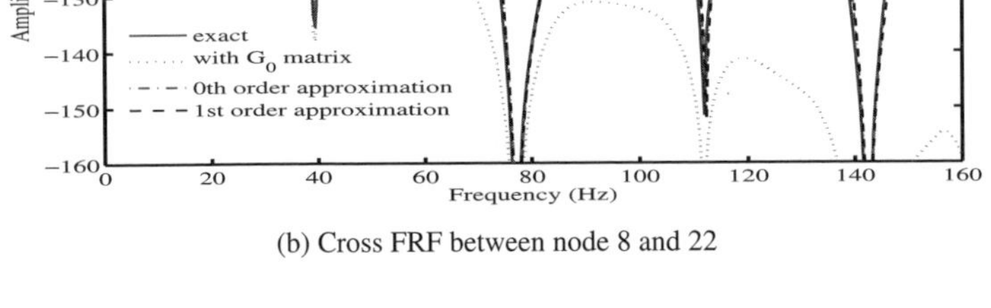

(b) Cross FRF between node 8 and 22

Figure 6.6. *Frequency response functions (FRF) for case (b)*

The order of the state-space model of an MDOF linear non-viscously damped system can be very large due to the additional dissipation coordinates. To reduce the computational cost, the model order reduction of non-viscously damped systems has been considered. In particular, the possibility of using an equivalent second-order system has been discussed. The main message is that an equivalent second-order model can be obtained by simply treating the non-viscously damped kernel function as a matrix function of a suitable argument matrix. This particular matrix can in turn be obtained using elementary matrix operations of the known system matrices. The idea behind this approximation arose from the concept of generalized proportional damping where any frequency-dependent damping can be incorporated within the scope of the proportional damping approximation. Here, generalized proportional damping is used together with the expressions of approximate eigenvalues of the non-viscously damped system. Like any approximate method, this approach has regions of acceptable accuracy. The error arising in the eigenvalue approximation depends on the validity of two assumptions, namely (1) small non-viscous damping and (2) small non-proportionality. As a result, the accuracy of the proposed second-order model depends on these two conditions. The proposed approach is valid for any general non-viscously damped kernel function (smooth functions of frequency), such as the GHM, ADF, fractional derivative and Biot model. In the numerical examples, a system with order 90 in the state space is reduced to a second-order system with dimension 25 using the proposed method. Acceptable agreements between the full system and the reduced system were obtained for the FRFs.

The eigenvalues and the eigenvectors of system are functions of the system parameters. In [ADH 14b], Chapter 1, we discuss how the eigensolutions of viscous and non-viscously damped systems change due to the changes in the system parameters.

Appendix

Symmetrizable Damped Systems

Theory of dynamics of multiple-degrees-of-freedom *symmetric* systems has been studied in this book so far. However, dynamical behavior of some systems encountered in practice cannot be expressed in terms of symmetric coefficient matrices or self-adjoint linear operators. Some examples are gyroscopic and circulatory systems [HUS 73], aircraft flutter [FAW 77b], ship motion in sea water [BIS 79], contact problems [SOO 83] and many actively controlled systems [CAU 93]. The aim of this appendix is to propose methods by which asymmetric dynamic systems can be transformed into symmetric systems. In this way, methods proposed in the book can, in turn, be applied to asymmetric systems as well.

Few authors have considered such general asymmetric dynamical systems. Fawzy and Bishop [FAW 76] presented several relationships satisfied by the eigenvectors and "eigenrows" of a damped asymmetric system and also presented a method to normalize them. Caughey and Ma [CAU 93] have derived conditions under which such systems can be diagonalized by a *similarity transformation*. In a subsequent chapter, Ma and Caughey [MA 95] utilized *equivalence transformation* to analyze asymmetric non-conservative systems and gave the condition under which they can be diagonalized. Adhikari [ADH 99a] proposed a method to obtain (complex) eigensolutions of general asymmetric non-conservative systems without converting the equations of motion into the first-order form.

The above-mentioned works have certainly enhanced the power of modal analysis in dealing with asymmetric systems. However, asymmetric systems are still not as well understood as symmetric systems. For example, rightly pointed out in [KLI 92], stability investigations become substantially easier if the system matrices are symmetric. Without doubt, it would be preferable if asymmetric systems could be transformed into equivalent symmetric systems so that we can take advantage of the well-developed theories for symmetric systems to analyze them. This is the primary reason to study "symmetrizability" of asymmetric systems. In linear dynamics literature, symmetrizability has been addressed by *similarity transformation* based on Taussky's [TAU 68] definition. We call Taussky's approach of symmetrization by similarity transformation as "symmetrization of first kind". In this appendix, symmetrizability of a matrix is redefined in the context of *equivalence transformation*. Such symmetrization will be called "symmetrization of the second kind".

Equivalence transformations are the most general class of non-singular linear transformations. This motivates us to utilize equivalence transformation rather than similarity transformation for symmetrization of asymmetric systems. It will be shown that much generality can be achieved by using symmetrization of the second kind compared to the first kind. Notations and definitions of some terminologies frequently used in this appendix are given in section A1.1. Taussky's version of symmetrization is presented in section A1.2. In section A1.3, we formally define the symmetrization of the second kind. For the sake of generality, some basic results on such symmetrization are presented on complex matrices. Numerical methods are outlined to carry out such a symmetrization procedure. In view of the undamped and viscously damped dynamical systems, simultaneous symmetrization of two and three matrices is considered in section A1.4 and section A1.5 respectively. Finally, section A1.6 summarizes the main results of this appendix. Throughout the appendix, suitable numerical examples are provided to illustrate the derived results.

A1.1. Notations, basic concepts and definitions

By $\mathbb{R}^{N\times N}$, we mean the space of $N \times N$ real matrices and $\mathbb{C}^{N\times N}$stands for the space of $N \times N$ complex matrices. An $N \times N$ matrix taken from either a real or complex number field will be denoted by $\mathbb{F}^{N\times N}$. A matrix $\mathbf{A}$ is called positive definite if all of its eigenvalues $\lambda_j > 0$ and will be denoted by $\mathbf{A} > 0$. A unit matrix $\mathbf{I}$ $\in \mathbb{R}^{N\times N}$is a diagonal matrix with all diagonal entries equal to one. Let $\mathbf{A} \in \mathbb{C}^{N\times N}$, then we denote $\mathbf{A}^T$, $\bar{\mathbf{A}}$, $\mathbf{A}^{-1}$, $\mathbf{A}^{-T}$ and $\mathbf{A}^*$ be the transpose, complex conjugate, inverse, inverse transposed and transposed conjugate. A matrix $\mathbf{A}$ is called *symmetric* if $\mathbf{A} = \mathbf{A}^T$, *Hermitian* if $\mathbf{A} = \mathbf{A}^*$ and *unitary* if $\mathbf{A}\mathbf{A}^* = \mathbf{I}$. If $\mathbf{A}$ is real, then a Hermitian matrix is equivalent to a symmetric matrix and a unitary matrix is equivalent to a real orthogonal matrix as $\mathbf{A}\mathbf{A}^T = \mathbf{I}$. For a matrix $\mathbf{A}$, by saying $\mathbf{A}^{-1}$ exists, we mean that none of its eigenvalues are equal to 0 and that $\mathbf{A}$ is non-singular. Two matrices $\mathbf{A}$ and $\mathbf{B}$, related by $\mathbf{B} = \mathbf{V}^T\mathbf{A}\mathbf{U}$ for some non-zero $\mathbf{U}$ and $\mathbf{V}$, are said to be related by an *equivalence transformation*. When $\mathbf{V}^T = \mathbf{U}^{-1}$, the equivalence transformation is called *similarity transformation* and we call $\mathbf{A}$ and $\mathbf{B}$ *similar*. In the event $\mathbf{V} = \mathbf{U}$, the equivalence transformation is a *congruence transformation*. Classical modal transformation in symmetric systems is an example of congruence transformation. When $\mathbf{V}^T = \mathbf{U}^{-1} = \mathbf{U}^T$, we call such transformation *orthogonal* *transformation*. Furthermore, if $\mathbf{V}^T = \mathbf{V}^{-1}$ and $\mathbf{U}^T = \mathbf{U}^{-1}$, such transformation is called *biorthogonal transformation*. A matrix $\mathbf{A}$ is said to be *diagonalizable* if it is similar to a diagonal matrix and that diagonal matrix contains all the eigenvalues of $\mathbf{A}$ with proper multiplicities. A matrix is diagonalizable if and only if it has N linearly independent eigenvectors.

A1.2. Symmetrizability of the first kind

The concept of symmetrizability of an asymmetric matrix originated from an excellent result by Taussky [TAU 59], which states that "for every $\mathbf{A} \in \mathbb{F}^{N\times N}$ there is a non-singular symmetric matrix transforming $\mathbf{A}$ into its transpose". Based on this general result, it was shown that every real square matrix can be expressed as a product of two real symmetric matrices, that is

$$\mathbf{A} = \mathbf{S}_1\mathbf{S}_2; \quad \mathbf{S}_j = \mathbf{S}_j^T \in \mathbb{R}^{N\times N} \qquad [A1.1]$$

always holds. Later, Taussky [TAU 68] proved that if one of the factors in this representation is positive definite, then $\mathbf{A}$ can be transformed into a symmetric matrix by a similarity transformation and vice versa. She formally defined *symmetrizability* of a matrix as follows:

DEFINITION A1.1.– *A matrix* $\mathbf{A}$ *is symmetrizable if and only if any one of the following conditions holds:*

1) $\mathbf{A}$ *is the product of two symmetric matrices, one of which is positive definite;*

2) $\mathbf{A}$ *is similar to a symmetric matrix;*

3) $\mathbf{A}^T = \mathbf{S}^{-1}\mathbf{A}\mathbf{S}$ *with* $\mathbf{S} = \mathbf{S}^T > 0$*;*

4) $\mathbf{A}$ *has real characteristic roots and a full set of characteristic vectors.*

Huseyin and Liepholz [HUS 73] were possibly the first to recognize the importance of Taussky's result in the context of structural dynamics. However, symmetrization was not properly exploited until a decade later when Inman's [INM 83] paper appeared. Inman's work inspired several authors to consider dynamics of symmetrizable asymmetric systems [AHM 84a, AHM 84b, AHM 87, SHA 89, CHE 93]. Next, a more general approach for symmetrization is proposed.

A1.3. Symmetrizability of the second kind

A1.3.1. *Definition*

Now, we will introduce a new definition of symmetrizability by utilizing equivalence transformation.

DEFINITION A1.2.– *A matrix* $\mathbf{A}$ *is symmetrizable of the second kind if and only if there exist two non-zero matrices* $\mathbf{L}$ *and* $\mathbf{R}$ *such that* $\widetilde{\mathbf{A}} = \mathbf{L}^T\mathbf{A}\mathbf{R}$ *is a symmetric matrix.*

This definition of symmetrizability is quite general and valid for both real and complex matrices. It also holds Taussky's definition as a special case when $\mathbf{L}^T = \mathbf{R}^{-1} = \mathbf{R}^{-T}$. In the above definition, even $\mathbf{A}$ is real $\mathbf{L}$ and $\mathbf{R}$ might be complex in order to make $\widetilde{\mathbf{A}}$ symmetric. For any $\mathbf{A}$, when $\mathbf{L}$ and $\mathbf{R}$ are real matrices, we call such $\mathbf{A}$ as *real symmetrizable* and *complex symmetrizable* when $\mathbf{L}$ and $\mathbf{R}$ are complex. Taussky [TAU 72] and Pommer and Kliem [POM 87] have discussed complex symmetrizable matrices in the context of similarity transformation. Applications of such complex symmetrizability in linear dynamical systems were given in [KLI 92].

A1.3.2. *Basic results*

Some basic results on symmetrizability of the second kind and a numerical method to carry out such symmetrization procedure will be developed. To achieve more generality, one of our main results is presented for non-square complex matrices in the following theorem.

THEOREM A1.1.– Every rectangular complex matrix $\mathbf{A} \in \mathbb{C}^{r\times s}$ can be transformed to a real symmetric (square) matrix by an equivalence transformation.

PROOF.– We use a result by Eckart and Young (theorem 1) that states that *for every* $\mathbf{A} \in \mathbb{C}^{r\times s}$ *there are two unitary matrices* $\mathbf{X} \in \mathbb{C}^{r\times r}$ *and* $\mathbf{Y} \in \mathbb{C}^{s\times s}$ *such that* $\mathbf{D} = \mathbf{X}^*\mathbf{A}\mathbf{Y}$ *is a diagonal matrix with real elements, none of which are negative.* If $rank(\mathbf{A}) = k$, then it was shown that $\mathbf{D} = \begin{bmatrix} \mathbf{\Lambda} & \mathbf{0}_1 \\ \mathbf{0}_2 & \mathbf{0}_3 \end{bmatrix}$ where $\mathbf{\Lambda}$=diag$(\lambda_1, \cdots, \lambda_k)$ is a real diagonal matrix and $\mathbf{0}_1$, $\mathbf{0}_2$ and $\mathbf{0}_3$ are $(k, s-k)$, $(r-k, k)$ and $(r-k, s-k)$ null matrices [WIL 39]. Without loss of generality, we select $\mathbf{L} = \overline{\mathbf{X}}\mathbf{Q}_1$ and $\mathbf{R} = \mathbf{Y}\mathbf{Q}_2$ for some non-zero $\mathbf{Q}_1 \in \mathbb{R}^{r\times n}$ and $\mathbf{Q}_2 \in \mathbb{R}^{s\times n}$ such that $\mathbf{Q}_1 = \begin{bmatrix} \mathbf{\Theta} & \mathbf{Q}_1^1 \\ \mathbf{Q}_2^1 & \mathbf{Q}_3^1 \end{bmatrix}$ and $\mathbf{Q}_2 = \begin{bmatrix} \mathbf{\Theta} & \mathbf{Q}_1^2 \\ \mathbf{Q}_2^2 & \mathbf{Q}_3^2 \end{bmatrix}$ where $\mathbf{\Theta} \in \mathbb{R}^{k\times k}$ and $\mathbf{Q}_j^l, l = 1,2; j = 1,2,3$ are all real matrices of proper orders. Using these, we have $\widetilde{\mathbf{A}} = \mathbf{L}^T\mathbf{A}\mathbf{R} = \mathbf{Q}_1^T\mathbf{X}^*\mathbf{A}\mathbf{Y}\mathbf{Q}_2 = \mathbf{Q}_1^T\mathbf{D}\mathbf{Q}_2 = \begin{bmatrix} \mathbf{\Theta}^T\mathbf{\Lambda}\mathbf{\Theta} & \widetilde{\mathbf{0}}_1 \\ \widetilde{\mathbf{0}}_2 & \widetilde{\mathbf{0}}_3 \end{bmatrix} \in \mathbb{R}^{n\times n}$ where $\widetilde{\mathbf{0}}_j; j = 1,2,3$ are all null matrices with proper orders. So, $\widetilde{\mathbf{A}}$ is a real symmetric matrix and this completes the proof.

REMARK A1.1.– The two matrices $\mathbf{Q}_1$ and $\mathbf{Q}_2$ used above can be selected in such a way that the symmetric form becomes non-singular and has the same rank as $\mathbf{A}$.

Observe that because $\mathbf{D}$ is a diagonal matrix with non-negative real elements, $\mathbf{Q}_1$ and $\mathbf{Q}_2$ can also be selected in a way such that the symmetric form becomes positive definitive. The symmetric form of $\mathbf{A}$ is, however, not unique because $\mathbf{Q}_1$ and $\mathbf{Q}_2$ can be chosen in several ways.

This result clearly demonstrates generality of the symmetrization of the second kind compared to the first kind that cannot be applied to rectangular matrices. Of course, theorem A1.1 is applicable for real matrices as a special case. When $\mathbf{A}$ is a real matrix, we have the following interesting special result.

THEOREM A1.2.– Every real matrix $\mathbf{A}$ is real symmetrizable of the second kind.

PROOF.– We use a result by Pearl [PEA 67] (theorem 1). This gives the condition under that the unitary matrices that reduces the matrix $\mathbf{A}$ to a real diagonal form by an equivalence transformation in Eckart and Young's [ECK 39] theorem are real orthogonal matrices. According to this result, *for* $\mathbf{A} \in \mathbb{C}^{r\times s}$ *there are two real orthogonal matrices* $\mathbf{O}_1 \in \mathbb{R}^{r\times r}$ *and* $\mathbf{O}_2 \in \mathbb{R}^{s\times s}$ *such that* $\mathbf{D} = \mathbf{O}_1^*\mathbf{A}\mathbf{O}_2$ *is a diagonal matrix with real non-negative elements if and only if* $\mathbf{A}\mathbf{A}^*$ *and* $\mathbf{A}^*\mathbf{A}$ *are both real matrices*. The theorem is proved since (trivially) this condition is always satisfied when $\mathbf{A}$ is real.

We have described two important results on symmetrization of the second kind. From a practical point of view, the obvious question is, for a given matrix $\mathbf{A}$, how do we calculate $\mathbf{L}$ and $\mathbf{R}$ so that $\widetilde{\mathbf{A}} = \mathbf{L}^T\mathbf{A}\mathbf{R}$ is a real symmetric matrix? In answer to this question, we observe that the proof given in [ECK 39] of their theorem described before can be used for obtaining $\mathbf{L}$ and $\mathbf{R}$. For computational purposes, the following steps may be followed to transform a general complex matrix $\mathbf{A} \in \mathbb{C}^{r\times s}$ to a non-singular real square matrix using an equivalence transformation.

1) Calculate the matrix $\mathcal{M} = \mathbf{A}\mathbf{A}^* \in \mathbb{C}^{r\times r}$. Solve the eigenvalue problem $\mathcal{M}\mathbf{q}_j = \lambda_j^2\mathbf{q}_j, \forall j = 1, \cdots r$. Since $\mathcal{M}$ is a non-negative Hermitian, its eigenvector matrix can be normalized to a unitary matrix. Suppose $rank(\mathbf{A}) = k$, arrange the numbering of the eigenvalues so that $\lambda_1 \geq \lambda_2 \geq \cdots \geq \lambda_k > 0; \lambda_{k+1} = \cdots = \lambda_r = 0$. Denote $\mathbf{X} = [\mathbf{q}_1, \mathbf{q}_2, \cdots, \mathbf{x}_k] \in \mathbb{C}^{r\times k}$ as the ordered collection of the eigenvectors.

2) Evaluate $\mathbf{y}_j = \mathbf{A}^*\mathbf{q}_j/\lambda_j, \forall j = 1, \cdots k$. Set $\mathbf{Y} = [\mathbf{y}_1\mathbf{y}_2, \cdots, \mathbf{y}_k] \in \mathbb{C}^{s\times k}$.

3) Consider any non-zero $\mathbf{Q} \in \mathbb{R}^{k\times k}$ and obtain $\mathbf{L} = \overline{\mathbf{X}}\mathbf{Q} \in \mathbb{C}^{r\times k}$ and $\mathbf{R} = \mathbf{Y}\mathbf{Q} \in \mathbb{C}^{s\times k}$. Finally, check that $\widetilde{\mathbf{A}} = \mathbf{L}^T\mathbf{A}\mathbf{R}$ is a $(k \times k)$ real non-singular symmetric matrix.

For the sake of generality, the method is proposed for rectangular complex matrices. This procedure is obviously applicable to real square matrices that we normally encounter in the equations of motion of linear systems. When $\mathbf{A}$ is real, $\mathcal{M}$

becomes a real symmetric matrix and consequently **X**, **Y**, **L** and **R** all become real matrices. For a further special case, when **A** is a symmetrizable matrix of the first kind, this procedure provides an alternative and easy way to find symmetric forms as calculation of Taussky's factorization can be avoided.

EXAMPLE A1.1.– The procedure outlined above can be illustrated by the following rectangular complex matrix

$$\mathbf{A} = \begin{bmatrix} 1.0 + 3.0\mathrm{i} & -2.0 + 1.0\mathrm{i} & 1.5 - 2.0\mathrm{i} \\ -2.0 - 3.0\mathrm{i} & 6.0 + 2.0\mathrm{i} & 7.0 + 2.0\mathrm{i} \end{bmatrix}. \quad \text{[A1.2]}$$

Step 1: We calculate $\mathcal{M} = \mathbf{A}\mathbf{A}^* = \begin{bmatrix} 21.25 & -14.50 - 10.00\mathrm{i} \\ -14.50 + 10.00\mathrm{i} & 106.00 \end{bmatrix}$. Solving the eigenvalue problem $\mathcal{M}\mathbf{q}_j = \lambda_j^2 \mathbf{x}_j$ and arranging accordingly, we have $\lambda_j^2 = \{109.5150,\ 17.7350\}$ and

$$\mathbf{X} = \begin{bmatrix} -0.1611 - 0.1111\mathrm{i} & 0.9807 \\ 0.9807 & 0.1611 - 0.1111\mathrm{i} \end{bmatrix}. \quad \text{[A1.3]}$$

Note that $rank(\mathbf{A}) = 2$.

Step 2: By calculating we obtain

$$\mathbf{Y} = \begin{bmatrix} -0.2347 + 0.3167\mathrm{i} & 0.2355 - 0.5311\mathrm{i} \\ 0.5824 - 0.1508\mathrm{i} & -0.2890 - 0.4677\mathrm{i} \\ 0.6541 - 0.2341\mathrm{i} & 0.5643 + 0.2045\mathrm{i} \end{bmatrix}.$$

Step 3: Now arbitrarily select the matrix $\mathbf{Q} = \begin{bmatrix} 0.1021 & 0.2605 \\ 1.4109 & 0.3519 \end{bmatrix}$ so that

$$\mathbf{L} = \overline{\mathbf{X}}\mathbf{Q} = \begin{bmatrix} 1.3672 + 0.0113\mathrm{i} & 0.3031 + 0.0289\mathrm{i} \\ 0.3274 + 0.1568\mathrm{i} & 0.3121 + 0.0391\mathrm{i} \end{bmatrix}$$

and

$$\mathbf{R} = \mathbf{Y}\mathbf{Q} = \begin{bmatrix} 0.3083 - 0.7170\mathrm{i} & 0.0218 - 0.1044\mathrm{i} \\ -0.3483 - 0.6752\mathrm{i} & 0.0500 - 0.2039\mathrm{i} \\ 0.8630 + 0.2647\mathrm{i} & 0.3690 + 0.0110\mathrm{i} \end{bmatrix}. \quad \text{[A1.4]}$$

Using this **L** and **R**, we have $\widetilde{\mathbf{A}} = \mathbf{L}^T\mathbf{A}\mathbf{R} = \begin{bmatrix} 8.4923 & 2.3692 \\ 2.3692 & 1.2315 \end{bmatrix}$, which is a real non-singular symmetric matrix.

EXAMPLE A1.2.– Consider a real asymmetric matrix

$$\mathbf{A} = \begin{bmatrix} 1.0 & -2.0 & 1.5 \\ 12.0 & 6.0 & 7.0 \\ -2.0 & 4.0 & 9.0 \end{bmatrix}. \qquad [A1.5]$$

Observe that $rank(\mathbf{A}) = 3$. Selecting the matrix

$$\mathbf{Q} = \begin{bmatrix} 0.0645 & -0.1647 & 0.1436 \\ -0.8923 & 0.2226 & 0.6155 \\ 0.1611 & 0.2298 & 0.3708 \end{bmatrix} \qquad [A1.6]$$

and following the procedure described before, we obtain

$$\mathbf{L} = \begin{bmatrix} 0.1735 & 0.2198 & 0.3701 \\ 0.3901 & -0.2469 & -0.1170 \\ -0.8026 & 0.1421 & 0.6215 \end{bmatrix}$$

$$\text{and} \qquad \mathbf{R} = \begin{bmatrix} 0.7299 & -0.2298 & -0.2942 \\ -0.2638 & -0.2377 & -0.1620 \\ -0.4733 & 0.1419 & 0.6513 \end{bmatrix} \qquad [A1.7]$$

are both real matrices. From these, we have

$$\widetilde{\mathbf{A}} = \mathbf{L}^T \mathbf{A} \mathbf{R} = \begin{bmatrix} 7.0392 & -1.7960 & -4.4600 \\ -1.7960 & 1.0004 & 1.0316 \\ -4.4600 & 1.0316 & 3.9722 \end{bmatrix}$$

a real non-singular symmetric matrix. Interestingly, note that $eig(\mathbf{A}) = \{12.8914, 1.5543 + 5.0507\mathrm{i}, 1.5543 - 5.0507\mathrm{i}\}$, i.e. $\mathbf{A}$ does not satisfy Taussky's condition of symmetrizability (definition A1.1). Thus, $\mathbf{A}$ in [A1.5] is a real symmetrizable matrix of the second kind but not real symmetrizable of the first kind. This illustrates generality of the proposed approach of symmetrization compared to the conventional approach.

Because the coefficient matrices in equations of motion of linear vibrations are real, next we consider only real square matrices.

A1.4. Simultaneous symmetrization of two matrices

The equation of motion of a linear undamped (non-gyroscopic) system can be expressed by

$$\mathbf{A}\ddot{\mathbf{q}}(t) + \mathbf{B}\mathbf{q}(t) = \mathbf{0}, \quad t > 0, \tag{A1.8}$$

where $\mathbf{A} \in \mathbb{R}^{N \times N}$ and $\mathbf{B} \in \mathbb{R}^{N \times N}$ and $\mathbf{q}(t) \in \mathbb{R}^{N}$ is vector of the *generalized coordinates*. Were **A** and **B** symmetric and positive definite matrices, they would, respectively, represent the mass and stiffness matrices. Here, however, no such restrictions are imposed on **A** and **B**. It was shown [HUS 78] that the classical modal analysis, originally developed for symmetric systems, can be extended to solve equation [A1.8]. There exist *left eigenvectors* or *normal modes* and *right eigenvectors* or *adjoint modes* that uncouple the equations of motion into N single degree-of-freedom oscillators. Although system [A1.8] can always be solved using normal modes and adjoint modes, it would be more useful if it could be transformed to a symmetric system. We consider the following two cases to investigate symmetrizability of system [A1.8].

A1.4.1. *Case 1: A and B have real roots and A is non-singular*

This is the most common case we encounter in practice. If system [A1.8] is "truly" undamped, then **A** or **B** does not possess any complex roots; otherwise a stable system will not have periodic motion for an infinitely long time but vibration (in some DOF) will decay due to the complex nature of the eigenvalues. As **A** is non-singular, rewriting equation [A1.8], we have

$$\mathbf{I}\ddot{\mathbf{q}}(t) + \mathbf{E}\mathbf{q}(t) = \mathbf{0} \tag{A1.9}$$

where $\mathbf{E} = \mathbf{A}^{-1}\mathbf{B} \in \mathbb{R}^{N \times N}$ has real roots as both **A** and **B** have real roots. Because this matrix satisfies condition 4 of Taussky's definition of symmetrizability, system [A1.9] can be transformed to a symmetric system by a similarity transformation. From this discussion, we have the following interesting result.

THEOREM A1.3.– All asymmetric undamped systems are similar to symmetric undamped systems.

Practical implementation of Taussky's factorization [A1.1] is not straightforward (see [AHM 87, SHA 89, CHE 93]). For this reason, it is required to develop efficient numerical methods for finding associated symmetric form(s) of system [A1.8]. Recall that as system [A1.8] is symmetrizable of the first kind, it is also symmetrizable of the second kind. We use the latter approach to avoid the calculation of Taussky's factorization.

It has been mentioned that there always exist two matrices $\mathbf{U} \in \mathbb{R}^{N \times N}$ (normal modal matrix) and $\mathbf{V} \in \mathbb{R}^{N \times N}$ (adjoint modal matrix) such that $\mathbf{V}^T\mathbf{A}\mathbf{U}$ and $\mathbf{V}^T\mathbf{B}\mathbf{U}$ are both real diagonal matrices. Numerical methods for obtaining $\mathbf{U}$ and $\mathbf{V}$ are well developed, see [HUS 78, MA 95] for further discussions. Select $\mathbf{L} = \mathbf{V}\mathbf{Q}$ and $\mathbf{R} = \mathbf{U}\mathbf{Q}$ for some non-zero $\mathbf{Q} \in \mathbb{R}^{N \times N}$. Clearly, $\mathbf{L}^T\mathbf{A}\mathbf{R}$ and $\mathbf{L}^T\mathbf{B}\mathbf{R}$ are both symmetric matrices now. This approach is much easier for finding symmetric forms than the approach via Taussky's factorization. Also note that symmetric forms are non-unique since we can select the matrix $\mathbf{Q}$ in many ways. Following example demonstrates this procedure.

EXAMPLE A1.3.– Suppose the coefficient matrices of an undamped system of the form [A1.8] are given by

$$\mathbf{A} = \begin{bmatrix} 0.5740 & 1.3858 & 1.3858 \\ 0.7070 & 0.7070 & -0.7070 \\ 0.4620 & -0.1914 & -0.1914 \end{bmatrix}$$

and
$$\mathbf{B} = \begin{bmatrix} 1.3748 & 10.9440 & 25.2975 \\ 1.2625 & 2.8770 & -17.4195 \\ 0.7455 & -4.1244 & 0.8625 \end{bmatrix}. \qquad [A1.10]$$

Numerical values of $\mathbf{A}$ and $\mathbf{B}$ are taken from [MA 95]. It may be verified that $\mathbf{A}$, $\mathbf{B}$ and consequently $\mathbf{E} = \mathbf{A}^{-1}\mathbf{B}$ are symmetrizable matrices of the first kind.

Solving the right and left eigenvalue problems $\mathbf{B}\mathbf{U} = \mathbf{\Lambda}\mathbf{A}\mathbf{U}$ and $\mathbf{V}^T\mathbf{B} = \mathbf{\Lambda}\mathbf{V}^T\mathbf{A}$, we obtain the undamped modal matrices

$$\mathbf{U} = \begin{bmatrix} 0.9996 & -0.5434 & 0.3951 \\ -0.0271 & 0.8391 & -0.4850 \\ -0.0044 & -0.0253 & 0.7802 \end{bmatrix}$$

and
$$\mathbf{V} = \begin{bmatrix} 0.2868 & 0.4247 & 0.4617 \\ 0.4677 & 0.2613 & -0.7124 \\ 0.8361 & -0.8668 & 0.5285 \end{bmatrix} \qquad [A1.11]$$

and the natural frequencies squared $diag(\mathbf{\Lambda}) = \{1.8241, 9.5561, 23.5486\}$. Selecting the matrix $\mathbf{Q}$ given by [A1.6], we obtain

$$\mathbf{L} = \mathbf{VQ} = \begin{bmatrix} -0.2860 & 0.1534 & 0.4737 \\ -0.3177 & -0.1826 & -0.0362 \\ 0.9126 & -0.2092 & -0.2176 \end{bmatrix}$$

$$\text{and} \qquad \mathbf{R} = \mathbf{UQ} = \begin{bmatrix} 0.6131 & -0.1948 & -0.0445 \\ -0.8286 & 0.0798 & 0.3328 \\ 0.1480 & 0.1744 & 0.2731 \end{bmatrix}. \qquad [A1.12]$$

Using these matrices, we have

$$\widetilde{\mathbf{A}} = \mathbf{L}^T\mathbf{AR} = \begin{bmatrix} 0.6282 & -0.1303 & -0.3608 \\ -0.1303 & 0.1032 & 0.1514 \\ -0.3608 & 0.1514 & 0.4150 \end{bmatrix}$$

$$\text{and} \qquad \widetilde{\mathbf{B}} = \mathbf{L}^T\mathbf{BR} = \begin{bmatrix} 6.2653 & -0.7600 & -2.8429 \\ -0.7600 & 1.3947 & 2.5574 \\ -2.8429 & 2.5574 & 5.3631 \end{bmatrix}. \qquad [A1.13]$$

Equation [A1.13] is a symmetric form of the asymmetric system [A1.10]. Note that symmetric form of this system can also be obtained alternatively by calculating Taussky's factorization of $\mathbf{E}$.

A1.4.2. *Case 2: A and B are general matrices*

We have shown that when $\mathbf{A}$ and $\mathbf{B}$ have real roots, system [A1.3] has an equivalent symmetric form via real linear transformations. However, if these matrices have complex roots, matrix $\mathbf{E}$ in equation [A1.9], in general, does not satisfy Taussky's definition of symmetrizability. In that case, can system [A1.8] be transformed into a real symmetric system using a real linear transformation? Our answer is the following:

LEMMA A1.1.– Linear undamped system [A1.8] is real symmetrizable if $\mathbf{A}^T\mathbf{B}$ and $\mathbf{B}\mathbf{A}^T$ are symmetric matrices.

PROOF.– According to Thompson [THO 69] "there exist two real orthogonal matrices $\mathbf{O}_1 \in \mathbb{R}^{N\times N}$ and $\mathbf{O}_2 \in \mathbb{R}^{N\times N}$ such that square matrices $\mathbf{O}_1^T\mathbf{A}\mathbf{O}_2 = \mathbf{\Lambda}_a$ and $\mathbf{O}_1^T\mathbf{B}\mathbf{O}_2 = \mathbf{\Lambda}_b$ are diagonal matrices if and only if $\mathbf{A}^T\mathbf{B}$ and $\mathbf{B}\mathbf{A}^T$ are symmetric matrices". Now construct $\mathbf{L} = \mathbf{O}_1\mathbf{Q}$ and $\mathbf{R} = \mathbf{O}_2\mathbf{Q}$ for some non-zero $\mathbf{Q} \in \mathbb{R}^{N\times N}$. Using these, we have $\mathbf{L}^T\mathbf{AR} = \mathbf{Q}^T\mathbf{\Lambda}_a\mathbf{Q}$ and $\mathbf{L}^T\mathbf{BR} = \mathbf{Q}^T\mathbf{\Lambda}_b\mathbf{Q}$ are both symmetric matrices. This completes the proof.

This result provides only a *sufficient* condition for simultaneous real symmetrizability of the general matrices **A** and **B**. Lack of it does not necessarily preclude existence of a real symmetric form of system [A1.8]. The following examples illustrate this fact.

EXAMPLE A1.4.– Consider an undamped system with **A** same as example A1.2 and

$$\mathbf{B} = \begin{bmatrix} 1.2044 & -5.4425 & 2.7013 \\ 2.1007 & 0.8393 & 0.1894 \\ -1.8393 & 0.8953 & 2.5087 \end{bmatrix}. \qquad [A1.14]$$

We may verify that **A** and **B** are not symmetrizable of the first kind. Observe that the matrices

$$\mathbf{A}^T\mathbf{B} = \begin{bmatrix} 30.0910 & 2.8380 & -0.0427 \\ 2.8380 & 19.5018 & 5.7688 \\ -0.0427 & 5.7688 & 27.9563 \end{bmatrix}$$

$$\text{and} \qquad \mathbf{B}\mathbf{A}^T = \begin{bmatrix} 16.1414 & 0.7063 & 0.1331 \\ 0.7063 & 31.5697 & 0.8607 \\ 0.1331 & 0.8607 & 29.8380 \end{bmatrix} \qquad [A1.15]$$

are both symmetric. Since the system matrices satisfy the condition of lemma A1.1, the transforming matrices **L** and **R** (calculated in equation [A1.7]), which symmetrizes **A** and also symmetrizes **B** as

$$\mathbf{L}^T\mathbf{B}\mathbf{R} = \begin{bmatrix} 2.8766 & -0.4670 & -1.4786 \\ -0.4670 & 0.5501 & 0.9468 \\ -1.4786 & 0.9468 & 2.1818 \end{bmatrix},$$

are symmetric matrix.

EXAMPLE A1.5.– Consider another undamped system with **A** in the above example and

$$\mathbf{B} = \begin{bmatrix} 0.1955 & 1.8857 & -3.2199 \\ -2.1312 & -0.2236 & 0.3609 \\ 1.0378 & 1.9501 & -1.5606 \end{bmatrix}. \qquad [A1.16]$$

Observe that the matrices

$$\mathbf{A}^T\mathbf{B} = \begin{bmatrix} -27.4545 & -4.6975 & 4.2325 \\ -9.0269 & 2.6874 & 2.3629 \\ -5.2845 & 18.8142 & -16.3493 \end{bmatrix}$$

and $$\mathbf{B}\mathbf{A}^T = \begin{bmatrix} -8.4058 & -8.8789 & -21.8278 \\ -1.1426 & -24.3894 & 6.6164 \\ -5.2033 & 13.2300 & -8.3212 \end{bmatrix} \quad [A1.17]$$

are not symmetric. However, the transforming matrices **L** and **R** obtained in equation [A1.5], which also symmetrizes **B** as

$$\mathbf{L}^T\mathbf{B}\mathbf{R} = \begin{bmatrix} -1.2355 & 0.8082 & 1.2380 \\ 0.8082 & -0.4868 & -0.9953 \\ 1.2380 & -0.9953 & -2.0335 \end{bmatrix}, \quad [A1.18]$$

are symmetric matrix. So the condition outlined in lemma A1.1 is sufficient.

A1.5. Simultaneous symmetrization of three matrices

The equations of motion describing free vibration of a viscously damped linear asymmetric system can be expressed by

$$\mathbf{A}\ddot{\mathbf{q}}(t) + \mathbf{C}\dot{\mathbf{q}}(t) + \mathbf{B}\mathbf{q}(t) = \mathbf{0}, \quad t > 0. \quad [A1.19]$$

We assume **A**, **B** and **C** are $N \times N$ arrays of real numbers but otherwise general. Were these matrices symmetric and positive definite, **A**, **B** and **C** would, respectively, be the mass, stiffness and viscous damping matrices. The equations of motion are now characterized by *three* real matrices, and this brings additional complication in the system dynamics. It is required to find a non-singular linear transformation which simultaneously symmetrizes **A**, **B** and **C**. Unlike the undamped case, where in absence of the **C** matrix conditions for obtaining such a transformation were easy to meet (theorem A1.3), admissible forms of these matrices get restricted here. We will investigate what forms of restrictions should be imposed on the system matrices so that the equations of motion can be transformed into a symmetric form. Two cases are considered here.

A1.5.1. *Case 1: A is non-singular and all the matrices have real roots*

When **A** is non-singular, equation [A1.19] can be rewritten as

$$\mathbf{I}\ddot{\mathbf{q}}(t) + \mathbf{F}\dot{\mathbf{q}}(t) + \mathbf{E}\mathbf{q}(t) = \mathbf{0} \quad [A1.20]$$

where $\mathbf{E} = \mathbf{A}^{-1}\mathbf{B} \in \mathbb{R}^{N \times N}$ and $\mathbf{F} = \mathbf{A}^{-1}\mathbf{C} \in \mathbb{R}^{N \times N}$. Since **A**, **B** and **C** have real roots, **E** and **F** also have real roots. Thus, **E** and **F** are individually symmetrizable of the first kind. But this does not imply that system [A1.20] is also symmetrizable, unless **I**, **E** and **F** are simultaneously symmetrizable. A basic result in this regard is the following:

THEOREM A1.4.– All diagonalizable damped asymmetric systems are similar to symmetric systems.

It is well known that system [A1.20] is diagonalizable by a similarity transformation if and only if **E** and **F** commute in multiplication, i.e. $\mathbf{EF} = \mathbf{FE}$. When this condition is satisfied, the system is symmetrizable of the first kind and hence also symmetrizable of the second kind. Following the approach outlined in section A1.4.1, symmetric form(s) may be obtained by utilizing equivalence transformation to avoid the calculation of Taussky's factorization. A result similar to theorem A1.4 was also established by Inman [INM 83] (theorem 1) and Shahruz and Ma [SHA 89] (theorem 3.1). Note that the opposite statement of the above theorem, i.e., "a damped asymmetric system similar to a symmetric system is diagonalizable", is in general not true. Thus, diagonalizability of a system is only a sufficient condition for symmetrizability and rather restrictive. A stronger and more liberal result is the following:

THEOREM A1.5.– [INM 83] (theorem A1.2) System [A1.20] is symmetrizable if and only if **E** and **F** have a common symmetric positive definite factor.

From Taussky's factorization [A1.1], one can write $\mathbf{E} = \mathbf{S}_1\mathbf{S}_2$, $\mathbf{S}_1 = \mathbf{S}_1^T > 0$, $\mathbf{S}_2 = \mathbf{S}_2^T$ and $\mathbf{F} = \mathbf{T}_1\mathbf{T}_2$, $\mathbf{T}_1 = \mathbf{T}_1^T > 0, \mathbf{T}_2 = \mathbf{T}_2^T$. According to the above theorem, system [A1.20] is symmetrizable if and only if $\mathbf{S}_1 = \mathbf{T}_1$. If this condition is satisfied, equation [A1.20] can be brought into symmetric form using a similarity transformation involving a *symmetric* matrix. For practical purposes, this condition is difficult to check. A more convenient result is the following:

THEOREM A1.6.– [KLI 92](theorem A1.1) System [A1.20] is real symmetrizable if and only if there exist modal matrices $\mathbf{U}_E \in \mathbb{R}^{N \times N}$ and $\mathbf{U}_F \in \mathbb{R}^{N \times N}$ (consisting of full eigenvector sets of **E** and **F**) such that $\mathbf{U}_F^{-1}\mathbf{U}_E$ is orthogonal, or equivalently $\mathbf{U}_F\mathbf{U}_F^T = \mathbf{U}_E\mathbf{U}_E^T$.

Further discussions on this theorem can be found in [POM 87]. When the condition of the above theorem is satisfied, equation [A1.20] can be brought into symmetric form by a similarity transformation. In this case, the symmetrizing matrix may be any non-zero matrix, unlike the symmetric matrix used in theorem A1.5. For this reason, Kliem's (1992) condition is more liberal and holds Inman's (1983) condition as a special case when the symmetrizing similarity transformation itself is a symmetric matrix. Thus, theorem A1.5 and, consequently, results obtained in [AHM 87] (theorem A1.2) and [SHA 89] (theorem 4.1) based on this theorem are only *sufficient* conditions for symmetrization of the first kind of system [A1.19].

Now, symmetrizability of system [A1.19] by means of equivalence transformation will be introduced. Our main result is the following:

THEOREM A1.7.– System [A1.19] is real symmetrizable of the second kind if and only if there exist $\mathbf{U} \in \mathbb{R}^{N\times N}$ and $\mathbf{V} \in \mathbb{R}^{N\times N}$, the matrices of undamped right and left eigenvectors, such that $\mathbf{V}^T\mathbf{C}\mathbf{U}$ is symmetric.

PROOF.– Because $\mathbf{U} \in \mathbb{R}^{N\times N}$ and $\mathbf{V} \in \mathbb{R}^{N\times N}$ are the matrices of right and left eigenvectors of the generalized eigenvalue problem involving $\mathbf{A}$ and $\mathbf{B}$, $\mathbf{V}^T\mathbf{A}\mathbf{U} = \mathbf{\Lambda}_a$ and $\mathbf{V}^T\mathbf{B}\mathbf{U} = \mathbf{\Lambda}_b$ are both real diagonal matrices. Select $\mathbf{L} = \mathbf{V}\mathbf{Q}$ and $\mathbf{R} = \mathbf{U}\mathbf{Q}$, where $\mathbf{Q} \in \mathbb{R}^{N\times N}$ is non-zero. Clearly, $\mathbf{L}^T\mathbf{A}\mathbf{R} = \mathbf{Q}^T\mathbf{\Lambda}_a\mathbf{Q}$ and $\mathbf{L}^T\mathbf{B}\mathbf{R} = \mathbf{Q}^T\mathbf{\Lambda}_b\mathbf{Q}$ are both real symmetric matrices. Using this transformation, the second term of equation [A1.19] is now $\mathbf{L}^T\mathbf{C}\mathbf{R} = \mathbf{Q}^T\left(\mathbf{V}^T\mathbf{C}\mathbf{U}\right)\mathbf{Q}$. Thus, the transformed system is symmetric if and only if $\mathbf{V}^T\mathbf{C}\mathbf{U}$ is a symmetric matrix.

Since equivalence transformations are more general than similarity transformations, it is expected that the condition of symmetrizability given by this theorem is less restrictive than the three previously established results discussed before. The following example illustrates this fact.

EXAMPLE A1.6.– Suppose for a damped dynamic system $\mathbf{A}$ and $\mathbf{B}$ are as in example A1.3 and

$$\mathbf{C} = \begin{bmatrix} 4.1429 & 9.4510 & 9.0014 \\ 4.6465 & 3.8206 & -3.2816 \\ 1.2006 & 0.5460 & -1.5001 \end{bmatrix}. \qquad \text{[A1.21]}$$

Transforming the system into the form of equation [A1.20], we may verify that both $\mathbf{E}$ and $\mathbf{F}$ are individually symmetrizable of first kind. However, none of the conditions of simultaneous symmetrizability outlined in theorems A1.4–A1.6 are satisfied by $\mathbf{E}$ and $\mathbf{F}$. Now, using the modal matrices $\mathbf{U}$ and $\mathbf{V}$ calculated before in equation [A1.11], we have

$$\mathbf{V}^T\mathbf{C}\mathbf{U} = \begin{bmatrix} 4.2300 & 1.7900 & -0.8400 \\ 1.7900 & 2.6500 & 1.8900 \\ -0.8400 & 1.8900 & 3.2100 \end{bmatrix}, \qquad \text{[A1.22]}$$

which is a symmetric matrix, i.e., the condition of theorem A1.7 is satisfied. The transformation matrices **L** and **R** given by equation [A1.12], which symmetrizes **A** and **B** also symmetrizes **C** as

$$\mathbf{L}^T\mathbf{C}\mathbf{R} = \begin{bmatrix} 1.4439 & -0.4735 & -1.8602 \\ -0.4735 & 0.5412 & 0.8591 \\ -1.8602 & 0.8591 & 2.6222 \end{bmatrix}, \qquad [A1.23]$$

are symmetric matrix. Thus, the system under consideration is real symmetrizable of the second kind but not of the first kind.

A1.5.2. *Case 2: A, B and C are general real matrices*

When the system matrices are general, some or all of them may have complex roots so that they do not satisfy Tausskky's condition of symmetrizability. For such systems, in general, Kliem's result (theorem A1.6) gives the condition of *complex symmetrizability* of the first kind and theorem A1.7 gives the condition of complex symmetrizability of the second kind. A *sufficient* condition for real symmetrizability of such systems is the following:

LEMMA A1.2.– System [A1.19] is real symmetrizable of the second kind if there exist **U** and **V**, the matrices of undamped right and left eigenvectors, such that $\mathbf{A}^T\mathbf{B}$, $\mathbf{B}\mathbf{A}^T$ and $\mathbf{V}^T\mathbf{C}\mathbf{U}$ are all symmetric matrices.

This lemma can be proved by easily following the results of lemma A1.1 and theorem A1.7. The example considered below illustrates this result.

EXAMPLE A1.7.– Suppose for a damped linear system **A** and **B** are given by example A1.4 and

$$\mathbf{C} = \begin{bmatrix} 4.3160 & -2.5771 & -1.4626 \\ 2.7122 & 1.8365 & -0.1999 \\ 1.3827 & -2.5631 & 4.3419 \end{bmatrix}. \qquad [A1.24]$$

We may easily verify that none of the system matrices are individually real symmetrizable of the first kind. We compute the undamped modal matrices

$$\mathbf{U} = \begin{bmatrix} 0.6498 & 0.7383 & 0.1807 \\ 0.4364 & -0.1676 & -0.8840 \\ 0.6224 & -0.6533 & 0.4311 \end{bmatrix}$$

$$\text{and} \qquad \mathbf{V} = \begin{bmatrix} 0.0445 & -0.0108 & 0.9990 \\ 0.9245 & -0.3784 & -0.0453 \\ 0.3785 & 0.9256 & -0.0068 \end{bmatrix} \qquad [A1.25]$$

and note that

$$\mathbf{V}^T\mathbf{CU} = \begin{bmatrix} 3.2290 & 1.3664 & 0.6412 \\ 1.3664 & -2.0229 & 4.4962 \\ 0.6412 & 4.4962 & 2.4504 \end{bmatrix} \qquad [A1.26]$$

is a symmetric matrix. So all the conditions outlined in lemma A1.2 are satisfied. Using the matrix **Q** given by [A1.6], we have

$$\mathbf{L} = \mathbf{VQ} = \begin{bmatrix} 0.1735 & 0.2198 & 0.3701 \\ 0.3901 & -0.2469 & -0.1170 \\ -0.8026 & 0.1421 & 0.6215 \end{bmatrix}$$

$$\text{and} \qquad \mathbf{R} = \mathbf{UQ} = \begin{bmatrix} -0.5878 & 0.0988 & 0.6148 \\ 0.0353 & -0.3123 & -0.3683 \\ 0.6926 & -0.1489 & -0.1529 \end{bmatrix}. \qquad [A1.27]$$

These real matrices transform the system into a real symmetric form.

A1.6. Summary

A method of symmetrization of asymmetric linear dynamical systems by means of equivalence transformation is introduced. This method of symmetrization is called "symmetrization of the second kind" and the existing symmetrization method based on similarity transformation is called "symmetrization of the first kind". Because the equivalence transformations are most generally linear transformations, symmetrization of the second kind holds that of the first kind as a special case and offers a generalized method to obtain symmetric forms of finite-dimensional asymmetric linear systems. Using this symmetrization, the methods proposed in the book can be used for viscously damped asymmetric systems.

Several results are provided on symmetrization of the second kind. We have discussed real and complex symmetrizability based on the nature of the symmetrizing matrices. It was proved that every rectangular complex matrix is complex symmetrizable of the second kind to a real square matrix. It is possible to perform the symmetrization in such a way that the resulting symmetric form becomes positive definite. As a special case, it was shown that all real matrices are real symmetrizable of the second kind. These results are more general than symmetrizability of the first kind which cannot be applied to rectangular matrices. An easy numerical method is presented to calculate symmetric forms of general matrices. This method is much easier than Taussky's factorization for calculating the symmetric form of the first kind.

In view of the undamped and damped dynamical systems, we have discussed simultaneous symmetrizability of two and three real square matrices. It was shown that (truly) undamped asymmetric systems are always real symmetrizable. A condition of real symmetrizability of general undamped systems is derived. We also have given a condition for real symmetrizability of the second kind for damped systems. Our result is more liberal than the existing results and also much easier to check from a numerical point of view.

Bibliography

[ABR 65] ABRAMOWITZ M., STEGUN I.A., *Handbook of Mathematical Functions, with Formulas, Graphs, and Mathematical Tables*, Dover Publications, New York, 1965.

[ABU 09] ABUAZOUM L.A., GARVEY S.D., "Eigenvalue and eigenvector derivatives of second-order systems using structure-preserving equivalences", *Journal of Sound and Vibration*, vol. 323, no. 3–5, pp. 1061–1076, 2009.

[ABU 12] ABU-MALLOUH R., ABU-ALSHAIKH I., ZIBDEH H.S., *et al.*, "Response of fractionally damped beams with general boundary conditions subjected to moving loads", *Shock and Vibration*, vol. 19, no. 3, pp. 333–347, 2012.

[ADE 86] ADELMAN H.M., HAFTKA R.T., "Sensitivity analysis of discrete structural system", *AIAA Journal*, vol. 24, no. 5, pp. 823–832, May 1986.

[ADH 98] ADHIKARI S., Energy dissipation in vibrating structures, Master's Thesis, Cambridge University Engineering Department, First Year Report, Cambridge, UK, May 1998.

[ADH 99a] ADHIKARI S., "Modal analysis of linear asymmetric non-conservative systems", *ASCE Journal of Engineering Mechanics*, vol. 125, no. 12, pp. 1372–1379, December 1999.

[ADH 99b] ADHIKARI S., "Rates of change of eigenvalues and eigenvectors in damped dynamic systems", *AIAA Journal*, vol. 37, no. 11, pp. 1452–1458, November 1999.

[ADH 99c] ADHIKARI S., MANOHAR C.S., "Dynamic analysis of framed structures with statistical uncertainties", *International Journal for Numerical Methods in Engineering*, vol. 44, no. 8, pp. 1157–1178, 1999.

[ADH 00a] ADHIKARI S., "Calculation of derivative of complex modes using classical normal modes", *Computer and Structures*, vol. 77, no. 6, pp. 625–633, August 2000.

[ADH 00b] ADHIKARI S., Damping models for structural vibration, PhD Thesis, Cambridge University Engineering Department, Cambridge, UK, September 2000.

[ADH 00c] ADHIKARI S., "On symmetrizable systems of second kind", *Transactions of ASME, Journal of Applied Mechanics*, vol. 67, no. 4, pp. 797–802, December 2000.

[ADH 01a] ADHIKARI S., "Classical normal modes in non-viscously damped linear systems", *AIAA Journal*, vol. 39, no. 5, pp. 978–980, May 2001.

[ADH 01b] ADHIKARI S., "Eigenrelations for non-viscously damped systems", *AIAA Journal*, vol. 39, no. 8, pp. 1624–1630, August 2001.

[ADH 01c] ADHIKARI S., WOODHOUSE J., "Identification of damping: part 1, viscous damping", *Journal of Sound and Vibration*, vol. 243, no. 1, pp. 43–61, May 2001.

[ADH 01d] ADHIKARI S., WOODHOUSE J., "Identification of damping: part 2, non-viscous damping", *Journal of Sound and Vibration*, vol. 243, no. 1, pp. 63–88, May 2001.

[ADH 02a] ADHIKARI S., "Derivative of eigensolutions of non-viscously damped linear systems", *AIAA Journal*, vol. 40, no. 10, pp. 2061–2069, October 2002.

[ADH 02b] ADHIKARI S., "Dynamics of non-viscously damped linear systems", *ASCE Journal of Engineering Mechanics*, vol. 128, no. 3, pp. 328–339, March 2002.

[ADH 02c] ADHIKARI S., "Lancaster's method of damping identification revisited", *Transactions of ASME, Journal of Vibration and Acoustics*, vol. 124, no. 4, pp. 617–627, October 2002.

[ADH 02d] ADHIKARI S., WOODHOUSE J., "Identification of damping: part 3, symmetry-preserving method", *Journal of Sound and Vibration*, vol. 251, no. 3, pp. 477–490, March 2002.

[ADH 02e] ADHIKARI S., WOODHOUSE J., "Identification of damping: part 4, error analysis", *Journal of Sound and Vibration*, vol. 251, no. 3, pp. 491–504, March 2002.

[ADH 03a] ADHIKARI S., WAGNER N., "Analysis of asymmetric non-viscously damped linear dynamic systems", *Transactions of ASME, Journal of Applied Mechanics*, vol. 70, no. 6, pp. 885–893, December 2003.

[ADH 03b] ADHIKARI S., WOODHOUSE J., "Quantification of non-viscous damping in discrete linear systems", *Journal of Sound and Vibration*, vol. 260, no. 3, pp. 499–518, February 2003.

[ADH 04a] ADHIKARI S., "Optimal complex modes and an index of damping non-proportionality", *Mechanical System and Signal Processing*, vol. 18, no. 1, pp. 1–27, January 2004.

[ADH 04b] ADHIKARI S., WAGNER N., "Direct time-domain approach for exponentially damped systems", *Computer and Structures*, vol. 82, no. 29–30, pp. 2453–2461, November 2004.

[ADH 05] ADHIKARI S., "Qualitative dynamic characteristics of a non-viscously damped oscillator", *Proceedings of the Royal Society of London, Series- A*, vol. 461, no. 2059, pp. 2269–2288, July 2005.

[ADH 06a] ADHIKARI S., "Damping modelling using generalized proportional damping", *Journal of Sound and Vibration*, vol. 293, no. 1–2, pp. 156–170, May 2006.

[ADH 06b] ADHIKARI S., FRISWELL M.I., "The calculation of eigensolution derivatives for non-viscously damped systems using Nelson's method", *AIAA Journal*, vol. 44, no. 8, pp. 1799–1806, August 2006.

[ADH 07a] ADHIKARI S., "On the quantification of damping model uncertainty", *Journal of Sound and Vibration*, vol. 305, no. 1–2, pp. 153–171, September 2007.

[ADH 07b] ADHIKARI S., LEI Y., FRISWELL M.I., "Modal analysis of non-viscously damped beams", *Transactions of ASME, Journal of Applied Mechanics*, vol. 74, no. 5, pp. 1026–1030, September 2007.

[ADH 08] ADHIKARI S., "Dynamic response characteristics of a non-viscously damped oscillator", *Transactions of ASME, Journal of Applied Mechanics*, vol. 75, no. 1, pp. 011003:1–011003:12, January 2008.

[ADH 09a] ADHIKARI S., PASCUAL B., "Eigenvalues of linear viscoelastic systems", *Journal of Sound and Vibration*, vol. 325, no. 4–5, pp. 1000–1011, September 2009.

[ADH 09b] ADHIKARI S., PHANI A.S., "Experimental identification of generalized proportional damping", *Transactions of ASME, Journal of Vibration and Acoustics*, vol. 131, no. 1, pp. 011008:1–011008:12, January 2009.

[ADH 10] ADHIKARI S., "A general derivation of dynamic response of viscoelastic structures", *International Journal of Aerospace Innovations*, vol. 2, no. 1–2, pp. 29–41, 2010.

[ADH 11a] ADHIKARI S., "An iterative approach for nonproportionally damped systems", *Mechanics Research Communications*, vol. 38, no. 3, pp. 226–230, 2011.

[ADH 11b] ADHIKARI S., PASCUAL B., "Iterative methods for eigenvalues of viscoelastic systems", *Transactions of ASME, Journal of Vibration and Acoustics*, vol. 133, no. 2, pp. 021002:1–021002:7, April 2011.

[ADH 14b] ADHIKARI S., *Structural Dynamic Analysis with Generalized Damping Models: Identification*, ISTE, London, John Wiley & Sons, New York, 2014.

[AHM 84a] AHMADIAN M., INMAN D.J., "Classical normal modes in asymmetric nonconservative dynamic systems", *AIAA Journal*, vol. 22, no. 7, pp. 1012–1015, July 1984.

[AHM 84b] AHMADIAN M., INMAN D.J., "On the nature of eigenvalues of general nonconservative systems", *Transactions of ASME, Journal of Applied Mechanics*, vol. 51, pp. 193–194, March 1984.

[AHM 87] AHMADIAN M., CHOU S.-H., "A new method for finding symmetric form of asymmetric finite-dimensional dynamic systems", *Transactions of ASME, Journal of Applied Mechanics*, vol. 54, pp. 700–705, September 1987.

[ALI 13] ALI S.F., ADHIKARI S., "Energy harvesting dynamic vibration absorbers", *Transactions of ASME, Journal of Applied Mechanics*, vol. 80, no. 4, pp. 041004:1–041004:9, 2013.

[ALL 82] ALLEMANG R.J., BROWN D.L., "A correlation coefficient for modal vector analysis", *Proceedings of the 1st International Modal Analysis Conference (IMAC)*, Orlando, FL, pp. 110–116, 1982.

[ALV 97] ALVIN K.F., PETERSON L.D., PARK K.C., "Extraction of normal modes and full modal damping from complex modal parameters", *AIAA Journal*, vol. 35, no. 7, pp. 1187–1194, July 1997.

[AND 12] ANDRIANNE T., DIMITRIADIS G., "Damping identification of lightly damped linear dynamic systems using common-base proper orthogonal decomposition", *Mechanical Systems and Signal Processing*, vol. 28, no. Special Issue, pp. 492–506, 2012.

[ARO 09a] ARORA V., SINGH S.P., KUNDRA T.K., "Comparative study of damped FE model updating methods", *Mechanical Systems and Signal Processing*, vol. 23, no. 7, pp. 2113–2129, 2009.

[ARO 09b] ARORA V., SINGH S.P., KUNDRA T.K., "Damped FE model updating using complex updating parameters: its use for dynamic design", *Journal of Sound and Vibration*, vol. 324, no. 1–2, pp. 350–364, 2009.

[ARO 09c] ARORA V., SINGH S.P., KUNDRA T.K., "Damped model updating using complex updating parameters", *Journal of Sound and Vibration*, vol. 320, no. 1–2, pp. 438–451, 2009.

[ARO 09d] ARORA V., SINGH S.P., KUNDRA T.K., "Finite element model updating with damping identification", *Journal of Sound and Vibration*, vol. 324, no. 3–5, pp. 1111–1123, 2009.

[ARO 10] ARORA V., SINGH S.P., KUNDRA T.K., "Further experience with model updating incorporating damping matrices", *Mechanical Systems and Signal Processing*, vol. 24, no. 5, Special Issue, pp. 1383–1390, 2010.

[BAB 94] BABURAJ V., MATSUKAI Y., "A study on the material damping of thin angle-ply laminated plates", *Journal of Sound and Vibration*, vol. 172, no. 3, pp. 415–419, 1994.

[BAG 83] BAGLEY R.L., TORVIK P.J., "Fractional calculus – a different approach to the analysis of viscoelastically damped structures", *AIAA Journal*, vol. 21, no. 5, pp. 741–748, May 1983.

[BAI 02] BAI Z., "Krylov subspace techniques for reduced-order modeling of large-scale dynamical systems", *Applied Numerical Mathematics*, vol. 43, no. 1–2, pp. 9–44, 2002.

[BAI 05] BAI Z., SU Y., "Soar: a second-order Arnoldi method for the solution of the quadratic eigenvalue problem", *SIAM Journal on Matrix Analysis and Applications*, vol. 26, no. 3, pp. 640–659, 2005.

[BAL 95] BALMÈS E., *Structural Dynamics Toolbox, User's Manual 2.0*, Sèrves, Scientific Software Group, France, 1995.

[BAL 96] BALMÈS E., "Frequency domain identification of structural dynamics using the pole/residue parametrization", *Proceedings of the 14th International Modal Analysis Conference (IMAC)*, vol. 1, Dearbon, MI, pp. 540–546, 1996.

[BAL 97] BALMÈS E., "New results on the identification of normal modes from experimental complex modes", *Mechanical Systems and Signal Processing*, vol. 11, no. 2, pp. 229–243, 1997.

[BAN 83] BANDSTRA J.P., "Comparison of equivalent viscous damping in discrete and continuous vibrating system", *Transactions of ASME, Journal of Vibration, Acoustics, Stress and Reliability in Design*, vol. 105, pp. 382–392, July 1983.

[BAN 91] BANKS H.T., INMAN D.J., "On damping mechanisms in beams", *Transactions of ASME, Journal of Applied Mechanics*, vol. 58, pp. 716–723, September 1991.

[BAR 97] BARUCH M., Identification of the damping matrix, Report No. TAE No.803, Technion, Israel, Faculty of Aerospace Engineering, Israel Institute of Technology, Haifa, May 1997.

[BAT 76] BATHE K.J., WILSON E.L., *Numerical Methods in Finite Element Analysis*, Prentice Hall, Englewood Cliffs, NJ, 1976.

[BAT 95] BATHE K.-J., *Finite Element Procedures*, Prentice Hall, Englewood Cliffs, NJ, 1995.

[BEA 77] BEARDS C.F., WILLIAMS J.L., "The damping of structural vibration by rotational slip in structural joint", *Journal of Sound and Vibration*, vol. 53, no. 3, pp. 333–340, 1977.

[BEA 96] BEARDS C., *Structural Vibration: Analysis and Damping*, ButterHeinem ST, Oxford, UK, 1996.

[BEL 60] BELLMAN R., *Introduction to Matrix Analysis*, McGraw-Hill, New York, 1960.

[BÉL 76] BÉLIVEAU J.-G., "Identification of viscous damping in structures from modal information", *Transactions of ASME, Journal of Applied Mechanics*, vol. 43, pp. 335–339, June 1976.

[BÉL 77] BÉLIVEAU J.-G., "Eigenrelations in structural dynamics", *AIAA Journal*, vol. 15, no. 7, pp. 1039–1041, July 1977.

[BEL 90] BELLOS J., INMAN D.J., "Frequency response of non-proportionally damped, lumped parameter, linear systems", *Transactions of ASME, Journal of Vibration and Acoustics*, vol. 112, pp. 194–201, April 1990.

[BER 73] BERT C.W., "Material damping: an introductory review of mathematical models, measure and experimental techniques", *Journal of Sound and Vibration*, vol. 29, no. 2, pp. 129–153, 1973.

[BHA 95] BHASKAR A., "Estimates of errors in the frequency response of non-classically damped systems", *Journal of Sound and Vibration*, vol. 184, no. 1, pp. 59–72, 1995.

[BHA 99] BHASKAR A., "Mode shapes during asynchronous motion and non-proportionality indices", *Journal of Sound and Vibration*, vol. 224, no. 1, pp. 1–16, 1999.

[BIL 06] BILBAO A., AVILES R., AGIRREBEITIA J., *et al.*, "Proportional damping approximation for structures with added viscoelastic dampers", *Finite Elements in Analysis and Design*, vol. 42, no. 6, pp. 492–502, 2006.

[BIO 55] BIOT M.A., "Variational principles in irreversible thermodynamics with application to viscoelasticity", *Physical Review*, vol. 97, no. 6, pp. 1463–1469, 1955.

[BIO 58] BIOT M.A., "Linear thermodynamics and the mechanics of solids", *Proceedings of the Third U. S. National Congress on Applied Mechanics*, New York, ASME, pp. 1–18, 1958.

[BIS 79] BISHOP R.E.D., PRICE W.G., "An investigation into the linear theory of ship response to waves", *Journal of Sound and Vibration*, vol. 62, no. 3, pp. 353–363, 1979.

[BIS 86] BISHOP R.E.D., PRICE W.G., "A note on hysteretic damping of transient motion", in ELISHAKOFF I., LYON R.H. (eds), *Random Vibration Status and Recent Developments*, Elsevier, Amsterdam, pp. 39–45, 1986.

[BLA 60] BLAND D.R., *Theory of Linear Viscoelasticity*, Pergamon Press, London, 1960.

[BOI 09] DU BOIS J.L., ADHIKARI S., LIEVEN N.A.J., "Mode veering in stressed framed structures", *Journal of Sound and Vibration*, vol. 322, no. 4–5, pp. 1117–1124, May 2009.

[BRA 84] BRANDON J.A., "Derivation and significance of second-order modal design sensitivities", *AIAA Journal*, vol. 22, no. 5, pp. 723–724, May 1984.

[BRA 91] BRANDON J.A., "Second-order design sensitivities to assess the applicability of sensitivity analysis", *AIAA Journal*, vol. 29, no. 1, pp. 135–139, January 1991.

[BUH 82] BUHARIWALA K.J., Dynamic response of viscoelastic fibre reinforced composite materials, Master's Thesis, Institute for Aerospace Studies, University of Toronto, 1982.

[BUR 09] BURCHETT B.T., "QZ-based algorithm for system pole, transmission zero, and residue derivatives", *AIAA Journal*, vol. 47, no. 8, pp. 1889–1901, August 2009.

[CAL 09] CALIM F.F., "Dynamic analysis of beams on viscoelastic foundation", *European Journal of Mechanics A-Solids*, vol. 28, no. 3, pp. 469–476, 2009.

[CAL 11] CALIM F.F., AKKURT F.G., "Static and free vibration analysis of straight and circular beams on elastic foundation", *Mechanics Research Communications*, vol. 38, no. 2, pp. 89–94, 2011.

[CAP 01] CAPECCHI D., DE FELICE G., "Hysteretic systems with internal variables", *ASCE Journal of Engineering Mechanics*, vol. 127, no. 9, pp. 891–898, September 2001.

[CAR 74] CARAVANI P., THOMSON W.T., "Identification of damping coefficients in multidimensional linear systems", *Transactions of ASME, Journal of Applied Mechanics*, vol. 41, pp. 379–382, June 1974.

[CAR 11] CARPINTERI A., PUGNO N., SAPORA A., "Dynamic response of damped von Koch antennas", *Journal of Sound and Vibration*, vol. 17, no. 5, pp. 733–740, 2011.

[CAU 60] CAUGHEY T.K., "Classical normal modes in damped linear dynamic systems", *Transactions of ASME, Journal of Applied Mechanics*, vol. 27, pp. 269–271, June 1960.

[CAU 65] CAUGHEY T.K., O'KELLY M.E.J., "Classical normal modes in damped linear dynamic systems", *Transactions of ASME, Journal of Applied Mechanics*, vol. 32, pp. 583–588, September 1965.

[CAU 93] CAUGHEY T.K., MA F., "Complex modes and solvability of non-classical linear systems", *Transactions of ASME, Journal of Applied Mechanics*, vol. 60, pp. 26–28, March 1993.

[CAV 09] CAVACECE M., VALENTINI P.P., VITA L., "Identification of modal damping ratios of four-flue chimney of a thermoelectrical plant using pseudo-inverse matrix method", *Structural Design of Tall and Special Buildings*, vol. 18, no. 2, pp. 203–216, 2009.

[CHA 78] CHAKRABARTI S., BHATTACHARYYA B.B., SWAMY M.N.S., "On simultaneous diagonalization of a collection of Hermitian matrices", *The Matrix and Tensor Quarterly*, vol. 29, no. 2, pp. 35–54, December 1978.

[CHA 98] CHASSIAKOS A.G., MASRI S.F., SMYTH A.W., *et al.*, "On-line identification of hysteretic systems", *Transactions of ASME, Journal of Applied Mechanics*, vol. 65, pp. 194–203, March 1998.

[CHA 10] CHAKRABORTY S., ROY S., "Element level identification of a viscously damped system", *International Journal of Acoustics and Vibration*, vol. 15, no. 3, pp. 113–120, 2010.

[CHA 11] CHA P.D., SABATER A.B., "Eigenvalue sensitivities of a linear structure carrying lumped attachments", *AIAA Journal*, vol. 49, no. 11, pp. 2470–2481, November 2011.

[CHA 13] CHANG S.-Y., "Nonlinear performance of classical damping", *Earthquake Engineering and Engineering Vibration*, vol. 12, no. 2, pp. 279–296, 2013.

[CHE 93] CHERNG A., ABDELHAMID M.K., "On the symmetrization of asymmetric finite dimensional linear dynamic systems", *Transactions of ASME, Journal of Vibration and Acoustics*, vol. 115, pp. 417–421, October 1993.

[CHE 94a] CHEN S.H., QIU Z., LIU Z., "Perturbation method for computing eigenvalue bounds in structural vibration systems with interval parameters", *Communications in Numerical Methods in Engineering*, vol. 10, no. 2, pp. 121–134, 1994.

[CHE 94b] CHEN S.-H., SONG D.-T., MA A.-J., "Eigensolution reanalysis of modified structures using perturbations and Rayleigh quotients", *Communications in Numerical Methods in Engineering*, vol. 10, no. 2, pp. 111–119, 1994.

[CHE 96a] CHEN S.Y., JU M.S., TSUEI Y.G., "Estimation of mass stiffness and damping matrices from frequency response function", *Transactions of ASME, Journal of Vibration and Acoustics*, vol. 118, pp. 78–82, January 1996.

[CHE 96b] CHEN S.Y., JU M.S., TSUEI Y.G., "Extraction of normal modes for highly coupled incomplete systems with general damping", *Mechanical Systems and Signal Processing*, vol. 10, no. 1, pp. 93–106, 1996.

[CHE 06] CHEN S.H., WU X.M., YANG Z.J., "Eigensolution reanalysis of modified structures using epsilon-algorithm", *International Journal for Numerical Methods in Engineering*, vol. 66, no. 13, pp. 2115–2130, June 25 2006.

[CHE 08] CHEN K.-F., ZHANG S.-W., "On the impulse response precursor of an ideal linear hysteretic damper", *Journal of Sound and Vibration*, vol. 312, no. 4–5, pp. 576–583, 2008.

[CHE 11] CHEN W.-R., "Bending vibration of axially loaded Timoshenko beams with locally distributed Kelvin-Voigt damping", *Journal of Sound and Vibration*, vol. 330, no. 13, pp. 3040–3056, 2011.

[CHI 09] CHIROIU V., MUNTEANU L., DONESCU S., "On the beams with external auxetic patches", *Advances in Mechanical Engineering*, 2009.

[CHO 04] CHOI K.M., JO H.K., KIM W.H., *et al.*, "Sensitivity analysis of non-conservative eigensystems", *Journal of Sound and Vibration*, vol. 274, no. 3–5, pp. 997–1011, 22 July 2004.

[CHO 07] CHOUCHANE M., GUEDRIA N., SMAOUI H., "Eigensensitivity computation of asymmetric damped systems using an algebraic approach", *Mechanical System and Signal Processing*, vol. 21, no. 7, pp. 2761–2776, October 2007.

[CHR 13] CHRONOPOULOS D., ICHCHOU M., TROCLET B., *et al.*, "Predicting the broadband vibroacoustic response of systems subject to aeroacoustic loads by a Krylov subspace reduction", *Applied Acoustics*, vol. 74, no. 12, pp. 1394–1405, 2013.

[CHU 86] CHUNG K.R., LEE C.W., "Dynamic reanalysis of weakly non-proportionally damped systems", *Journal of Sound and Vibration*, vol. 111, pp. 37–50, 1986.

[CLO 76] CLOUGH R.W., MOJTAHEDI S., "Earthquake response analysis considering non-proportional damping", *Earthquake Engineering and Structural Dynamics*, vol. 4, pp. 489–496, 1976.

[CLO 93] CLOUGH R.W., PENZIEN J., *Dynamics of Structures*, McGraw-Hill, New York, 1993.

[COO 01] COOK R.D., MALKUS D.S., PLESHA M.E. *et al.*, *Concepts and Applications of Finite Element Analysis*, 4th ed., John Wiley & Sons, New York, 2001.

[COR 06a] CORTES F., ELEJABARRIETA M.J., "An approximate numerical method for the complex eigenproblem in systems characterised by a structural damping matrix", *Journal of Sound and Vibration*, vol. 296, no. 1–2, pp. 166–182, 2006.

[COR 06b] CORTES F., ELEJABARRIETA M.J., "Longitudinal vibration of a damped rod. Part 1: complex natural frequencies and mode shapes", *International Journal of Mechanical Sciences*, vol. 48, no. 9, pp. 969–975, 2006.

[COR 06c] CORTES F., ELEJABARRIETA M.J., "Computational methods for complex eigenproblems in finite element analysis of structural systems with viscoelastic damping treatments", *Computer Methods in Applied Mechanics and Engineering*, vol. 195, no. 44–47, pp. 6448–6462, 2006.

[COR 07a] CORTES F., ELEJABARRIETA M.J., "Finite element formulations for transient dynamic analysis in structural systems with viscoelastic treatments containing fractional derivative models", *International Journal for Numerical Methods in Engineering*, vol. 69, no. 10, pp. 2173–2195, 2007.

[COR 07b] CORTES F., ELEJABARRIETA M.J., "Viscoelastic materials characterisation using the seismic response", *Materials & Design*, vol. 28, no. 7, pp. 2054–2062, 2007.

[COR 08] CORTES F., ELEJABARRIETA M.J., "Forced response of a viscoelastically damped rod using the superposition of modal contribution functions", *Journal of Sound and Vibration*, vol. 315, no. 1–2, pp. 58–64, 2008.

[COR 09] CORTES F., MATEOS M., ELEJABARRIETA M.J., "A direct integration formulation for exponentially damped structural systems", *Computers & Structures*, vol. 87, no. 5–6, pp. 391–394, 2009.

[COU 89] COURANT R., HILBERT D., *Methods of Mathematical Physics: Volume 1*, 1st ed., John Wiley & Sons, New York, 1989.

[CRA 70] CRANDALL S.H., "The role of damping in vibration theory", *Journal of Sound and Vibration*, vol. 11, no. 1, pp. 3–18, 1970.

[CRA 91] CRANDALL S.H., "The hysteretic damping model in vibration theory", *Journal of Mechanical Engineering Science*, vol. 205, pp. 23–28, 1991.

[CRE 73] CREMER L., HECKL M., *Structure-Borne Sound*, 2nd ed., E.E. UNGAR (trans.) Springer-Verlag Berlin, Germany, 1973.

[CRI 85] CRIGHTON D.G., "Acoustics of a stiff locally reacting structure", *Proceedings of the Royal Society of London, Series- A*, vol. 397, pp. 99–120, 1985.

[CRO 76] CRONIN D.L., "Approximation for determining harmonically excited response of non-classically damped system", *ASME Journal of Engineering for Industry*, vol. 98, no. 1, pp. 43–47, 1976.

[CRO 90] CRONIN D.L., "Eigenvalue and eigenvector determination of non-classically damped dynamic systems", *Computer and Structures*, vol. 36, no. 1, pp. 133–138, 1990.

[DAL 99] DALENBRING M., "Damping function estimation based on measured vibration frequency responses and finite-element displacement modes", *Mechanical Systems and Signal Processing*, vol. 13, no. 4, pp. 547–569, 1999.

[DAW 84] DAWE D., *Matrix and Finite Element Displacement Analysis of Structures*, Oxford University Press, Oxford, UK, 1984.

[DAY 01] DAYA E.M., POTIER-FERRY M., "A numerical method for nonlinear eigenvalue problems and application to vibrations of viscoelastic structures", *Computers & Structures*, vol. 79, pp. 533–541, 2001.

[DE 09] DE LIMA A., RADE D., NETO F.L., "An efficient modeling methodology of structural systems containing viscoelastic dampers based on frequency response function substructuring", *Mechanical Systems and Signal Processing*, vol. 23, no. 4, pp. 1272–1281, 2009.

[DE 10] DE LIMA A., DA SILVA A., RADE D., *et al.*, "Component mode synthesis combining robust enriched Ritz approach for viscoelastically damped structures", *Engineering Structures*, vol. 32, no. 5, pp. 1479–1488, 2010.

[DI 11] DI PAOLA M., FAILLA G., SOFI A., *et al.*, "A mechanically based approach to non-local beam theories", *International Journal of Mechanical Sciences*, vol. 53, no. 9, pp. 676–687, 2011.

[DI 13] DI PAOLA M., FAILLA G., ZINGALES M., "Non-local stiffness and damping models for shear-deformable beams", *European Journal of Mechanics A-Solids*, vol. 40, pp. 69–83, 2013.

[DIC 98] DICKSON L.E., "A new solution of the cubic equation", *American Mathematical Monthly*, vol. 5, pp. 38–39, 1898.

[DIN 11] DING Y., LAW S.S., "Structural damping identification based on an iterative regularization method", *Journal of Sound and Vibration*, vol. 330, no. 10, pp. 2281–2298, 2011.

[DU 11] DU BOIS J.L., ADHIKARI S., LIEVEN N.A.J., "On the quantification of eigenvalue curve veering: a veering index", *Transactions of ASME, Journal of Applied Mechanics*, vol. 78, no. 4, pp. 041007:1–8, 2011.

[DUN 79] DUNCAN P.E., TAYLOR R.E., "A note on the dynamic analysis of non-proportionally damped systems", *Earthquake Engineering and Structural Dynamics*, vol. 7, pp. 99–105, 1979.

[EAR 66] EARLS S.W.E., "Theoretical estimation of frictional energy dissipation in a simple lap joint", *Journal of Mechanical Engineering Science*, vol. 8, no. 2, pp. 207–214, 1966.

[ECK 39] ECKART C., YOUNG G., "A principal axis transformation for non-Hermitian matrices", *Bulletin of the American Mathematical Society*, vol. 45, pp. 118–121, February 1939.

[ELB 09] ELBEHEIRY E., "On eigenproblem solution of damped vibrations associated with gyroscopic moments", *Journal of Sound and Vibration*, vol. 320, no. 3, pp. 691–704, 2009.

[ELV 13] ELVIN N., ERTURK A., *Advances in Energy Harvesting Methods*, Springer, New York, 2013.

[ERL 07] ERLICHER S., ARGOUL P., "Modal identification of linear non-proportionally damped systems by wavelet transform", *Mechanical Systems and Signal Processing*, vol. 21, no. 3, pp. 1386–1421, 2007.

[ERT 11] ERTURK A., INMAN D.J., *Piezoelectric Energy Harvesting: Modelling and Application*, Wiley-Blackwell, Sussex, UK, 2011.

[EWI 84] EWINS D.J., *Modal Testing: Theory and Practice*, Research Studies Press, Taunton, 1984.

[EWI 00] EWINS D.J., *Modal Testing: Theory and Practice*, 2nd ed. Research Studies Press, Baldock, England, 2000.

[FAB 88] FABUNMI J., CHANG P., VORWALD J., "Damping matrix identification using the spectral basis technique", *Transactions of ASME, Journal of Vibration, Acoustics, Stress and Reliability in Design*, vol. 110, pp. 332–337, July 1988.

[FAI 13] FAILLA G., SANTINI A., ZINGALES M., "A non-local two-dimensional foundation model", *Archive of Applied Mechanics*, vol. 83, no. 2, pp. 253–272, 2013.

[FAN 94] FANG J., LYONS G.J., "Material damping of free hanging pipes: theoretical and experimental studies", *Journal of Sound and Vibration*, vol. 172, no. 3, pp. 371–389, 1994.

[FAW 76] FAWZY I., BISHOP R.E.D., "On the dynamics of linear non-conservative systems", *Proceedings of the Royal Society of London, Series-A*, vol. 352, pp. 25–40, 1976.

[FAW 77a] FAWZY I., "Orthogonality of generally normalized eigenvectors and eigenrows", *AIAA Journal*, vol. 15, no. 2, pp. 276–278, February 1977.

[FAW 77b] FAWZY I., BISHOP R.E.D., "On the nature of resonance in non-conservative systems", *Journal of Sound and Vibration*, vol. 55, no. 4, pp. 475–485, 1977.

[FEL 93] FELSZEGHY S.F., "On uncoupling and solving the equations of motion of vibrating linear discrete systems", *Transactions of ASME, Journal of Applied Mechanics*, vol. 60, pp. 456–462, June 1993.

[FEL 94] FELSZEGHY S.F., "Development of biorthonormal eigenvectors for modal analysis of linear discrete non-classically damped systems", *Journal of Sound and Vibration*, vol. 176, no. 2, pp. 225–269, 1994.

[FOS 58] FOSS K.A., "Co-ordinates which uncouple the equations of motion of damped linear dynamic systems", *Transactions of ASME, Journal of Applied Mechanics*, vol. 25, pp. 361–364, September 1958.

[FOX 68] FOX R.L., KAPOOR M.P., "Rates of change of eigenvalues and eigenvectors", *AIAA Journal*, vol. 6, no. 12, pp. 2426–2429, December 1968.

[FRI 86] FRITZEN C.-P., "Identification of mass, damping and stiffness matrices of mechanical systems", *Transactions of ASME, Journal of Vibration, Acoustics, Stress and Reliability in Design*, vol. 108, pp. 9–16, January 1986.

[FRI 95a] FRISWELL M.I., "Calculation of second and higher order eigenvector derivatives", *Journal of Guidance, Control and Dynamics*, vol. 18, no. 4, pp. 919–921, 1995.

[FRI 95b] FRISWELL M.I., MOTTERSHEAD J.E., *Finite Element Model Updating in Structural Dynamics*, Kluwer Academic Publishers, Netherlands, 1995.

[FRI 97] FRISWELL M.I., INMAN D.J., LAM M.J., "On the realisation of GHM models in viscoelasticity", *Journal of Intelligent Material Systems and Structures*, vol. 8, no. 11, pp. 986–993, November 1997.

[FRI 99] FRISWELL M.I., INMAN D.J., "Reduced-order models of structures with viscoelastic components", *AIAA Journal*, vol. 37, no. 10, pp. 1318–1325, October 1999.

[FRI 00] FRISWELL M.I., ADHIKARI S., "Derivatives of complex eigenvectors using Nelson's method", *AIAA Journal*, vol. 38, no. 12, pp. 2355–2357, December 2000.

[FRI 01] FRISWELL M.I., MOTTERSHEAD J.E., AHMADIAN H., "Finite element model updating using experimental test data: parametrization and regularization", *Philosophical Transactions of the Royal Society of London Series A - Mathematical Physical and Engineering Sciences*, vol. 359, no. 1778, pp. 169–186, 2001.

[FRI 07a] FRISWELL M.I., ADHIKARI S., LEI Y., "Nonlocal finite element analysis of damped beams", *International Journal of Solids and Structures*, vol. 44, no. 22–23, pp. 7564–7576, November 2007.

[FRI 07b] FRISWELL M.I., ADHIKARI S., LEI Y., "Vibration analysis of beams with non-local foundations using the finite element method", *International Journal for Numerical Methods in Engineering*, vol. 71, no. 11, pp. 1365–1386, September 2007.

[FRI 10a] FRISWELL M.I., DUTT J.K., ADHIKARI S., *et al.*, "Time domain analysis of a viscoelastic rotor using internal variable models", *International Journal of Mechanical Sciences*, vol. 52, no. 10, pp. 1319–1324, 2010.

[FRI 10b] FRISWELL M.I., PENNY J.E.T., GARVEY S.D., *et al.*, *Dynamics of Rotating Machines*, Cambridge University Press, Cambridge, UK, 2010.

[FUN 65] FUNG Y.C., *Foundations of Solid Mechanics*, Prentice Hall, Englewood Cliffs, NJ, 1965.

[GAL 03] GALLINA P., "Effect of damping on asymmetric systems", *Journal of Vibration and Acoustics – Transactions of the ASME*, vol. 125, no. 3, pp. 359–364, 2003.

[GAR 73] GARG S., "Derivatives of eigensolutions for a general matrix", *AIAA Journal*, vol. 11, no. 8, pp. 1191–1194, August 1973.

[GAR 95] GARVEY S.D., PENNY J.E.T., FRISWELL M.I., "The relationship between the real and imaginary parts of complex modes", *Journal of Sound and Vibration*, vol. 212, no. 1, pp. 75–83, 1995.

[GAR 02a] GARVEY S.D., FRISWELL M.I., PRELLS U., "Co-ordinate transformations for second order systems. Part I: general transformations", *Journal of Sound and Vibration*, vol. 258, no. 5, pp. 885–909, 2002.

[GAR 02b] GARVEY S.D., FRISWELL M.I., PRELLS U., "Co-ordinate transformations for, second order systems. Part II: elementary structure-preserving transformations", *Journal of Sound and Vibration*, vol. 258, no. 5, pp. 911–930, 2002.

[GAR 04] GARVEY S.D., PRELLS U., FRISWELL M.I., *et al.*, "General isospectral flows for linear dynamic systems", *Linear Algebra and Its Applications*, vol. 385, pp. 335–368, 2004.

[GAR 11] GARCIA-BARRUETABENA J., CORTES F., ABETE J.M., "Influence of nonviscous modes on transient response of lumped parameter systems with exponential damping", *Journal of Vibration and Acoustics – Transactions of the ASME*, vol. 133, no. 6, pp. 064502:1–8, 2011.

[GAR 12] GARCIA-BARRUETABENA J., CORTES F., ABETE J.M., "Dynamics of an exponentially damped solid rod: analytic solution and finite element formulations", *International Journal of Solids and Structures*, vol. 49, no. 3–4, pp. 590–598, 2012.

[GAR 13] GARCÍA-PALENCIA A., SANTINI-BELL E., "A two-step model updating algorithm for parameter identification of linear elastic damped structures", *Computer-Aided Civil and Infrastructure Engineering*, vol. 28, no. 7, pp. 509–521, 2013.

[GAU 91] GAUL L., KLEIN P., KEMPLE S., "Damping description involving fractional operators", *Mechanical Systems and Signal Processing*, vol. 5, no. 2, pp. 81–88, 1991.

[GAU 99] GAUL L., "The influence of damping on waves and vibrations", *Mechanical Systems and Signal Processing*, vol. 13, no. 1, pp. 1–30, 1999.

[GAW 97] GAWRONSKI W., SAWICKI J.T., "Response errors of non-proportionally lightly damped structures", *Journal of Sound and Vibration*, vol. 200, no. 4, pp. 543–550, 1997.

[GEN 07] GENG Z., POPOV A.A., COLE D.J., "Measurement, identification and modelling of damping in pneumatic tyres", *International Journal of Mechanical Sciences*, vol. 49, no. 10, pp. 1077–1094, 2007.

[GEN 10] GENTA G., AMATI N., "Hysteretic damping in rotordynamics: an equivalent formulation", *Journal of Sound and Vibration*, vol. 329, no. 22, pp. 4772–4784, 2010.

[GÉR 97] GÉRADIN M., RIXEN D., *Mechanical Vibrations*, 2nd ed. John Wiley & Sons, New York, 1997, [Translation of *Théorie des Vibrations*].

[GOL 85] GOLLA D.F., HUGHES P.C., "Dynamics of viscoelastic structures - a time domain finite element formulation", *Transactions of ASME, Journal of Applied Mechanics*, vol. 52, pp. 897–906, December 1985.

[GON 12] GONZALEZ-LOPEZ S., FERNANDEZ-SAEZ J., "Vibrations in Euler-Bernoulli beams treated with non-local damping patches", *Computers & Structures*, vol. 108, no. Special Issue, pp. 125–134, 2012.

[GUE 06] GUEDRIA N., SMAOUI H., CHOUCHANE M., "A direct algebraic method for eigensolution sensitivity computation of damped asymmetric systems", *International Journal for Numerical Methods in Engineering*, vol. 68, no. 6, pp. 674–689, 5 November 2006.

[GUE 07] GUEDRIA N., CHOUCHANE M., SMAOUI H., "Second-order eigensensitivity analysis of asymmetric damped systems using Nelson's method", *Journal of Sound and Vibration*, vol. 300, no. 3–5, pp. 974–992, 6 March 2007.

[HAL 99] HALEVI Y., KENIGSBUCH R., "Model updating of the complex modeshapes and the damping matrix", in FRISWELL M.I., MOTTERSHEAD J.E., LEES A.W. (eds), *Inverse Problems in Engineering*, Swansea, Wales, UK, pp. 251–260, 1999.

[HAR 98] HARVILLE D.A., *Matrix Algebra from a Statistician's Perspective*, Springer-Verlag, New York, 1998.

[HAS 72] HASSELSMAN T.K., "A method of constructing a full modal damping matrix from experimental measurements", *AIAA Journal*, vol. 10, no. 4, pp. 526–527, 1972.

[HAS 76] HASSELSMAN T.K., "Modal coupling in lightly damped structures", *AIAA Journal*, vol. 14, pp. 1627–1628, 1976.

[HOL 12a] HOLLAND D.E., RODGERS J.R., EPUREANU B.I., "Measurement point selection and modal damping identification for bladed disks", *Mechanical Systems and Signal Processing*, vol. 33, pp. 97–108, 2012.

[HOL 12b] HOLLAND D.E., EPUREANU B.I., FILIPPI S., "Structural damping identification for mistuned bladed disks and blisks", *Transactions of ASME, Journal of Vibration and Acoustics*, vol. 134, no. 2, pp. 024504:1–3, 2012.

[HOL 13] HOLLAND D.E., EPUREANU B.I., "A component damping identification method for mistuned blisks", *Mechanical Systems and Signal Processing*, 2013, in press.

[HUG 00] HUGHES T.J.R., *The Finite Element Method: Linear Static and Dynamic Finite Element Analysis*, Dover Publications, New York, 2000.

[HUI 07] HUI-QING X., HUA D., "Derivatives of repeated eigenvalues and corresponding eigenvectors of damped systems", *Applied Mathematics and Mechanics*, vol. 28, no. 6, pp. 837–845, June 2007.

[HUS 73] HUSEYIN K., LIEPHOLZ H.H.E., "Divergence instability of multiple parameter circulatory systems", *Quarterly Journal of Applied Mathematics*, vol. 31, pp. 185–197, July 1973.

[HUS 78] HUSEYIN K., *Vibration and Stability of Multiple Parameter Systems*, Sijthoff & Noordhoff, Netherlands, 1978.

[HWA 93] HWANG J.H., MA F., "On the approximate solution of non-classically damped linear systems", *Transactions of ASME, Journal of Applied Mechanics*, vol. 60, pp. 695–701, September 1993.

[IBR 83a] IBRAHIM S.R., "Computation of normal modes from identified complex modes", *AIAA Journal*, vol. 21, no. 3, pp. 446–451, March 1983.

[IBR 83b] IBRAHIM S.R., "Dynamic modeling of structures from measured complex modes", *AIAA Journal*, vol. 21, no. 6, pp. 898–901, June 1983.

[IBR 89] IBRAHIMBEGOVIC A., WILSON E.L., "Simple numerical algorithms for the mode superposition analysis of linear structural systems with non-proportional damping", *Computer and Structures*, vol. 33, no. 2, pp. 523–531, 1989.

[IBR 95] IBRAHIM S.R., SESTIERI A., "Existence and normalization of complex modes in post experimental use in modal analysis", *Proceedings of the 13th International Modal Analysis Conference (IMAC)*, vol. 1, Nashville, TN, pp. 483–489, 1995.

[IMR 95] IMREGUN M., EWINS D.J., "Complex modes - origin and limits", *Proceedings of the 13th International Modal Analysis Conference (IMAC)*, Nashville, TN, pp. 496–506, 1995.

[INM 83] INMAN D.J., "Dynamics of asymmetric non-conservative systems", *Transactions of ASME, Journal of Applied Mechanics*, vol. 50, pp. 199–203, March 1983.

[INM 89] INMAN D.J., "Vibration analysis of viscoelastic beams by separation of variables and modal analysis", *Mechanics Research Communications*, vol. 16, no. 4, pp. 213–218, 1989.

[INM 03] INMAN D.J., *Engineering Vibration*, Prentice Hall PTR, NJ, 2003.

[JON 01] JONES D.I.G., *Handbook of Viscoelastic Vibration Damping*, Wiley-Blackwell, Sussex, UK, 2001.

[KAW 11] KAWANO D.T., MORZFELD M., MA F., "The decoupling of defective linear dynamical systems in free motion", *Journal of Sound and Vibration*, vol. 330, no. 21, pp. 5165–5183, 2011.

[KHA 07] KHALIL M., ADHIKARI S., SARKAR A., "Linear system identification using proper orthogonal decomposition", *Mechanical System and Signal Processing*, vol. 21, no. 8, pp. 3123–3145, November 2007.

[KHA 09] KHALIL M., SARKAR A., ADHIKARI S., "Nonlinear filters for chaotic oscillatory systems", *Nonlinear Dynamics*, vol. 55, no. 1–2, pp. 113–137, January 2009.

[KHA 10] KHALIL M., SARKAR A., ADHIKARI S., "Tracking noisy limit cycle oscillation with nonlinear filters", *Journal of Sound and Vibration*, vol. 329, no. 2, pp. 150–170, January 2010.

[KLI 92] KLIEM W.R., "Symmetrizable systems in mechanics and control theory", *Transactions of ASME, Journal of Applied Mechanics*, vol. 69, pp. 454–456, June 1992.

[KOR 13] KORUK H., SANLITURK K.Y., "A novel definition for quantification of mode shape complexity", *Journal of Sound and Vibration*, vol. 332, no. 14, pp. 3390–3403, 2013.

[KRE 99] KREYSZIG E., *Advanced Engineering Mathematics*, 8th ed., John Wiley & Sons, New York, 1999.

[KRE 06] KREYSZIG E., *Advanced Engineering Mathematics*, 9th ed., John Wiley & Sons, New York, 2006.

[LAZ 12] LÁZARO M., PÉREZ-APARICIO J.L., EPSTEIN M., "Computation of eigenvalues in proportionally damped viscoelastic structures based on the fixed-point iteration", *Applied Mathematics and Computation*, vol. 219, no. 8, pp. 3511–3529, 2012.

[LAZ 13a] LÁZARO M., PEREZ-APARICIO J.L., "Multiparametric computation of eigenvalues for linear viscoelastic structures", *Computers & Structures*, vol. 117, pp. 67–81, 2013.

[LAZ 13b] LÁZARO M., PÉREZ-APARICIO J.L., EPSTEIN M., "A viscous approach based on oscillatory eigensolutions for viscoelastically damped vibrating systems", *Mechanical Systems and Signal Processing*, vol. 40, no. 2, pp. 767–782, 2013.

[LAL 95] LALLEMENT G., INMAN D.J., "A tutorial on complex eigenvalues", *Proceedings of the 13th International Modal Analysis Conference (IMAC)*, Nashville, TN, pp. 490–495, 1995.

[LAN 61] LANCASTER P., "Expression of damping matrices in linear vibration problems", *Journal of Aerospace Sciences*, vol. 28, p. 256, 1961.

[LAN 66] LANCASTER P., *Lambda-Matrices and Vibrating System*, Pergamon Press, London, 1966.

[LAZ 59] LAZAN B.J., "Energy dissipation mechanisms in structures with particular reference to material damping", in RUZCKA J.E. (ed.), *Structural Dynamics*, Atlantic City, NJ, ASME Annual Meeting, 1959.

[LAZ 68] LAZAN B.J., *Damping of Materials and Members in Structural Mechanics*, Pergamon Press, Oxford, 1968.

[LEE 99a] LEE I.-W., KIM D.-O., JUNG H.-H., "Natural frequency and mode shape sensitivities of damped systems: part I, distinct natural frequencies", *Journal of Sound and Vibration*, vol. 223, no. 3, pp. 399–412, 1999.

[LEE 99b] LEE I.-W., KIM D.-O., JUNG H.-H., "Natural frequency and mode shape sensitivities of damped systems: part II, multiple natural frequencies", *Journal of Sound and Vibration*, vol. 223, no. 3, pp. 413–424, 1999.

[LEI 74] LEISSA A.W., "On a curve veering aberration", *Journal of Applied Mathematics and Physics (ZAMP)*, vol. 25, pp. 99–111, 1974.

[LEI 06] LEI Y., FRISWELL M.I., ADHIKARI S., "A Galerkin method for distributed systems with non-local damping", *International Journal of Solids and Structures*, vol. 43, no. 11–12, pp. 3381–3400, 2006.

[LEI 13a] LEI Y., FRISWELL M.I., ADHIKARI S., "Vibration of nonlocal Kelvin-Voigt viscoelastic damped Timoshenko beams", *International Journal of Engineering Science*, vol. 66–67, no. 1, pp. 1–13, 2013.

[LEI 13b] LEI Y., MURMU T., ADHIKARI S. *et al.*, "Dynamic characteristics of damped viscoelastic nonlocal Euler-Bernoulli beams", *European Journal of Mechanics - A/Solids*, vol. 42, no. 12, pp. 125–136, 2013.

[LES 90] LESIEUTRE G.A., MINGORI D.L., "Finite element modeling of frequency-dependent material properties using augmented thermodynamic fields", *AIAA Journal of Guidance, Control and Dynamics*, vol. 13, pp. 1040–1050, 1990.

[LES 92] LESIEUTRE G.A., "Finite elements for dynamic modeling of uniaxial rods with frequency dependent material properties", *International Journal of Solids and Structures*, vol. 29, no. 12, pp. 1567–1579, 1992.

[LES 95] LESIEUTRE G.A., BIANCHINI E., "Time-domain modeling of linear viscoelasticity using anelastic displacement fields", *Transactions of ASME, Journal of Vibration and Acoustics*, vol. 117, no. 4, pp. 425–430, October 1995.

[LES 96a] LESIEUTRE G.A., GOVINDSWAMY K., "Finite element modeling of frequency-dependent and temperature-dependent dynamic behaviour of viscoelastic material in simple shear", *International Journal of Solids an Structures*, vol. 33, pp. 419–432, 1996.

[LES 96b] LESIEUTRE G.A., LEE U., "A finite element for beams having segmented active constrained layers with frequency-dependent viscoelastics", *Smart Materials and Structures*, vol. 5, no. 5, pp. 615–627, 1996.

[LES 04] LESIEUTRE G.A., OTTMAN G.K., HOFMANN H.F., "Damping as a result of piezoelectric energy harvesting", *Journal of Sound and Vibration*, vol. 269, no. 3–5, pp. 991–1001, 2004.

[LI 05] LI W., "Evaluation of the damping ratio for a base-excited system by the modulations of responses", *Journal of Sound and Vibration*, vol. 279, no. 3–5, pp. 1181–1194, 2005.

[LI 09] LI X.Y., LAW S.S., "Identification of structural damping in time domain", *Journal of Sound and Vibration*, vol. 328, no. 1–2, pp. 71–84, 2009.

[LI 12a] LI L., HU Y., WANG X., LING L., "Computation of eigensolution derivatives for nonviscously damped systems using the algebraic method", *AIAA Journal*, vol. 50, no. 10, pp. 2282–2284, October 2012.

[LI 12b] LI L., HU Y., WANG X., "A parallel way for computing eigenvector sensitivity of asymmetric damped systems with distinct and repeated eigenvalues", *Mechanical System and Signal Processing*, vol. 30, pp. 61–77, July 2012.

[LI 13a] LI L., HU Y., WANG X., "Improved approximate methods for calculating frequency response function matrix and response of MDOF systems with viscoelastic hereditary terms", *Journal of Sound and Vibration*, vol. 332, no. 15, pp. 3945–3956, 2013.

[LI 13b] LI L., HU Y., WANG X., "Numerical methods for evaluating the sensitivity of element modal strain energy", *Finite Elements in Analysis and Design*, vol. 64, pp. 13–23, February 2013.

[LI 13c] LI L., HU Y., WANG X., "A study on design sensitivity analysis for general nonlinear eigenproblems", *Mechanical Systems and Signal Processing*, vol. 34, no. 1–2, pp. 88–105, 2013.

[LI 13d] LI L., HU Y., WANG X., *et al.*, "Eigensensitivity analysis for asymmetric nonviscous systems", *AIAA Journal*, vol. 51, no. 3, pp. 738–741, 2013.

[LI 13e] LI L., HU Y., WANG X., *et al.*, "Eigensensitivity analysis of damped systems with distinct and repeated eigenvalues", *Finite Elements in Analysis and Design*, vol. 72, pp. 21–34, 15 September 2013.

[LIA 92] LIANG Z., TONG M., LEE C.G., "Complex modes in damped linear systems", *International Journal of Analytical and Experimental Modal Analysis*, vol. 7, no. 1, pp. 1–20, January 1992.

[LIA 11] LIANG Z., LEE G.C., DARGUSH G.F., *et al.*, *Structural Damping: Applications in Seismic Response Modification*, Advances in Earthquake Engineering, CRC Press, Boca Raton, FL, 2011.

[LIN 06] LIN R.M., ZHU J., "Model updating of damped structures using FRF data", *Mechanical Systems and Signal Processing*, vol. 20, no. 8, pp. 2200–2218, 2006.

[LIN 09] LIN R.M., ZHU J., "On the relationship between viscous and hysteretic damping models and the importance of correct interpretation for system identification", *Journal of Sound and Vibration*, vol. 325, no. 1–2, pp. 14–33, 2009.

[LIU 94] LIU Q.K., SNECKENBERGER J.E., "Free vibration analysis of linear vibrating deficient systems with arbitrary damping", *Journal of Sound and Vibration*, vol. 177, no. 1, pp. 43–55, 1994.

[LIU 00] LIU K., KUJATH M.R., ZHENG W., "Quantification of non-proportionality of damping in discrete vibratory systems", *Computer and Structures*, vol. 77, no. 5, pp. 557–569, 2000.

[LIU 05] LIU C.Q., "A method of symmetrisation of asymmetric dynamical systems", *Shock and Vibration*, vol. 12, no. 4, pp. 309–315, 2005.

[LIU 07] LIU H., LIU B., YUAN D. *et al.*, "Identification for sucker-rod pumping system's damping coefficients based on chain code method of pattern recognition", *Transactions of ASME, Journal of Vibration and Acoustics*, vol. 129, no. 4, pp. 434–440, 2007.

[LIU 08a] LIU C.-S., "Identifying time-dependent damping and stiffness functions by a simple and yet accurate method", *Journal of Sound and Vibration*, vol. 318, no. 1–2, pp. 148–165, 2008.

[LIU 08b] LIU C.-S., "A Lie-group shooting method estimating nonlinear restoring forces in mechanical systems", *CMES-Computer Modeling in Engineering & Sciences*, vol. 35, no. 2, pp. 157–180, 2008.

[LIU 08c] LIU C.-S., "A lie-group shooting method for simultaneously estimating the time-dependent damping and stiffness coefficients", *CMES-Computer Modeling in Engineering & Sciences*, vol. 27, no. 3, pp. 137–149, 2008.

[LIU 08d] LIU C.-S., CHANG J.-R., CHANG K.-H., *et al.*, "Simultaneously estimating the time-dependent damping and stiffness coefficients with the aid of vibrational data", *CMC-Computers Materials & Continua*, vol. 7, no. 2, pp. 97–107, 2008.

[LIU 09] LIU C.-S., KUO C.-L., "New methods for the computations of direct and inverse mechanical vibration problems", in SAPRI R. (ed.), *Mechanical Vibrations: Measurement, Effects and Control*, Nova Science Publishers, pp. 953–993, 2009.

[LIU 10] LIU M.H., ZHENG G.T., "Component synthesis method for transient response of nonproportionally damped structures", *AIAA Journal*, vol. 48, no. 11, pp. 2556–2563, 2010.

[LUO 89] LUO D.Z., "A graphic explanation of undamped and damped mode shapes and its application", *Journal of Sound and Vibration*, vol. 135, pp. 351–356, 1989.

[MA 95] MA F., CAUGHEY T.K., "Analysis of linear conservative vibrations", *Transactions of ASME, Journal of Applied Mechanics*, vol. 62, pp. 685–691, September 1995.

[MA 04] MA F., NG C.H., "On the orthogonality of natural modes of vibration", *Mechanics Research Communications*, vol. 31, no. 3, pp. 295–299, 2004.

[MA 09] MA F., IMAM A., MORZFELD M., "The decoupling of damped linear systems in oscillatory free vibration", *Journal of Sound and Vibration*, vol. 324, no. 1–2, pp. 408–428, 2009.

[MA 10] MA F., MORZFELD M., IMAM A., "The decoupling of damped linear systems in free or forced vibration", *Journal of Sound and Vibration*, vol. 329, no. 15, pp. 3182–3202, 2010.

[MAI 97a] MAIA N.M.M., SILVA J.M.M. (eds), *Theoretical and Experimental Modal Analysis*, Engineering Dynamics Series, J.B. ROBETRS (ser. ed.) Research Studies Press, Taunton, England, 1997.

[MAI 97b] MAIA N.M.M., SILVA J.M.M., RIBEIRO A.M.R., "A contribution to understanding the identification of systems with general damping", *Proceedings of the 17th International Modal Analysis Conference (IMAC)*, vol. 2, pp. 1145–1151, 1997.

[MAI 98] MAIA N.M.M., SILVA J.M.M., RIBEIRO A.M.R., "On a general model for damping", *Journal of Sound and Vibration*, vol. 218, no. 5, pp. 749–767, 1998.

[MAK 99] MAKRIS N., "Frequency-independent dissipation and causality", in SPANOS P. (ed.), *Computational Stochastic Mechanics*, Rotterdam, Balkema, pp. 435–441, 1999.

[MAL 97] MALONE D.P., CRONIN D.L., RANDOLPH T.W., "Eigenvalue and eigenvector determination for damped gyroscopic systems", *Transactions of ASME, Journal of Applied Mechanics*, vol. 64, pp. 710–712, September 1997.

[MAN 98] MANOHAR C.S., ADHIKARI S., "Dynamic stiffness of randomly parametered beams", *Probabilistic Engineering Mechanics*, vol. 13, no. 1, pp. 39–51, January 1998.

[MCI 78] MCINTYRE M.E., WOODHOUSE J., "The influence of geometry on damping", *Acustica*, vol. 39, no. 4, pp. 210–224, 1978.

[MCI 88] MCINTYRE M.E., WOODHOUSE J., "On measuring the elastic and damping constants of orthotropic sheet materials", *Acta Metallurgica*, vol. 36, no. 6, pp. 1397–1416, 1988.

[MCT 93] MCTAVISH D.J., HUGHES P.C., "Modeling of linear viscoelastic space structures", *Transactions of ASME, Journal of Vibration and Acoustics*, vol. 115, pp. 103–110, January 1993.

[MEA 02] MEAD D.J., "Structural damping and damped vibration", *Applied Mechanics Reviews*, vol. 55, no. 6, pp. R45–R54, 2002.

[MEI 67] MEIROVITCH L., *Analytical Methods in Vibrations*, Macmillan Publishing, New York, 1967.

[MEI 80] MEIROVITCH L., *Computational Methods in Structural Dynamics*, Sijthoff & Noordohoff, Netherlands, 1980.

[MEI 85] MEIROVITCH L., RYLAND G., "A perturbation technique for gyroscopic systems with small internal and external damping", *Journal of Sound and Vibration*, vol. 100, no. 3, pp. 393–408, 1985.

[MEI 97] MEIROVITCH L., *Principles and Techniques of Vibrations*, Prentice Hall International, NJ, 1997.

[MEN 12a] MENDIGUREN J., CORTES F., GALDOS L., "A generalised fractional derivative model to represent elastoplastic behaviour of metals", *International Journal of Mechanical Sciences*, vol. 65, no. 1, pp. 12–17, 2012.

[MEN 12b] MENTRASTI L., "Exact deflation in the complex modal analysis of low-rank non-classically damped structures", *Engineering Structures*, vol. 45, pp. 496–508, December 2012.

[MIN 91] MINAS C., INMAN D.J., "Identification of a nonproportional damping matrix from incomplete modal information", *Transactions of ASME, Journal of Vibration and Acoustics*, vol. 113, pp. 219–224, April 1991.

[MIN 12] MIN C., PARK H., PARK S., "Direct identification of non-proportional modal damping matrix for lumped mass system using modal parameters", *Journal of Mechanical Science and Technology*, vol. 26, no. 4, pp. 993–1002, 2012.

[MIT 90] MITCHELL L., "Complex modes: a review", *Proceedings of the 8th International Modal Analysis Conference (IMAC)*, vol. 2, Kissimmee, FL, pp. 891–899, 29 January–1 February 1990.

[MOO 04] MOON Y.J., KIM B.W., KO M.G., *et al.*, "Modified modal methods for calculating eigenpair sensitivity of asymmetric damped system", *International Journal for Numerical Methods in Engineering*, vol. 60, no. 11, pp. 1847–1860, 2004.

[MOR 08a] MORZFELD M., AJAVAKOM N., MA F., "A remark about the decoupling approximation of damped linear systems", *Mechanics Research Communications*, vol. 35, no. 7, pp. 439–446, 2008.

[MOR 08b] MORZFELD M., MA F., AJAVAKOM N., "On the decoupling approximation in damped linear systems", *Journal of Vibration and Control*, vol. 14, no. 12, pp. 1869–1884, 2008.

[MOR 09] MORZFELD M., AJAVAKOM N., MA F., "Diagonal dominance of damping and the decoupling approximation in linear vibratory systems", *Journal of Sound and Vibration*, vol. 320, no. 1–2, pp. 406–420, 2009.

[MOR 11a] MORZFELD M., MA F., "The decoupling of damped linear systems in configuration and state spaces", *Journal of Sound and Vibration*, vol. 330, no. 2, pp. 155–161, 2011.

[MOR 11b] MORZFELD M., MA F., PARLETT B.N., "The transformation of second-order linear systems into independent equations", *SIAM Journal on Applied Mathematics*, vol. 71, no. 4, pp. 1026–1043, 2011.

[MOT 90] MOTTERSHEAD J.E., "Theory for the estimation of structural vibration parameters from incomplete data", *AIAA Journal*, vol. 28, no. 3, pp. 559–561, 1990.

[MOT 93] MOTTERSHEAD J.E., FRISWELL M.I., "Model updating in structural dynamics – a survey", *Journal of Sound and Vibration*, vol. 167, no. 2, pp. 347–375, 1993.

[MUL 05] MULLER P., "Are the eigensolutions of a 1-d.o.f. system with viscoelastic damping oscillatory or not?", *Journal of Sound and Vibration*, vol. 285, no. 1–2, pp. 501–509, July 2005.

[MUR 88] MURTHY D.V., HAFTKA R.T., "Derivatives of eigenvalues and eigenvectors of a general complex matrix", *International Journal for Numerical Methods in Engineering*, vol. 26, no. 2, pp. 293–311, 1988.

[MUR 97a] MURAVYOV A., "Analytical solutions in the time domain for vibration problems of discrete viscoelastic systems", *Journal of Sound and Vibration*, vol. 199, no. 2, pp. 337–348, 1997.

[MUR 97b] MURAVYOV A., HUTTON S.G., "Closed-form solutions and the eigenvalue problem for vibration of discrete viscoelastic systems", *Transactions of ASME, Journal of Applied Mechanics*, vol. 64, pp. 684–691, September 1997.

[MUR 98a] MURAVYOV A., "Forced vibration responses of a viscoelastic structure", *Journal of Sound and Vibration*, vol. 218, no. 5, pp. 892–907, December 1998.

[MUR 98b] MURAVYOV A., HUTTON S.G., "Free vibration response characteristics of a simple elasto-hereditary system", *Transactions of ASME, Journal of Vibration and Acoustics*, vol. 120, no. 2, pp. 628–632, April 1998.

[MUS 05] MUSCOLINO G., PALMERI A., RICCIARDELLI F., "Time-domain response of linear hysteretic systems to deterministic and random excitations", *Earthquake Engineering and Structural Dynamics*, vol. 34, pp. 1129–1147, 2005.

[NAI 86] NAIR S.S., SING R., "Examination of the validity of proportional damping approximations with two further numerical indices", *Journal of Sound and Vibration*, vol. 104, pp. 348–350, 1986.

[NAS 85] NASHIF A.D., JONES D.I.G., HENDERSON J.P., *Vibration Damping*, John Wiley, New York, 1985.

[NAY 70] NAYLOR V.D., "Some fallacies in modern damping theory", *Journal of Sound and Vibration*, vol. 11, pp. 278–280, 1970.

[NEL 76] NELSON R.B., "Simplified calculation of eigenvector derivatives", *AIAA Journal*, vol. 14, no. 9, pp. 1201–1205, September 1976.

[NEL 79] NELSON H.D., GLASGOW D.A., "Eigenrelations for general second-order systems", *AIAA Journal*, vol. 17, no. 7, pp. 795–797, July 1979.

[NEW 87] NEWLAND D.E., "On the modal analysis of nonconservative linear systems", *Journal of Sound and Vibration*, vol. 112, no. 1, pp. 69–96, 1987.

[NEW 89] NEWLAND D.E., *Mechanical Vibration Analysis and Computation*, Longman, Harlow and John Wiley, New York, 1989.

[NIC 87a] NICHOLSON D.W., "Response bounds for non-classically damped mechanical systems under transient loads", *Transactions of ASME, Journal of Applied Mechanics*, vol. 54, pp. 430–433, 1987.

[NIC 87b] NICHOLSON D.W., "Stable response of non-classically damped mechanical systems", *Applied Mechanics Reviews, ASME*, vol. 40, no. 6, pp. 733–740, June 1987.

[NIC 96] NICHOLSON D.W., BAOJIU L., "Stable response of non-classically damped mechanical systems - II", *Applied Mechanics Reviews, ASME*, vol. 49, no. 10, Part-2, pp. S49–S54, October 1996.

[NIG 83] NIGAM N.C., *Introduction to Random Vibration*, MIT Press, Cambridge, MA, 1983.

[OLI 96] OLIVETO G., SANTINI A., "Complex modal analysis of a continuous model with radiation damping", *Journal of Sound and Vibration*, vol. 192, no. 1, pp. 15–33, 1996.

[OME 12] OMENZETTER P., "Sensitivity analysis of the eigenvalue problem for general dynamic systems with application to bridge deck flutter", *Journal of Engineering Mechanics-ASCE*, vol. 138, no. 6, pp. 675–682, June 2012.

[OTR 07a] OTRIN M., BOLTEZAR M., "Damped lateral vibrations of straight and curved cables with no axial pre-load", *Journal of Sound and Vibration*, vol. 300, no. 3–5, pp. 676–694, 6 March 2007.

[OTR 07b] OTRIN M., BOLTEZAR M., "The vibration over a spatially curved steel wire with an outer band", *Strojniski Vestnik-Journal of Mechanical Engineering*, vol. 53, no. 10, pp. 635–656, 2007.

[OTR 09a] OTRIN M., BOLTEZAR M., "On the modeling of vibration transmission over a spatially curved cable with casing", *Journal of Sound and Vibration*, vol. 325, no. 4–5, pp. 798–815, 2009.

[OTR 09b] OTRIN M., BOLTEZAR M., "On the modeling of vibration transmission over a spatially curved cable with casing", *Journal of Sound and Vibration*, vol. 325, no. 4–5, pp. 798–815, 2009.

[PAL 03] PALMERI A., RICCIARDELLI F., LUCA A.D., *et al.*, "State space formulation for linear viscoelastic dynamic systems with memory", *ASCE Journal of Engineering Mechanics*, vol. 129, no. 7, pp. 715–724, July 2003.

[PAL 04] PALMERI A., RICCIARDELLI F., MUSCOLINO G., *et al.*, "Random vibration of systems with viscoelastic memory", *ASCE Journal of Engineering Mechanics*, vol. 130, no. 9, pp. 1052–1061, September 2004.

[PAL 11a] PALMERI A., ADHIKARI S., "A Galerkin-type state-space approach for transverse vibrations of slender double-beam systems with viscoelastic inner layer", *Journal of Sound and Vibration*, vol. 330, no. 26, pp. 6372–6386, 2011.

[PAL 11b] PALMERI A., MUSCOLINO G., "A numerical method for the time-domain dynamic analysis of buildings equipped with viscoelastic dampers", *Structural Control and Health Monitoring*, vol. 18, no. 5, pp. 519–539, 2011.

[PAN 13] PAN Y., WANG Y., "Frequency-domain analysis of exponentially damped linear systems", *Journal of Sound and Vibration*, vol. 332, no. 7, pp. 1754–1765, 2013.

[PAP 12] PAPAI F., ADHIKARI S., WANG B., "Estimation of modal dampings for unmeasured modes", *Slovak Journal of Civil Engineering*, vol. 20, no. 4, pp. 17–27, 2012.

[PAR 86] PARTER G., SING R., "Quantification of the extent on non-proportional damping discrete vibratory systems", *Journal of Sound and Vibration*, vol. 104, no. 1, pp. 109–125, 1986.

[PAR 92a] PARK I.W., KIM J.S., MA F., "On the modal coupling in non-classically damped linear systems", *Mechanics Research Communications*, vol. 13, pp. 407–413, 1992.

[PAR 92b] PARK S., PARK I.W., MA F., "Decoupling approximation of non-classically damped structures", *AIAA Journal*, vol. 30, no. 9, pp. 2348–2351, September 1992.

[PAR 94] PARK I.W., KIM J.S., MA F., "Characteristics of modal coupling in non-classically damped systems under harmonic excitation", *Transactions of ASME, Journal of Applied Mechanics*, vol. 61, pp. 77–83, March 1994.

[PAR 99] PARK C.H., INMAN D.J., LAM M.J., "Model reduction of viscoelastic finite element models", *Journal of Sound and Vibration*, vol. 219, no. 4, pp. 619–637, 1999.

[PAW 13] PAWLAK Z., LEWANDOWSKI R., "The continuation method for the eigenvalue problem of structures with viscoelastic dampers", *Computers & Structures*, vol. 125, no. 9, pp. 53–61, 2013.

[PAZ 80] PAZ M., *Structural Dynamics: Theory and Computation*, 2nd ed., Van Nostrand, Reinhold, 1980.

[PEA 67] PEARL M.H., "A decomposition theorem for matrices", *Canadian Journal of Mathematics*, vol. 19, pp. 344–349, 1967.

[PER 95] PERES-DA-SILVA S.S., CRONIN D.L., RANDOLPH T.W., "Computation of eigenvalues and eigenvectors of nonclassically damped systems", *Computer and Structures*, vol. 57, no. 5, pp. 883–891, 1995.

[PET 98] PETYT M., *Introduction to Finite Element Vibration Analysis*, Cambridge University Press, Cambridge, UK, 1998.

[PHA 03] PHANI A.S., "On the necessary and sufficient conditions for the existence of classical normal modes in damped linear dynamic systems", *Journal of Sound and Vibration*, vol. 264, no. 3, pp. 741–745, 2003.

[PHA 04] PHANI A.S., Damping identification in linear vibrations, PhD Thesis, Cambridge University Engineering Department, Cambridge, CB2 1PZ, January 2004.

[PHA 07] PHANI A.S., WOODHOUSE J., "Viscous damping identification in linear vibration", *Journal of Sound and Vibration*, vol. 303, no. 3–5, pp. 475–500, 2007.

[PHA 08] PHANI A.S., WOODHOUSE J., "Rayleigh quotient and dissipative systems", *Transactions of ASME, Journal of Applied Mechanics*, vol. 75, no. 6, pp. 061005:1–6, November 2008.

[PHA 09] PHANI A.S., WOODHOUSE J., "Experimental identification of viscous damping in linear vibration", *Journal of Sound and Vibration*, vol. 319, no. 3–5, pp. 832–849, 2009.

[PIL 97] PILKEY D.F., INMAN D.J., "Iterative approach to viscous damping matrix identification", *Proceedings of the 15th International Modal Analysis Conference (IMAC)*, vol. 2, pp. 1152–1157, 1997.

[PIL 98] PILKEY D.F., INMAN D.J., "Survey of damping matrix identification", *Proceedings of the 16th International Modal Analysis Conference (IMAC)*, vol. 1, pp. 104–110, 1998.

[PLA 73] PLAUT R.H., HUSEYIN K., "Derivative of eigenvalues and eigenvectors in non-self adjoint systems", *AIAA Journal*, vol. 11, no. 2, pp. 250–251, February 1973.

[PLA 91] PLACIDI F., POGGI F., SESTIERI A., "Real modes computation from identified modal parameters with estimate of general damping", *Proceedings of the 9th International Modal Analysis Conference (IMAC)*, London, 1991.

[POM 87] POMMER C., KLIEM W.R., "Simultaneously normalizable matrices", *Linear Algebra and Its Applications*, vol. 94, pp. 113–125, 1987.

[POT 13] POTAPOV V.D., "Stability via nonlocal continuum mechanics", *International Journal of Solids and Structures*, vol. 50, no. 5, pp. 637–641, 1 March 2013.

[PRA 09] PRANDINA M., MOTTERSHEAD J.E., BONISOLI E., "An assessment of damping identification methods", *Journal of Sound and Vibration*, vol. 323, no. 3–5, pp. 662–676, 2009.

[PRA 12a] PRADHAN S., MODAK S.V., "A method for damping matrix identification using frequency response data", *Mechanical Systems and Signal Processing*, vol. 33, pp. 69–82, 2012.

[PRA 12b] PRADHAN S., MODAK S.V., "Normal response function method for mass and stiffness matrix updating using complex FRFs", *Mechanical Systems and Signal Processing*, vol. 32, no. Special Issue, pp. 232–250, 2012.

[PRE 92] PRESS W.H., TEUKOLSKY S.A., VETTERLING W.T. *et al.*, *Numerical Recipes in C*, Cambridge University Press, Cambridge, 1992.

[PRE 00] PRELLS U., FRISWELL M.I., "A measure of non-proportional damping", *Mechanical Systems and Signal Processing*, vol. 14, no. 2, pp. 125–137, 2000.

[PRE 09] PRELLS U., GARVEY S.D., "On the class of strictly isospectral systems", *Mechanical Systems and Signal Processing*, vol. 23, no. 6, pp. 2000–2007, 2009.

[QIA 95] QIAN D., HANSEN J.S., "A time domain substructure synthesis method for viscoelastic structures", *Journal of Applied Mechanics*, vol. 62, pp. 407–413, 1995.

[RAO 71] RAO C.R., MITRA S.K., *Generalized Inverse of Matrices and its Applications*, John Wiley & Sons, New York, 1971.

[RAO 11] RAO S.S., *Mechanical Vibrations*, Prentice Hall PTR, NJ, 2011.

[RAY 77] RAYLEIGH J.W., *Theory of Sound (Two Volumes)*, 2nd ed., Dover Publications, New York, 1945 re-issue, 1877.

[RAY 85] RAYLEIGH J.W., A theorem relating to the time-moduli ofFirst Year Report. dissipative systems, Report of the British Association, pp. 911–912, 1885.

[REN 97] REN G., ZHENG Z., "A quasi-decoupling approach for nonclassical linear systems in state-space", *Transactions of ASME, Journal of Applied Mechanics*, vol. 64, pp. 946–950, December 1997.

[ROE 92] ROEMER M.J., MOOK D.J., "Mass, stiffness and damping matrix identification: an integrated approach", *Transactions of ASME, Journal of Vibration and Acoustics*, vol. 114, pp. 358–363, July 1992.

[ROG 70] ROGERS L.C., "Derivatives of eigenvalues and eigenvectors", *AIAA Journal*, vol. 8, no. 5, pp. 943–944, May 1970.

[ROS 97] ROSSIKHIN Y.A., SHITIKOVA M.V., "Applications of fractional calculus to dynamic problems of linear and nonlinear hereditary mechanics of solids", *Applied Mechanics Reviews, ASME*, vol. 50, no. 1, pp. 15–67, 1997.

[ROY 09] ROY S., CHAKRABORTY S., "Identification of damping parameters of linear dynamic system", *International Journal of Structural Stability and Dynamics*, vol. 9, no. 3, pp. 473–487, 2009.

[RUD 74] RUDISILL C.S., "Derivatives of eigenvalues and eigenvectors for a general matrix", *AIAA Journal*, vol. 12, no. 5, pp. 721–722, May 1974.

[RUG 00] RUGE P., WAGNER N., "Quadratic time-interpolation versus discontinuous Hamiltonian approach", *Acta technica Academiae Scientiarum Hungaricae*, vol. 108, no. 1–2, pp. 161–175, 2000.

[RUG 01] RUGE P., "Restricted Padé scheme in computational structural dynamics", *Computer and Structures*, vol. 79, no. 20–21, pp. 1913–1921, August 2001.

[SCA 70] SCANLAN R.H., "Linear damping models and causality in vibrations", *Journal of Sound and Vibration*, vol. 13, no. 4, pp. 499–503, 1970.

[SEG 87] SEGALMAN D.J., "Calculation of damping matrices for linearly viscoelastic structures", *Journal of Applied Mechanics*, vol. 54, pp. 585–588, 1987.

[SEG 06] SEGALMAN D.J., "Modelling joint friction in structural dynamics", *Structural Control & Health Monitoring*, vol. 13, no. 1, pp. 430–453, 2006.

[SES 94] SESTIERI A., IBRAHIM S.R., "Analysis of errors and approximations in the use of modal coordinates", *Journal of Sound and Vibration*, vol. 177, no. 2, pp. 145–157, 1994.

[SHA 88] SHAHRUZ S.M., MA F., "Approximate decoupling of the equations of motion of linear underdamped system", *Transactions of ASME, Journal of Applied Mechanics*, vol. 55, pp. 716–720, September 1988.

[SHA 89] SHAHRUZ S.M., MA F., "On symmetrizability of asymmetric non-conservative systems", *Transactions of ASME, Journal of Applied Mechanics*, vol. 56, pp. 474–476, June 1989.

[SHA 90] SHAHRUZ S.M., "Approximate decoupling of the equations of motion of damped linear systems", *Journal of Sound and Vibration*, vol. 136, no. 1, pp. 51–64, 1990.

[SHA 95] SHAHRUZ S.M., "Comments on 'an index of damping non-proportionality for discrete vibrating systems'", *Journal of Sound and Vibration*, vol. 186, no. 3, pp. 535–542, 1995.

[SHA 97] SHAHRUZ S.M., SRIMATSYA P.A., "Approximate solutions of non-classically damped linear systems in normalized and physical coordinates", *Journal of Sound and Vibration*, vol. 201, no. 2, pp. 262–271, 1997.

[SHE 09] SHEN H., DUAN Z., "A time-history analysis algorithm of a non-viscously damped system using gauss precise integration", in DENG H., WANG L., WANG F., LEI J. (eds), *Artificial Intelligence and Computational Intelligence*, Lecture Notes in Computer Science, Springer, Berlin Heidelberg, vol. 5855, pp. 221–230, 2009.

[SIE 08] SIEBER J., WAGG D.J., ADHIKARI S., "On the interaction of exponential non-viscous damping with symmetric nonlinearities", *Journal of Sound and Vibration*, vol. 314, no. 1–2, pp. 1–11, 2008.

[SIL 98] SILVA J.M.M., MAIA N.M.M. (eds), *Modal Analysis and Testing: Proceedings of the NATO Advanced Study Institute*, NATO Science Series: E: Applied Science, Sesimbra, Portugal, 3–15 May 1998.

[SIV 09] SIVAPRASAD R., VENKATESHA S., MANOHAR C.S., "Identification of dynamical systems with fractional derivative damping models using inverse sensitivity analysis", *Computers, Materials & Continua*, vol. 9, no. 9, pp. 179–207, 2009.

[SKA 12] SKAROLEK A., "Exponential damping as an approach to internal hysteretic damping of rotor systems: FEM model of Timoshenko rotating beam with Maxwell-Weichert damping model", in BERAN J., BÍLEK M., HEJNOVA M., ZABKA P. (eds), *Advances in Mechanisms Design*, Mechanisms and Machine Science, Springer, Netherlands, vol. 5855, pp. 253–261, 2012.

[SLA 93] SLATER J.C., BELVIN W.K., INMAN D.J., "A survey of modern methods for modeling frequency dependent damping in finite element models", *Proceedings of the 11th International Modal Analysis Conference (IMAC)*, pp. 1508–1512, 1993.

[SMI 06] SMITH D.E., SIDDHI V., "Generalized approach for incorporating normalization conditions in design sensitivity analysis of eigenvectors", *AIAA Journal*, vol. 44, no. 11, pp. 2552–2561, November 2006.

[SOO 83] SOOM A., KIM C., "Roughness-induced dynamic loading at dry boundary-lubricated sliding contacts", *ASME Journal of Lubrication Technology*, vol. 105, pp. 514–517, 1983.

[SRA 11] SRACIC M.W., ALLEN M.S., "Method for identifying models of nonlinear systems using linear time periodic approximations", *Mechanical Systems and Signal Processing*, vol. 25, no. 7, pp. 2705–2721, 2011.

[STA 97] STAREK L., INMAN D.J., "A symmetric inverse vibration problem for nonproportional underdamped systems", *Transactions of ASME, Journal of Applied Mechanics*, vol. 64, pp. 601–605, September 1997.

[STR 88] STRANG G., *Linear Algebra and Its Applications*, 3rd ed. Hardcourt Brace Jovanovich, Orlando, FL, 1988.

[SUA 87] SUAREZ L.E., SING M.P., "Eigenproperties of nonclassically damped primary structure and oscillator systems", *Transactions of ASME, Journal of Applied Mechanics*, vol. 54, pp. 668–673, September 1987.

[SUA 89] SUAREZ L.E., SING M.P., "Dynamic synthesis of nonclassically damped substructures", *ASCE Journal of Engineering Mechanics*, vol. 115, no. 1, pp. 52–70, 1989.

[SUL 10] SULTAN C., "Proportional damping approximation using the energy gain and simultaneous perturbation stochastic approximation", *Mechanical Systems and Signal Processing*, vol. 24, no. 7, pp. 2210–2224, October 2010.

[SUL 13] SULTAN C., "Decoupling approximation design using the peak to peak gain", *Mechanical Systems and Signal Processing*, vol. 36, no. 2, pp. 582–603, April 2013.

[SUN 95] SUN C.T., LU Y.P., *Vibration Damping of Structural Elements*, Prentice Hall PTR, NJ, 1995.

[TAN 97] TAN G.B., Non-linear vibration of cable-deployed space structures, PhD Thesis, Cambridge University Engineering Department, Cambridge, UK, 1997.

[TAU 59] TAUSSKY O., ZASSENHAUS H., "On the similarity transformation between a matrix and its transpose", *Pacific Journal of Mathematics*, vol. 9, pp. 893–896, 1959.

[TAU 68] TAUSSKY O., "Positive definite matrices and their role in the study of the characteristic roots of general matrices", *Advances in Mathematics*, vol. 2, pp. 175–186, 1968.

[TAU 72] TAUSSKY O., "The role of symmetric matrices in the study of general matrices", *Linear Algebra and its Applications*, vol. 5, pp. 147–154, 1972.

[THO 69] THOMPSON R.C., "On Pearl's paper 'a decomposition theorem for matrices'", *Canadian Mathematical Bulletin*, vol. 12, pp. 805–808, 1969.

[THO 74] THOMSON W.T., CALKINS C., CARAVANI P., "A numerical study of damping", *Earthquake Engineering and Structural Dynamics*, vol. 3, pp. 97–103, 1974.

[TIS 01] TISSEUR F., MEERBERGEN K., "The quadratic eigenvalue problem", *SIAM Review*, vol. 43, no. 2, pp. 235–286, 2001.

[TIS 11] TISSEUR F., GARVEY S.D., MUNRO C., "Deflating quadratic matrix polynomials with structure preserving transformations", *Linear Algebra and Its Applications*, vol. 435, no. 3, Special Issue, pp. 464–479, 2011.

[TON 92] TONG M., LIANG Z., LEE G.C., "On non-proportionality of generally damped systems", *Proceedings of the 10th International Modal Analysis Conference (IMAC)*, San Diego, CA, pp. 1302–1308, 1992.

[TON 94] TONG M., LIANG Z., LEE G.C., "An index of damping non-proportionality for discrete vibrating systems", *Journal of Sound and Vibration*, vol. 174, pp. 37–55, 1994.

[TOR 87] TORVIK P.J., BAGLEY R.L., "Fractional derivatives in the description of damping: materials and phenomena", *The Role of Damping in Vibration and Noise Control*, ASME DE-5, pp. 125–135, 1987.

[TRI 00] TRINDADE M.A., BENJEDDOU A., OHAYON R., "Modeling of frequency-dependent viscoelastic materials for active-passive vibration damping", *Journal of Vibration and Acoustics - Transactions of the ASME*, vol. 122, no. 2, pp. 169–174, 2000.

[TRI 02] TRINKS C., RUGE P., "Treatment of dynamic systems with fractional derivatives without evaluating memory-integrals", *Computational Mechanics*, vol. 29, no. 6, pp. 471–476, 2002.

[TRI 06] TRINDADE M.A., "Reduced-order finite element models of viscoelastically damped beams through internal variables projection", *Journal of Vibration and Acoustics - Transactions of the ASME*, vol. 128, no. 4, pp. 501–508, 2006.

[TSA 09] TSAI T.-C., TSAU J.-H., CHEN C.-S., "Vibration analysis of a beam with partially distributed internal viscous damping", *International Journal of Mechanical Sciences*, vol. 51, no. 11–12, pp. 907–914, 2009.

[UDW 90] UDWADIA F.E., ESFANDIARI R.S., "Non-classically damped dynamic systems: an iterative approach", *Transactions of ASME, Journal of Applied Mechanics*, vol. 57, pp. 423–433, June 1990.

[UDW 09] UDWADIA F.E., "A note on nonproportional damping", *Journal of Engineering Mechanics-ASCE*, vol. 135, no. 11, pp. 1248–1256, 2009.

[UNG 62] UNGAR E.E., KERWIN E.M., "Loss factors of viscoelastic systems in terms of energy concepts", *The Journal of the Acoustical Society of America*, vol. 34, no. 7, pp. 954–957, July 1962.

[UNG 73] UNGAR E.E., "The status of engineering knowledge concerning the damping of built-up structures", *Journal of Sound and Vibration*, vol. 26, pp. 141–154, 1973.

[UNG 00] UNGAR E.E., "Damping by viscoelastic layers", *Applied Mechanics Reviews, ASME*, vol. 53, no. 6, pp. R33–R38, June 2000.

[VAS 06] VASQUES C.M.A., MACE B.R., GARDONIO P., *et al.*, "Arbitrary active constrained layer damping treatments on beams: finite element modelling and experimental validation", *Computers & Structures*, vol. 84, no. 22–23, pp. 1384–1401, 2006.

[VAS 10a] VASQUES C.M.A., MOREIRA R.A.S., RODRIGUES J.D., "Viscoelastic damping technologies - part I: modeling and finite element implementation", *Journal of Advanced Research in Mechanical Engineering*, vol. 1, no. 2, pp. 76–95, 2010.

[VAS 10b] VASQUES C.M.A., MOREIRA R.A.S., RODRIGUES J.D., "Viscoelastic damping technologies - part II: experimental identification procedure and validation", *Journal of Advanced Research in Mechanical Engineering*, vol. 1, no. 2, pp. 76–95, 2010.

[VAS 11] VASQUES C.M.A., CARDOSO L.C., "Viscoelastic damping technologies: finite element modeling and application to circular saw blades", in VASQUES C., DIAS RODRIGUES J. (eds), *Vibration and Structural Acoustics Analysis*, pp. 207–264, Springer, Netherlands, 2011.

[VEL 86] VELETSOS A.S., VENTURA C.E., "Modal analysis of non-classically damped linear systems", *Earthquake Engineering and Structural Dynamics*, vol. 14, pp. 217–243, 1986.

[VES 11] VESELIC K., *Damped Oscillations of Linear Systems: A Mathematical Introduction*, Lecture Notes in Mathematics, Springer, New York, 2011.

[VIG 86] VIGNERON F.R., "A natural modes model and modal identities for damped linear structures", *Transactions of ASME, Journal of Applied Mechanics*, vol. 53, pp. 33–38, March 1986.

[VIN 03] VINOKUR R., "The relationship between the resonant and natural frequency for non-viscous systems", *Journal of Sound and Vibration*, vol. 267, pp. 187–189, 2003.

[WAG 03] WAGNER N., ADHIKARI S., "Symmetric state-space formulation for a class of non-viscously damped systems", *AIAA Journal*, vol. 41, no. 5, pp. 951–956, 2003.

[WAN 13] WANG Y., INMAN D.J., "Finite element analysis and experimental study on dynamic properties of a composite beam with viscoelastic damping", *Journal of Sound and Vibration*, 2013, in press.

[WAR 77] WARBURTON G.B., SONI S.R., "Errors in response calculations for non-classically damped structures", *Earthquake Engineering and Structural Dynamics*, vol. 5, pp. 365–376, 1977.

[WIL 39] WILLIAMSON J., "Note on a principal axis transformation for non-Hermitian matrices", *Bulletin of American Mathematical Society*, vol. 45, pp. 920–922, 1939.

[WIL 88] WILKINSON J.H., *The Algebraic Eigenvalue Problem*, Oxford University Press, Oxford, UK, 1988.

[WOO 98] WOODHOUSE J., "Linear damping models for structural vibration", *Journal of Sound and Vibration*, vol. 215, no. 3, pp. 547–569, 1998.

[XIE 08] XIE H., DAI H., "Calculation of derivatives of multiple eigenpairs of unsymmetrical quadratic eigenvalue problems", *International Journal of Computer Mathematics*, vol. 85, no. 12, pp. 1815–1831, 2008.

[XIE 12] XIE H., "An iterative method for partial derivatives of eigenvectors of quadratic eigenvalue problems", *BIT Numerical Mathematics*, vol. 52, no. 2, pp. 525–536, June 2012.

[XIE 13] XIE H., "Simultaneous iterative method for eigenpair derivatives of damped systems", *AIAA Journal*, vol. 51, no. 1, pp. 236–239, January 2013.

[XU 10] XU T., XU T., ZUO W., *et al.*, "Fast sensitivity analysis of defective system", *Applied Mathematics and Computation*, vol. 217, no. 7, pp. 3248–3256, 1 December 2010.

[XUE 08] XUE-CHUAN Z., YONG-JUN L., JIAN-PING Z., "Strain analysis of nonlocal viscoelastic Kelvin bar in-tension", *Applied Mathematics and Mechanics-English Edition*, vol. 29, no. 1, pp. 67–74, 2008.

[YAE 87] YAE K.H., INMAN D.J., "Response bounds for linear underdamped systems", *Transactions of ASME, Journal of Applied Mechanics*, vol. 54, pp. 419–423, June 1987.

[YAN 93] YANG B., "Exact receptances of non-proportionally damped dynamic systems", *Transactions of ASME, Journal of Vibration and Acoustics*, vol. 115, pp. 47–52, January 1993.

[YAN 98] YANG B., WU X., "Modal expansion of structural systems with time delays", *AIAA Journal*, vol. 36, no. 12, pp. 2218–2224, 1998.

[YAN 12] YANG S., SRACIC M.W., ALLEN M.S., "Two algorithms for mass normalizing mode shapes from impact excited continuous-scan laser Doppler vibrometry", *Transactions of ASME, Journal of Vibration and Acoustics*, vol. 134, no. 2, pp. 021004:1–8, 2012.

[YIU 93] YIU Y.C., "Finite element analysis of structures with classical viscoelastic materials", *34th AIAA/ASME/ASCE/AHS/ASC Structures, Structural Dynamics & Materials Conference*, vol. 4, La Jolla, CA, pp. 2110–2119, April 1993.

[YIU 94] YIU Y.C., "Substructure and finite element formulation for linear viscoelastic materials", *35th AIAA/ASME/ASCE/AHS/ASC Structures, Structural Dynamics & Materials Conference*, vol. 3, Hilton Head, SC, pp. 1585–1594, April 1994.

[YUK 05] YUKSEL S., DALLI U., "Longitudinally vibrating elastic rods with locally and non-locally reacting viscous dampers", *Shock and Vibration*, vol. 12, no. 2, pp. 109–118, 2005.

[ZEN 95] ZENG Q.H., "Highly accurate modal method for calculating eigenvector derivatives in viscous damping systems", *AIAA Journal*, vol. 33, no. 4, pp. 746–751, April 1995.

[ZHA 04] ZHANG Z.Y., ZHANG H.S., "Eigensensitivity analysis of a defective matrix with zero first-order eigenvalue derivatives", *AIAA Journal*, vol. 42, no. 1, pp. 114–123, 2004.

[ZHA 07a] ZHANG J., ZHENG G.T., "The Biot model and its application in viscoelastic composite structures", *Journal of Vibration and Acoustics*, vol. 129, pp. 533–540, 2007.

[ZHA 07b] ZHANGPING L., JINWU X., "Novel modal method for efficient calculation of complex eigenvector derivatives", *AIAA Journal*, vol. 45, no. 6, pp. 1406–1414, June 2007.

[ZHA 11] ZHANG Z.-Y., "Approximate method for eigensensitivity analysis of a defective matrix", *Journal of Computational and Applied Mathematics*, vol. 235, no. 9, pp. 2913–2927, 1 March 2011.

[ZHI 06] ZHI J.Y., SU H.C., WU X.M., "A method for modal reanalysis of topological modifications of structures", *International Journal for Numerical Methods in Engineering*, vol. 65, no. 13, pp. 2203–2220, 2006.

[ZHO 95] ZHOU K., DOYLE J.C., GLOVER K., *Robust and Optimal Control*, Prentice Hall, Upper Saddle River, NJ, 1995.

[ZIE 91] ZIENKIEWICZ O.C., TAYLOR R.L., *The Finite Element Method*, 4th ed., McGraw-Hill, London, 1991.

Author Index

Index

D

E

O

P

Q, R

S

T